Biocatalysts for Industry

TOPICS IN APPLIED CHEMISTRY

Series Editors: Alan R. Katritzky, FRS
Kenan Professor of Chemistry
University of Florida, Gainesville, Florida

Gebran J. Sabongi
Laboratory Manager, Encapsulation Technology Center
3M, St. Paul, Minnesota

BIOCATALYSTS FOR INDUSTRY
Edited by Jonathan S. Dordick

CHEMICAL TRIGGERING
Reactions of Potential Utility in Industrial Processes
Gebran J. Sabongi

THE CHEMISTRY AND APPLICATION OF DYES
Edited by David R. Waring and Geoffrey Hallas

HIGH-TECHNOLOGY APPLICATIONS OF ORGANIC COLORANTS
Peter Gregory

INFRARED ABSORBING DYES
Edited by Masaru Matsuoka

STRUCTURAL ADHESIVES
Edited by S. R. Hartshorn

TARGET SITES FOR HERBICIDE ACTION
Edited by Ralph C. Kirkwood

A Continuation Order Plan is available for this series. A continuation order will bring delivery of each new volume immediately upon publication. Volumes are billed only upon actual shipment. For further information please contact the publisher.

Biocatalysts for Industry

Edited by
JONATHAN S. DORDICK
The University of Iowa
Iowa City, Iowa

Plenum Press • New York and London

Library of Congress Cataloging-in-Publication Data

Biocatalysts for industry / edited by Jonathan S. Dordick.
p. cm. -- (Topics in applied chemistry)
Includes bibliographical references and index.
ISBN 0-306-43652-3
1. Enzymes--Industrial applications. 2. Enzymes--Biotechnology.
I. Dordick, Jonathan S. II. Series.
TP248.E5B57 1991
660'.634--dc20 91-9029
CIP

ISBN 0-306-43652-3

A Division of Plenum Publishing Corporation
233 Spring Street, New York, N.Y. 10013

Printed in the United States of America

Contributors

George Abraham, US Department of Agriculture, Southern Regional Research Center, New Orleans, LA 70124

Ali Al-Hakim, Department of Medicinal and Natural Products Chemistry, College of Pharmacy, Iowa City, IA 52242

Harvey W. Blanch, Department of Chemical Engineering, University of California, Berkeley, CA 95720

Douglas S. Clark, Department of Chemical Engineering, University of California, Berkeley, CA 94720

Jeffrey S. Deetz, Genetics Institute, Inc., 87 Cambridge Park Drive, Cambridge, MA 02140

Jonathan S. Dordick, Department of Chemical and Biochemical Engineering, University of Iowa, Iowa City, IA 52242

Patrick R. Dugan, Center for Bioprocessing and Biotechnology Group, Idaho National Engineering Laboratory, EG & G Idaho, Inc., P.O. Box 1625, Idaho Falls, ID 83415

Thomas P. Graycar, Genencor, International, 180 Kimball Way, South San Francisco, CA 94080

David E. Hansen, Department of Chemistry, Amherst College, Amherst, MA 01002

Vijay T. John, Department of Chemical Engineering, Tulane University, New Orleans, LA 70118

Michael E. McIlwain, Center for Bioprocessing and Biotechnology Group, Idaho National Engineering Laboratory, EG & G Idaho, Inc., P.O. Box 1625, Idaho Falls, ID 83415

Saul L. Neidleman, Biosource Genetics Corporation, 3333 Vaca Valley Parkway, Vacaville, CA 95688

Zivko L. Nikolov, Department of Food Technology, Iowa State University, Ames, IA 50011

Elizabeth A. Pease, Department of Molecular and Cell Biology, The Pennsylvania State University, University Park, PA 16802

Theodore W. Randolph, Department of Chemical Engineering, Yale University, New Haven, CT 06520

Peter J. Reilly, Department of Chemical Engineering, Iowa State University, Ames, IA 50011

David G. Rethwisch, Department of Chemical and Biochemical Engineering, University of Iowa, Iowa City, IA 52242

J. David Rozzell, Celgene Corporation, 7 Powder Horn Drive, Warren, NJ 07060

Ming Tien, Department of Molecular and Cell Biology, The Pennsylvania State University, University Park, PA 16802

Susan J. Tomazic, Abbott Laboratories, Department 90U, Building AP-8B, Abbott Park, IL 60064

Aleksey Zaks, Enzytech, Inc., 64 Sidney St., Cambridge, MA 02139

Preface

The application of enzymes or whole cells (fermentatively active or resting; microbial, plant, or animal) to carry out selective transformations of commercial importance is the central theme of industrial biocatalysis. Traditionally, biocatalysis has been in the domain of the life scientist or biochemical engineer. However, recent advances in this field have enabled biocatalytic processes to compete head on with, and in some cases outperform, conventional chemical processing. Chemo-biocatalytic systems are being developed thereby combining the most attractive features of biocatalysts, namely high specificity, with those of chemical catalysts, such as high reactivities and wide substrate specificities. Hence, synthetic chemists and chemical engineers are now beginning to use biocatalysts as highly selective reagents in chemical synthesis and processing.

This book is about biocatalysts and their past, present, and potential applications in the food, pharmaceutical, and chemical industries. The concept of the book did not emanate from a meeting. Rather, it is a compilation of selected examples where biocatalysis either has already made a significant impact in the aforementioned industries, or has the potential to make a substantial contribution. I have been fortunate to have assembled contributions from world-class researchers in the field of biocatalysis. Their timely contributions are sincerely appreciated.

The book mainly deals with enzymes, but whole resting cells and catalytic antibodies are also discussed. It is aimed at a reader who has some background in chemistry and who has a need to understand how biocatalysts are used, where they have been employed, and what the future of biocatalysis is likely to be. The book is divided into four sections. The first functions as background material for the rest of the text, including a historical overview of biocatalysis prior to 1950. The second section deals with examples of biocatalytic systems that have no direct competition from chemical catalysts. For the most part such examples have arisen from the inability of chemical catalysts to catalyze a desired transformation, either due to a lack in selectivity or high potential for causing unacceptable health hazards. Such examples have been typically in the food industry, most notably carbohydrate modification. However, the area of biopolymer synthesis and modification has also become increasingly important for the development of polymeric drugs, peptides and hormones, novel carbohydrate derivatives, lignin-derived materials, and biopulping. Key aspects of the third section include the use of biocatalysts in areas conventionally believed to be restricted to chemical processing; particularly, the application of enzymes in organic solvents both in general terms and with specific examples including stereoselective biotransformations, modification of commercially important fats, and the use of enzymes in supercritical fluids. The final section of this book deals with several key future biocatalytic developments. This includes the application of protein-stabilizing techniques to increase the lifetime of biocatalysts and thereby enhance productivities, the use of site-directed mutagenesis to alter systematically the catalytic properties of enzymes to make them more suitable as catalysts for selected transformations, the development of antibodies with high specificity coupled with catalytic activity, and the use of enzyme-containing reactors for selective bioseparations.

In each chapter, basic information is given to familiarize the general reader with the specific topic. In each case, the author(s) have speculated on the future applications of the specific technology or process. It is hoped that the reader will learn enough from the specific examples provided in the book to consider applying biocatalysts to fit their own catalytic requirements. Finally, the gracious support of my wife, Vera, is affectionately acknowledged.

Jonathan S. Dordick
Iowa City, Iowa

Contents

Part 1. The General Uses of Biocatalysts

1. An Introduction to Industrial Biocatalysis

J. S. Dordick

2. Historical Perspective on the Industrial Uses of Biocatalysts

S. L. Neidleman

Part 2. Biocatalysts without Chemical Competition

3. Enzymatic Depolymerization of Starch

Z. L. Nikolov and P. J. Reilly

4. Biocatalysis in the Production of Carbohydrates for Food Uses

A. J. Hacking

10. Lipase Catalysis and Its Applications

V. T. John and G. Abraham

11. Biocatalysis in Supercritical Fluids

T. W. Randolph, H. W. Blanch, and D. S. Clark

I

THE GENERAL USES OF BIOCATALYSTS

1

An Introduction to Industrial Biocatalysis

JONATHAN S. DORDICK

1.1. SCOPE OF BIOCATALYSTS

Nature is extremely diverse in terms of the large number and many types of organic molecules required for life. This diversity is made possible solely due to the wide catalytic scope of enzymes. It is the ability to harness the catalytic power of enzymes and use it for the synthesis of commercially important products that represents the core technology of applied biocatalysis. This chapter provides an introduction into the field of applied biocatalysis. Specifically, the commercial use of biocatalysts in the food, medical (diagnostic), chemical, and pharmaceutical industries will be discussed along with the future potential for biocatalyst application in such areas. In addition, the advantages and disadvantages of using biocatalysts vs. chemical catalysts will be examined in order to rationalize the use of biocatalysts for processes that have traditionally been slow to utilize such a novel approach.

1.1.1. Advantages of Enzymes vs. Chemical Catalysts

Enzymatic catalysis is inherently more selective than conventional chemical synthesis.[1,2] This selectivity is often positional (i.e., regioselectivity) or chiral (i.e., stereoselectivity). While a great deal of work has been

---

JONATHAN S. DORDICK • Department of Chemical and Biochemical Engineering, University of Iowa, Iowa City, Iowa 52242.

carried out to make organic catalysts highly selective,[3] enzymes have evolved over billions of years to their high degree of perfection. High selectivity has some key benefits including reduced side reactions, easier separations, and potentially fewer negative environmental effects. In addition to the great selectivity of enzymes, other factors make these catalysts potentially useful for commercial applications, including the wide variety of reactions catalyzed, the ability to operate optimally under mild reaction conditions (i.e., ambient temperature and pressure), and the high turnover numbers found in many enzyme-catalyzed reactions that are often higher than those of conventional chemical catalysts, which typically operate at much higher temperatures.[4]

1.1.2. Classification of Enzymes

Approximately 3000 enzymes have been characterized to some degree,[5] while over 300 are available commercially from enzyme suppliers.[6] All enzymes are grouped into one of six distinct classes depending on the reactions they catalyze (Table 1). Many enzymes require cofactors to aid in the catalytic event. These low-molecular-weight, non-protein molecules are often enzyme-bound; however, many are not, and cofactor recycling is expensive. This often limits the commercial potential for cofactor-requiring enzymes.

The basic chemistries of each enzyme class are given below.[7] Note that reactions listed are those that take place *in vivo*:

1. *Oxidoreductases.* These enzymes are divided into several classes. The dehydrogenases and oxidases catalyze oxidation and reduction reac-

Table 1. Classification of Enzymes

Enzyme classification	Representative subclasses
Oxidoreductases	Oxidases, oxygenases, peroxidases, dehydrogenases
Transferases	Glycosyltransferases, methyltransferases, transaldolases, transketolases, acyltransferases, alkyltransferases, transaminases, sulfotransferases, phosphotransferases, nucleotidyltransferases
Hydrolases	Esterases, lipases, glycosidases, proteases, sulfatases, phosphatases, aminoacylases, endo- and exo-nucleases, halohydrolases
Lyases	Decarboxylases, aldolases, ketolases, hydratases, dehydratases, polysaccharide lyases, ammonia lyases
Isomerases	Racemases, epimerases, isomerases
Ligases	Synthetases, carboxylases

tions on the following functional groups: hydroxyl, aldehyde, ketone, carboxyl, amino, $-CH=CH-$ (oxidation), $-CH_2-CH_2-$ (reduction), NAD(P)(H). Peroxidases oxidize reduced compounds with H_2O_2 as oxidant, and oxygenases incorporate either one or both atoms of molecular oxygen into the oxidation product.

2. *Transferases.* Transferases catalyze the transfer of a functional group of one compound (donor) to a second molecule (acceptor). Specifically, these enzymes transfer methyl, hydroxymethyl, formyl, glycosyl, acyl, alkyl, phosphate, and sulfate groups.

3. *Hydrolases.* These are the most readily commercially obtainable and well-studied enzymes. Their primary function is to catalyze the hydrolysis of a variety of compounds including esters and lipids, thiolesters, phosphates and pyrophosphates, sulfates, glycosides, peptides and proteins.

4. *Lyases.* Lyases cleave $C-C$, $C-N$, and $C-O$ bonds by elimination to produce double bonds, or add groups to double bonds. The most common lyases are decarboxylases, polysaccharide lyases (e.g., heparinase), ammonia lyases, aldehyde lyases, oxo-acid lyases, and hydro-lyases (e.g., hydratases, fumarase, aspartase).

5. *Isomerases.* These enzymes catalyze intramolecular isomerization. Common subclasses include racemases, epimerases, intramolecular oxidoreductases, mutases, cis-trans isomerases, and intramolecular lyases.

6. *Ligases.* Ligases catalyze the linking together of two or more molecules, simultaneous with the hydrolysis of ATP. Typical bonds formed include $C-O$, $C-N$, $C-S$, and $C-C$.

In addition to the formal classification of enzymes, one may classify enzymes into practical classes in terms of the synthetic chemistries involved. For example, hydrolytic enzymes are capable not only of breaking down esters, amides, lactones, glycosides, and anhydrides, often stereo- and regioselectively,[8, 9] but of catalyzing reverse reactions such as ester and peptide syntheses in organic media.[10, 11] Lyases are highly selective hydration and dehydration catalysts, while isomerases carry out highly selective and often complex molecular alterations; of particular commercial interest is glucose isomerization to fructose catalyzed by glucose isomerase (formally known as a xylose isomerase with specificity for glucose, as well). Hydrolases, lyases, and isomerases are optimal industrial enzymes, as cofactors are generally not required.

Oxidoreductases may also be highly useful practical catalysts and can be cofactor-independent. For example, polyphenol oxidase (tyrosinase) catalyzes the oxidation of phenols[12] and, under certain conditions, phenolic hydroxylations.[13] Tyrosinase requires two copper metal ions as cofactors; however, the metals are tightly bound to the enzyme and no complex and expensive cofactors are required.[12] Generally, though, oxidoreductases require cofactors or expensive cosubstrates. Peroxidases require hydrogen peroxide, dehydrogenases use nicotinamide cofactors (NAD^+) or flavins (FAD). The practical application of enzymes will be considered in Section 1.2 of this chapter.

1.1.3. Sources of Biocatalysts

Enzymes are isolated from microbial fermentations and animal and plant by-products. Microbial enzymes from bacteria, fungi, and yeasts are by far the most common and account for about 80% of the total industrial enzyme production.[14] Enzymes from animals are highly dependent on the availability of by-products from livestock slaughter.[5] Similarly, plant enzymes are obtained as a by-product from the fruit and vegetable industries.[5] Often the animal and plant enzymes are available only as unpurified powders or suspensions, and this can limit their use, for example, as diagnostics, which require highly purified enzyme preparations. Furthermore, as animal and plant enzymes are isolated through a by-products industry, their availabilities can fluctuate widely.

Microbial enzymes do not suffer the same limitations as animal or plant enzymes. They can be produced in large fermentation processes, and the scope of isolatable enzymes is tremendous; the starting point being the vast number of microorganisms that can be grown in fermentors.[15] Most commercial enzymes are produced extracellularly. The major exceptions are glucose (xylose), isomerase, invertase, uricase, and penicillin acylase.[16,17] Therefore, enzyme isolation is simplified by the removal of cells and cell debris prior to enzyme isolation and purification.

One great problem with enzyme isolation, however, is that cells tightly regulate the synthesis of their enzymes.[18] Hence, optimal production often requires the derepression or induction of the genetic expression systems of the cell. Bacterial cells are the most amenable to this metabolic manipulation, and in most cases commercial enzymes are produced via batch fermentations under carbon or nitrogen limitation to elicit induction of the enzyme of interest.[19] Media, therefore, must be carefully controlled, and there is often a trade-off between high specific productivities and the need for high cell densities.

Enzymes that are used as commodity products must be prepared cheaply. The fermentations cannot be expensive, and fairly simple raw materials such as starches, whey, and cereal meals are used as energy sources.[20] On the other hand, specialty enzymes that are employed as diagnostics or for highly selective biotransformations often must be prepared in a high-purity manner. However, complex media can compromise the ability to produce a pure enzyme, since complex and expensive purification procedures are needed.

1.1.4. Enzymes vs. Fermentation

Enzymes are the catalysts within living cells. When, then should enzymes be used as biocatalysts in lieu of fermentation? Generally speaking, enzymes are useful if single or few reactions are required. On the other hand, if a specific biotransformation requires multiple steps, enzymes may not be suitable for the biocatalytic process, simply because of process economics. For each specific biotransformation a separate enzyme is required, thereby necessitating separate enzyme isolations, often separate operating conditions (pH, temperature, substrate concentration, solvent, etc.), and the likelihood of elaborate purifications between successive enzyme steps.

Microorganisms, on the other hand, can carry out selective synthesis starting from a cheap and abundant raw material. If a microbial pathway exists for a particular product, it is highly probable that a fermentation process can be devised to overproduce the product. A prime example is the use of microorganisms to produce antibiotics.[21, 22] These molecules are highly complex and often require dozens of synthetic steps from glucose to the final product. Microbial transformations do suffer from disadvantages not found with isolated enzymes. One of the greatest problems is the complex regulatory control of metabolism in living cells which often complicates the effectiveness of fairly simple biotransformations. Often the biotransformation of interest is a single step that lies in the middle of a complex fermentative pathway. Significant genetic manipulation of cellular control may be necessary to isolate the product in high yields.

1.2. INDUSTRIAL APPLICATIONS OF BIOCATALYSTS

1.2.1. Specific Applications

Enzymes have found numerous applications in the food, medical (diagnostic), chemical, and pharmaceutical industries.[14, 23–25]. Tables 2–4 highlight specific applications of biocatalysts in these areas. By far the greatest commercial use of biocatalysts has been in the food and beverage industry.[14, 23, 24] The reasons for the preeminence of enzymes in the food and beverage industry are largely historical. Centuries of producing foods through biological processes (e.g., cheese, yogurt, beer, wine, soy, vinegar, bread, etc.) have left an indelible mark on food processing. Furthermore, biological processing of foods and food additives eliminates the use of harsh, degradative processes and potential toxicities associated with chemical catalysis for food applications. Hence strong acids or bases, solvents, and otherwise poisonous chemical catalysts have generally been off-limits to large-scale food processing. Where chemical and enzymatic catalysis could be directly compared (e.g., starch hydrolysis, invert sugar, cellulose hydrolysis, and glucose removal), biocatalysis has maintained an overwhelming advantage.[24]

Diagnostics has also been largely a biocatalyst driven market and enzymes have a proven record of accomplishment in this area.[25] This industry requires highly selective catalysis in order to develop the highly accurate and sensitive sensors that are necessary for diagnostics. Such selectivity can be obtained only by enzymes (Table 3).

Unlike the food and diagnostics industries, the use of enzymes in the chemical and pharmaceutical industries has fallen short of expectations.[14, 26] While a number of processes use enzymes (Table 4), the scope is insignificant compared to the use of organic catalysts for synthetic reac-

Table 2. Application of Enzymes in the Food and Beverage Industry[a]

Enzyme	Source	Application	Commercial suppliers[b]
Bacterial α-amylase	*Bacillus subtilis*, *B. licheniformis*	Starch conversion, alcohol fermentation, glucose production	A, B, C, D, E, F, G, H, I
Fungal α-amylase	*Aspergillus niger*, *A. oryzae*	Maltogenic saccharification, acid-resistant saccharification	A, B, E, G, H, I, J, K
β-Amylase (diastase)	Malt, *A. oryzae*	Maltogenic saccharification, supplement to bread	B, D, L, K
Amyloglucosidase (glucoamylase)	*Rhizopus niveus*, *A. niger*, *Endomycopsis fibuliger*	Starch syrups, glucose production, animal feeds, baking, brewing	I, K

[a] Data adapted from ref. 14.

[b] A – Alltech, Inc., Kentucky;
B – Biocon, Ltd., Ireland;
C – Corning Biosystems, New York;
D – Fermco Biochemics, Illinois;
E – Gist Brocades, The Netherlands;
F – Godo Shusei Co., Japan;
G – Miles Kali—Chemie GmbH, W. Germany;
H – Novo Industries A/S, Denmark;
I – Rohm GmbH, W. Germany;
J – Miles Laboratories, Inc., Indiana;
K – John & E. Sturge, Ltd., United Kingdom;
L – A.B.M.C. Food Division, United Kingdom;
M – Hokkaido Sugar Co., Japan;
N – Chr. Hansen Laboratories A/S, Denmark;
O – Dairyland Food Laboratories, Inc., Wisconsin;
P – Amano Pharmaceuticals Co., Ltd., Japan;
Q – Toyobo Enzymes, Japan;
R – Merck Co., New Jersey;
S – Nagase Co., Japan;
T – Pfizer, Inc., New York;
U – Kingsbridge Industrial, Inc., Taiwan;
V – Biozyme Laboratories, Ltd., Belgium;
W – Boehringer Mannheim GmbH, W. Germany;
X – Calbiochem, California;
Y – Genzyme Corp., Connecticut;
Z – Sigma Chemical Co., Missouri.

Table continued

Table 2. (continued)

Enzyme	Source	Application	Commercial suppliers[b]
Pullulanase	*Klebsiella aerogenes*	Debranching of starch	L
β-Glucanase	*B. subtilis, Penicillium emersoni, A. niger*	Brewing	B, H, J, L
β-Galactosidase	*A. oryzae, A. niger, Saccharomyces lactis, Kluyveromyces fragilis*	Removal of lactose from milk	E, F, G, H, K
Cellulase	*Tichoderma viride, Penicillium funiculosum, A. niger*	Cellulose hydrolysis, wine industry	B, D, K
Invertase	*S. cerevisiae, S. diastaticus, A. niger, A. oryzae*	Sucrose hydrolysis for confectionaries, high-test molasses	
Fructosyl transferase (levansucrase, inulase)	*B. subtilis, Candida* spp.	Sweeteners	H
α-Galactosidase (melibiase)	*A. niger, S. cerevisiae, Mortierella vivacea*	Raffinose removal from sugar beets	H, M
Glucose isomerase	*Bacillus coagulans, Actinomyces missouriensis, Lactobacillus brevis,* several *Streptomyces* spp.	High-fructose corn syrups	D, E, F, G, J, H, J
Glucose oxidase	*A. niger, P. glaucum, P. notatum, Phanerochaete chrysosporium*	Removal of oxygen or glucose from foods and wine, dried egg manufacture, butter	E, G, K
Lipases and esterases	Pancreas of porcine, bovine, calf, kid, lamb, *Candida cylindracea, Rhizopus arrhizus, R. delemar, A. niger, Mucor miehei,* wheat germ	Flavoring of milk products, piccante flavor, fat modification, cocoa butter, fat emulsions	E, H, J, K, N, O, P, Q
Catalase	*A. niger, Penicillium* spp., *Micrococcus lysodeikticus,* bovine liver	Sterilization of milk, butter, often used in conjunction with glucose oxidase	J, K, O
Pectinases	*A. niger, A. ochraceus, A. oryzae, Arthrobacter globiformis*	Juice and wine industries for clarification, pectin removal	B, C, D, E, G, H, I, J, K

Table continued

Table 2. (continued)

Enzyme	Source	Application	Commercial suppliers[b]
Anthocyanase	*A. niger*	Decolorizing grapes	C
Hemicellulase	*A. niger, A. oryzae, A. saitoi, B. subtilis*	Hydrolysis of pentosans, xylans, galactans and manans; baking, fruit juice, gums	H, I, J
Animal/plant proteases	Bovine/porcine pancreas (trypsin, chymotrypsin), papaya (papain), figs (ficin), kid, lamb (chymosin), porcine stomach (pepsin)	Protein hydrolysis, brewing (haze removal), fish processing, animal feeds, meat tenderizing, baking	B, E, H, I, J, L, P, Q, V
Microbial proteases	*B. subtilis, B. amyloliquifaciens, B. licheniformis, A. flavus, A. oryzae, A. saitoi, M. miehei, Rhizopus* spp., *Streptomyces griseus, B. thermoproteolyticus*	Same as above	B, C, F, G, H, I, L
Rennets	Calf, kid, lamb (stomach), *Endothia parasitica, M. miehei*	Cheese manufacturing	E, J, H, O

tions performed today on an industrial scale.[27] The major exception is the use of penicillin acylase in the synthesis of semisynthetic penicillins. It is instructive to understand why this has been the case.

1.2.2. Actual and Perceived Problems Associated with Enzymes

Below, I list several problems associated with biocatalytic systems for chemical synthesis. Several of these problems have been overcome by novel biocatalytic techniques (some of which are described in other chapters in this book). However, some of the problems may be due to certain misperceptions of biocatalytic processes and their potential.

1. *Enzymes can only function in aqueous solutions, thereby catalyzing reactions on nonpolar compounds only in dilute solutions.* While conventional biocatalysis has held the long-standing belief that enzymes function only in water, this notion has been proved to be untenable. Enzymes can now be employed in a wide variety of nonaqueous media.[28–32] Several

Table 3. Enzymes Used as Diagnostic Agents

Analyte	Enzymes	Detection
Acetic acid and other free fatty acids	Acyl-CoA synthetase, acyl-CoA oxidase, peroxidase	Colorimetric
Citric acid	Citrate lyase, lactate dehydrogenase	Disappearance of NADH
Lactic acid	Lactate dehydrogenase, lactate oxidase, peroxidase	Formation of NADH, colorimetric
Pyruvic acid	Lactate dehydrogenase, pyruvate oxidase, peroxidase	Disappearance of NADH, colorimetric
Ascorbic acid	Ascorbate oxidase	UV
Formaldehyde	Formaldehyde dehydrogenase	Formation of NADH
Urea	Urease, glutamate dehydrogenase	Disappearance of NADPH
Uric acid	Uric oxidase, peroxidase	Colorimetric
Ethanol, methanol	Alcohol dehydrogenase, alcohol oxidase, peroxidase	Formation of NADH, colorimetric
Glycerol	Glycerol dehydrogenase, glycerol kinase, glyceraldehyde-3-phosphate dehydrogenase	Formation of NADH
Acetaldehyde	Alcohol dehydrogenase	Disappearance of NADH
Malic acid	Malate dehydrogenase	Formation of NADH
Gluconic acid	Gluconate kinase, 6-phosphogluconate dehydrogenase	Formation of NADPH
Glucose	Hexokinase, glucose-6-phosphate dehydrogenase	Formation of NADH
Fructose	Hexokinase, phosphoglucose isomerase, glucose-6-phosphate dehydrogenase	Formation of NADH
Galactose	Galactose oxidase, peroxidase	Colorimetric
Sorbitol	Sorbitol dehydrogenase	Disappearance of NADH
Lactose	β-Galactosidase, galactose oxidase, peroxidase	Colorimetric
Phenols	Peroxidase	Colorimetric
	Polyphenol oxidase	Colorimetric
Maltose	μ-Glucosidase, glucose oxidase, peroxidase	Colorimetric
Raffinose	μ-Galactosidase, galactose oxidase, peroxidase	Colorimetric
Sucrose	Invertase, glucose oxidase, peroxidase	Colorimetric
Cholesterol	Cholesterol oxidase, peroxidase	Colorimetric

Table continued

Table 3. (continued)

Analyte	Enzymes	Detection
Cholesterol esters	Cholesterol esterase, cholesterol oxidase, peroxidase	Colorimetric
Hydrogen peroxide	Peroxidase	Colorimetric
Creatine	Creatine kinase, hexokinase, glucose-6-phosphate dehydrogenase	Formation of NADPH
Creatinine	Creatinine amidohydrolase, creatine amidinohydrolase	Colorimetric
Triglycerides	Lipase, glycerol kinase, glycerol-1-phosphate dehydrogenase	Formation of NADH
L-Aspartic acid	Aspartate aminotransferase	Colorimetric
L-Alanine	Alanine aminotransferase	Colorimetric
Bile acids	3-μ-Hydroxysteroid dehydrogenase	Formation of NADH
L-Asparagine	Asparagine, aspartate aminotransferase	Colorimetric
L-Glutamate	Glutamate dehydrogenase	Formation of NADH
General L-amino acids	L-Amino acid oxidase, peroxidase	Colorimetric
General D-amino acids	D-Amino acid oxidase, peroxidase	Colorimetric
Phospholipids	Phospholipase D	Colorimetric
Ketone bodies (D-3-hydroxybutyrate and acetoacetate)	D-3-hydroxybutyrate dehydrogenase	Formation of NADH
L-Leucine	Leucine dehydrogenase	Formation of NADH
NAD(P)H	Luciferase	Generation of light
Inorganic phosphate	Purine nucleoside phosphorylase, xanthine oxidase, peroxidase	Colorimetric
ATP	Adenosine triphosphatase, pyruvate kinase, lactate dehydrogenase	Disappearance of NADH
L-Arginine	Arginase, urease, glutamate dehydrogenase	Disappearance of NADPH

chapters in this book highlight this fact. Therefore, enzymes have been employed with water-insoluble substrates in organic solvents to yield high conversions and productivities in concentrated solutions.

2. *Enzymes are too fragile for commercial chemistry.* While enzymes do inactivate under high temperatures (most denature above 65 °C),[33] there exist highly stable enzymes from thermophilic organisms that can operate stably at temperatures exceeding 80 °C.[34] Furthermore, the rates of enzyme

catalysis at ambient temperatures are high, often higher than those of chemical catalysts under similar conditions.[4] Hence, high temperatures are generally not needed for biotransformation to occur. However, in many cases higher temperatures and pressures may be necessary for highly viscous reactions, and enzymes are not too amenable to such situations.

Enzymes may also be subjected to denaturation by surfactants, some

Table 4. Application of Biocatalysts in the Chemical and Pharmaceutical Industries[a]

1. Biotransformations involving enzymes or immobilized cells		
Product	**Starting material**	**Biocatalyst (source)**
L-Dopa	L-Tyrosine	Tyrosinase (mushroom)
L-Aspartate	Fumarate	Aspartase *(E. coli)*
L-Lysine	DL-Aminocaprolactam	L-Aminocaprolactam hydrolase *(Candida humicola)*
P-Hydroxyphenylglycine	DL-5-(*p*-Hydroxyphenyl)-hydantoin	DL-Hydantoinase (*Bacillus brevis* cells)
L-Cysteine	DL-2-Aminothiazoline-carboxylic acid	*Pseudomonas thiazoline-philum* cells
L-Alanine	L-Aspartate	L-Aspartate-β-decarboxylase *(P. dacunhae)*
L-Tryptophan	Indole + L-serine	Tryptophan synthetase *(E. coli)*
L-Phenylalanine	Trans-cinnamic acid	Phenylalanine ammonia lyase *(Rhodotorula rubra)*
	L-Aspartate + phenylpyruvate	Transaminase
L-Malate	Fumarate	Fumarase *(Brevibacterium ammoniagenes)*
Acrylamide	Acrylonitrite	Nitrilase (*Corynebacterium* sp.)
Prednisolone	11-Deoxycortisone	11-β-Hydroxylase + D′-dehydrogenase *(Curalasis lunatic)* + dehydrogenase
Cholestenone	Cholesterol	Cholesterol oxidase (*Nocardia* sp.)
6-Aminopenicillanic acid (precursor to semisynthetic antibiotics)	Penicillin G or V	Penicillin amidase (*Basodiomycete* spp.)
D-(+)-Mandelonitrile	Benzaldehyde	D-Oxynitrilase (*Corynebacterium* sp.)
Propylene glycol	Propylene	Chloroperoxidase
Inosine monophosphate	Adenosine monophosphate	Adenylic acid deaminase *(A. oryzae)*

[a] Data adapted from refs. 14, 59.

Table continued

Table 4. (continued)

2. Specific processes	
Application	**Biocatalyst(s)**
Leather manufacture — dehairing — bating (removal of protein fibers) — tanning	Keratinase (alkaline proteases), ficin, bacterial proteases, tannases, pectinases, cellulases, hemicellulases
Paper manufacture — starch modification — biopulping — wood preservation	α-Amylases (fungal and bacterial), cellulases, hemicellulases, peroxidases, pectinases
Detergents	Alkaline proteases, lipases
Silver recovery for photography	Microbial proteases
Textile desizing	Bacterial ∝-amylases
Digestive aids	Trypsin, papain, ficin, bromelain, fungal proteases, pancreatic lipases
Removal of necrotic tissue	Trypsin
Removal of penicillin	Penicillin acylase (*B. subtilis*, cereus)
Phenolic polymerizations	Peroxidases
Oligosaccharides	Fructosyltransferases, dextransucrases
Drying of lacquer	Laccase (fungal)
Resolution of racemic amino acids	Aminoacylases *(A. oryzae)*
Membrane cleaning	Amicrobial proteases, cellulases, pectinases, lipases
Oil recovery—removal of proteins from xantham gum fermentation	Proteases
Peptide synthesis	Proteases
Plastein formation	Bacterial and fungal proteases
Inosinic and guanylic acid manufacture	5′-Phosphodiesterase *(Penicillium citrinum, B. subtilis)*
Resolution of organic acids via esterification in organic media	Lipases
Contact lens cleaners	Papain

organic solvents, extremes of pH, and oxidants or reductants.[35] Sometimes, however, this is advantageous when it is desirable to eliminate catalysis following the reaction. It is much easier to selectively inactivate enzymes than chemical catalysts.

3. *Enzymes are slow catalysts on a per gram turnover basis.* Enzymes are large molecules. Molecular weights range from ca. 5000 for the smallest xylanases to well over a million for cellulolytic complexes.[36] The average

molecular weight is ca. 50,000, compared to chemical catalysts with molecular weights in the hundreds of daltons. The large enzyme molecules take up space, reducing the density of active sites in a reactor. This results in a lower productivity for enzymes as compared to chemical catalysts, assuming similar turnover numbers.

In general, assuming that both an enzymatic and a chemical catalyst exist for a given transformation, chemical catalysis is indicated if speed is of primary importance. However, if selectivity at moderate to high speeds is required, then enzymatic catalysis is often a favorable alternative.

4. *Enzymes are too difficult to isolate.* Most enzymes of commercial use are extracellular and come from microbial sources. Fermentation of these organisms is typically not difficult and uses cheap starting materials. Separation of the enzymes into rather crude mixtures is often sufficient for a particular reaction.[15] The advent of recombinant DNA technology will certainly increase the yield and reduce the costs for many enzymes, especially those from animal, plant, and difficult-to-grow microbial sources.[37]

5. *Enzymes are inhibited by relatively low concentrations of end products.* In general this is the case, and catalyst poisoning is more of a problem than with chemical catalysts. However, several ingenious methods of product removal (e.g., membrane reactors),[38, 39] application of non-aqueous enzymology,[28] and novel enzymes obtained from recombinant DNA technology[37] have enabled enzymes and enzyme-based systems to drastically improve their susceptibilities to inhibitors.

6. *Enzymes are too specific for general applicability.* While the highly selective nature of enzymatic catalysis is generally desired, especially in the area of specialty and chiral chemicals and diagnostics, not all enzymes are highly selective. Many enzymes catalyze a wide range of reactions with a large number of substrates. For example, lipases and proteases are capable of hydrolyzing a wide variety of fatty acid esters and proteins,[40–42] respectively, while peroxidases catalyze the oxidation of phenols and aromatic amines among other organic compounds.[43, 44] Literally hundreds of peroxidase substrates have been identified.[45]

In addition to these factors (some of which, as discussed, are perceived problems), the unfamiliarity of enzymes to the synthetic chemist can be a major stumbling block to the general acceptance of enzymes. Mechanistic enzymology is poorly understood compared to chemical catalysis. In many cases, the scope of substrate specificities has not been identified. These factors often discourage the chemist from employing an enzyme as an alternative to a chemical catalyst for which mechanistic and kinetic information is readily available.

1.2.3. Choosing an Enzyme for a Particular Process

The use of enzymes in commercial processes does presuppose that there is a need for the advantages inherent in biocatalytic processes. Simple chemical transformations of bulk commodity chemicals that require acid–base chemistries or very cheap starting materials are generally poor candidates for enzymic processes. In these cases, selectivity is not often sought, rather high yields and productivities are desired. Assuming that a need for a biocatalyst has been ascertained, what steps should be taken to choose the proper biocatalyst(s)? It should be noted that many potential applications may require more than one enzyme, and a choice of many enzymes may exist. The following steps attempt to provide a guideline to follow for enzyme selection.

1. *Identify a chemical reaction or process that is not currently satisfactory.* Is there a need for biocatalysis? To answer this question, the levels of process productivity and reaction selectivity must be ascertained. As previously mentioned, commodity chemicals production requires high rates of synthesis. Reaction selectivity is often not applicable, as many of these chemicals are simple and do not contain discrete regio- or stereoselective functional groups. Specialty chemicals (bulk and fine) as well as biologicals, however, are typically produced with reaction selectivity as the overriding process variable. In such cases, biocatalysis is a viable option.

2. *Is there a known enzymatic reaction to the catalysis of interest?* If an enzyme exists that can carry out the desired transformation, clearly there is a way to make the desired product biologically. However, the enzyme may be expensive and not amenable to scale-up, especially if expensive cofactors are required. Substitute enzymes within the same enzyme class should be considered, as well as different reaction mechanisms that will lead to the same product. In the latter case, enzymes from different classes may be more useful for obtaining the desired product.

3. *If necessary, begin a screening and/or genetic/protein engineering program to obtain an enzyme of optimal activity and function.* This is not a trivial task. However, with the advent of protein engineering it may be feasible to "synthesize" an enzyme with highly specific characteristics and substrate specificities.

4. *Are the substrates and products compatible with enzymes?* Numerous substrate functional groups are highly reactive and can cause severe inactivation of the enzyme. This is particularly common with anhydrides, acid halides, activated aldehydes, alkyl halides, and epoxides.[46] Furthermore, many substrates and products are unstable in aqueous reaction media, and organic solvents may be required. In this case, is there a precedence for a particular enzyme of interest to function in nonaqueous

media? Several chapters in this book highlight the advantages of non-aqueous enzymology and its gaining widespread applications.

This strategy can be applied to any enzyme-catalyzed process. As an example, we shall consider the synthesis of sugar esters for use as intermediate blocking groups during oligosaccharide synthesis. Chemical synthesis lacks selectivity. For example, glucose contains five free hydroxyl groups. Nearly all react equally with anhydrides or acyl halides,[47] conventional acylating agents used in chemical synthesis. There do exist synthetic procedures for blocking and deblocking, but they are cumbersome and expensive.[48] The high selectivity of enzymatic catalysis is attractive in this case. Hence, step 1 is satisfied.

Sugar esters are found in nature[49] with uses ranging from surfactants and emulsifiers[50] to antimicrobials.[51] Clearly enzyme systems are available that will catalyze this reaction selectively. However, the reaction is not thermodynamically favorable in aqueous solution as water is a coproduct. In the cell, such reactions are typically driven by ATP hydrolysis, and this would be expensive to recycle in an enzymic system *in vitro*.[2] Perhaps, a different class of enzymes might be employed. Hydrolytic enzymes are capable of functioning in reverse in organic solvents, and lipases and esterases are known to synthesize esters in addition to their normal function of ester hydrolysis.[10, 28] Thus, step 2 is satisfied. Step 4 is satisfied if glucose and a fatty acid or a fatty acid ester are employed as the acyl donor. Neither the sugar nor the fat is harmful to enzymes. Specificity can be obtained by a screen of commercially available hydrolases. As these enzymes are readily available from a variety of sources, it is likely that a commercial source of enzyme can be found. In fact, lipases from porcine pancreas and several microbial sources have been used to synthesize sugar esters in organic solvents.[52–54]

1.3. FUTURE DEVELOPMENTS

Biocatalysis is not a stagnant industry. While the use of biocatalysts is secure in the food, beverage, and diagnostics industries, several new technologies will increase the utility of enzymes in the chemical and pharmaceutical industries. Specifically, the most significant advances in applied biocatalysis are in the area of enzymes in organic solvents, in the potential for using highly selective antibodies as catalysts, and in recombinant DNA technology for enzyme production and protein engineering. With regard to the latter, several genetically engineered organisms have already been developed. For example, the gene for rennin from calf stomach has been cloned into *Aspergillus niger*,[55] and the gene for lignin peroxidase from

Phanerochaete chrysosporium has been cloned into *E. coli.*[56,57] Protein engineering of the bacterial protease subtilisin has progressed to the point of commercialization for use in detergents,[58] and it is only a matter of time before protein-engineered enzymes with improved stabilities (both chemical and thermal) and activities are used commercially. Biocatalytic improvement obtained from recombinant DNA technology and protein engineering will enable many enzymes that have been isolated and studied to be prepared in sufficient quantities for industrial development. The following chapters highlight the wide scope of enzyme technology, its past, current applications, and projected future directions.

REFERENCES

1. G. M. Whitesides and C.-H. Wong, *Aldrichimica Acta 16*, 27 (1983).
2. C. Walsh, *Enzymatic Reaction Mechanisms*, W. H. Freeman & Co., San Francisco (1979).
3. H. S. Mosher and J. D. Morrison, *Science 221*, 1013 (1983).
4. R. W. Maatman, *Catal. Rev. 8*, 1 (1973).
5. A. J. Hacking, *Economic Aspects of Biotechnology*, p. 147, Cambridge University Press, Cambridge (1986).
6. Sigma Chemical Catalogue (1989).
7. M. Dixon and E. C. Webb, *Enzymes*, 3rd ed., Academic Press, New York (1979).
8. B. Borgstrom and H. L. Brockman, eds., *Lipases*, Elsevier, Amsterdam (1984).
9. E. M. Meijer, W. H. J. Boesten, A. E. Schoemaker, and J. A. M. Van Balken, in: *Biocatalysis in Organic Synthesis* (J. Tramper, H. C. Van der Plas, and P. Linko, eds.), Elsevier, Amsterdam (1985).
10. A. Zaks and A. M. Klibanov, *Proc. Natl. Acad. Sci. USA 82*, 3192 (1985).
11. H. Ooshima, H. Mori, and Y. Harano, *Biotechnol. Lett. 7*, 789 (1985).
12. H. S. Mason, *Adv. Enzymol. Relat. Subj. Biochem. 19*, 79 (1957).
13. R. Z. Kazandjian and A. M. Klibanov, *J. Am. Chem. Soc. 107*, 5448 (1985).
14. T. Godfrey and J. Reichelt, *Industrial Enzymology*, The Nature Press, New York (1983).
15. W. E. Goldstein, in: *Basic Biotechnology* (J. Bu'Lock and B. Kristiansen, eds.), p. 385, Academic Press, London (1987).
16. S. H. Hemmingsen, *Appl. Biochem. Biotechnol. 2*, 157 (1979).
17. M. D. Lilly, *Appl. Biochem. Biotechnol. 2*, 1 (1979).
18. J. H. Miller and W. S. Reznikoff, eds., *The Operon*, 2nd ed., Cold Spring Harbor, New York (1980).
19. N. D. Davis and W. T. Blevins, in: *Microbial Technology: Fermentation Technology* (H. J. Peppler and D. Perlman, eds.), Vol. 2, 2nd ed., p. 303, Academic Press, New York (1979).
20. F. G. Priest, *Aspects of Microbiology*, Vol. 9, American Society for Microbiology, Washington (1984).
21. W. Crueger and A. Grueger, *Biotechnology, A Textbook of Industrial Microbiology*, Sinauer Assoc. Inc., Sunderland, MA (1984).
22. J. Berdy, *Adv. Appl. Microbiol. 18*, 309 (1974).
23. A. T. Andrews, ed., *Chemical Aspects of Food Enzymes*, Royal Society of Chemistry, London (1987).

24. W. M. Fogarty, ed., *Microbial Enzymes and Biotechnology*, Applied Science, London (1983).
25. G. G. Guilbault, *Analytical Uses of Immobilized Enzymes*, Pekker, New York (1984).
26. J. B. Jones, in: *Asymmetric Synthesis* (J. D. Morrison, ed.), Academic Press, New York (1983).
27. J. D. Morrison, ed., *Asymmetric Synthesis*, Vols. 1–5, Academic Press, New York, (1983–1985).
28. J. S. Drdick, *Enzyme Microb. Technol. 11*, 194 (1989).
29. A. M. Klibanov, *Chemtech. 16*, 354 (1986).
30. L. G. Butler, *Enzyme Microb. Technol. 1*, 253 (1979).
31. M. D. Lilly, *J. Chem. Tech. Biotechnol. 32*, 162 (1982).
32. P. J. Halling, *Biotechnol. Adv. 5*, 47 (1987).
33. T. J. Ahern and A. M. Klibanov, in: *Protein Engineering* (D. L. Oxender and C. F. Fox, eds.), p. 213, Alan R. Liss, New York (1987).
34. R. M. Kelly and J. W. Deming, *Biotechnol. Prog. 4*, 47 (1988).
35. R. D. Schmid, in: *Advances in Biochemical Engineering* (T. K. Ghose, A. Fiechter, and N. Blakebrough, eds.), Vol. 12, p. 41, Springer-Verlag, Berlin (1979).
36. T. E. Barman, *Enzyme Handbook*, Springer-Verlag, New York (1985).
37. M. V. Arbige and W. H. Pitcher, *Trends Biotechnol. 7*, 330 (1989).
38. S. L. Matson and J. A. Quinn, *Ann. NY Acad. Sci. 469*, 152 (1986).
39. D. G. Rethwisch, A. Subramanian, G. Yi, and J. S. Dordick, *J. Am. Chem. Soc. 112*, 1649 (1990).
40. L. H. Posorske, *J. Am. Oil Chem. Soc. 61*, 1758 (1984).
41. A. Neuberger and K. Brocklehurst, eds., *Hydrolytic Enzymes*, Elsevier, Amsterdam (1987).
42. K. Inouye, K. Watanabe, K. Morihara, Y. Tochino, T. Kanaya, J. Emura, and S. Sakakibara, *J. Am. Chem. Soc. 101*, 751 (1979).
43. B. C. Saunders, A. G. Holmes-Siedle, and B. P. Stark, *Peroxidase*, Butterworths, Washington (1964).
44. J. S. Dordick, M. A. Marletta, and A. M. Klibanov, *Biotechnol. Bioeng. 30*, 31 (1987).
45. H. Musso, in: *Oxidative Couplity of Phenols* (W. I. Taylor and A. R. Battersby, eds.), p. 1 (1967).
46. F. Notheisz, M. Bartok, and V. Remport, *Acta Phys. Chem. 18*, 89 (1972).
47. F. Desgupta, G. W. Hay, N. A. Szarek, and W. L. Schilling, *Carbohydr. Res. 114*, 153 (1983).
48. A. H. Haines, *Adv. Carbohydr. Chem. Biochem. 39*, 13 (1981).
49. A. Zuckerberg, A. Diver, Z. Peeri, D. L. Gutnick, and E. Rosenberg, *Appl. Environ. Microbiol. 37*, 414 (1979).
50. D. L. Marshall and L. B. Bullerman, *J. Food Sci. 51*, 468 (1986).
51. A. Kato and K. Arima, *Biochem. Biophys. Res. Commun. 42*, 596 (1971).
52. M. Therisod and A. M. Klibanov, *J. Am. Chem. Soc. 108*, 5638 (1986).
53. M. Therisod and A. M. Klibanov, *J. Am. Chem. Soc. 109*, 3977 (1987).
54. J. S. Dordick, A. J. Hacking, and R. A. Khan, UK Patent Application No. 8822673 (1988).
55. D. Cullen, G. L. Gray, L. J. Wilson, K. J. Hayenga, M. H. Lamsa, M. W. Rey, S. Norton, and R. M. Berka, *Bio/Technology 5*, 369 (1987).
56. M. Tien and C.-P. D. Tu, *Nature 326*, 520 (1987).
57. E. L. F. Holzbaur and M. Tien, *Biochem. Biophys. Res. Commun. 155*, 626 (1988).
58. J. A. Wells and D. A. Estell, *Trends Biochem. Sci. 13*, 291 (1988).
59. K. Arima, in: *Global Impacts of Applied Microbiology* (M. P. Starr, ed.), pp. 278–279, Wiley (1965).

2

Historical Perspective on the Industrial Uses of Biocatalysts

S. L. NEIDLEMAN

2.1. INTRODUCTION

The purpose of this chapter is to offer a short glance back in the development of commercial applications of enzyme technology. It is not a matter of paying homage to the pioneers of biocatalysis, but rather an attempt to achieve a sense of historical perspective as an adjunct to understanding the present and anticipating the future. This goal will be approached by illustrating many examples of enzyme applications in the period 1890–1960, spiced by selected vignettes to inject humanity into this view of the past.

This is not to suggest that the use of enzymes to produce goods for human consumption began in the late 19th century. For example, the origins of biocatalysis in the dairy and brewing industries lie veiled in the mists of prehistory. Consider the amazement of the accidental inventor in pre-patent days who carried milk in the stomachs of animals and discovered cheese, or agitated milk in the heat of the midday sun and obtained butter, or the innocent taster of aged crushed fruit who grew tipsy with alcohol and elation. Not much has changed, although technology,

S. L. NEIDLEMAN • Biosource Genetics Corporation, 3333 Vaca Valley Parkway, Vacaville, CA 95688.

commercialization, and government regulations have civilized these and other similarly ancient biocatalytic processes.

In selecting early heroes of modern industrial biocatalysis—a dangerous sport—significant names are bound not to be mentioned—but no one would argue with a list that began with Takamine, Wallerstein, Röhm, Boidin and Effront. This becomes evident when early patents are considered.

2.2. EARLY PATENTS

Table 1 presents selected early patents (1894–1938) related to the use of enzymes for commercial purposes. Scrutiny of the list begins to establish the idea that a creditable amount of interest existed in the period under dis-

Table 1. Selected Early Enzyme Patents

Inventors	Patent number	Year	Enzyme	Title
J. Takamine	US 525,823	1894	Amylases	Process of making diastatic enzyme
J. Takamine	US 826,699	1906	Amylases	Diastatic substance and method of making same
O. Röhm	US 886,411	1908	Trypsin, steapsin	Preparation of hides for the manufacture of leather
J. Takamine	US 991,560	1911	Amylases	Enzyme
J. Takamine	US 991,561	1911	Amylases	Amylolytic enzyme
L. Wallerstein	US 995,820	1911	Malt protease	Beer and method of preparing same
L. Wallerstein	US 995,823	1911	Proteases	Preparation of use in brewing
L. Wallerstein	US 995,824	1911	Pepsin	Method of treating beer or ale
L. Wallerstein	US 995,825	1911	Papain	Method of treating beer or ale
L. Wallerstein	US 997,826	1911	Bromelin	Method of treating beer or ale
L. Wallerstein	US 997,873	1911	Yeast protease	Method of treating beer or ale
O. Röhm	GER 283,923	1915	Pancreatin	Process for cleaning laundry of all types
S. Frankel	US 1,129,387	1915	Amylases	Manufacture of diastase
I. Pollak	US 1,153,640	1915	Amylases	Diastase preparations and method of making same
I. Pollak	US 1,153,641	1915	Amylases	Malt extract and method of making same
A. Boidin and J. Effront	US 1,227,374	1917	Amylases	Process for treating amylaceous substances

Table continued

Table 1. (continued)

Inventors	Patent number	Year	Enzyme	Title
A. Boidin and J. Effront	US 1,227,525	1917	Amylases	Process of manufacturing diastases and toxins by oxidizing ferments
V. G. Bloede	US 1,257,307	1918	Amylase	Process of manufacturing vegetable glue
V. G. Bloede	US 1,273,571	1918	Amylase	Process of manufacturing vegetable glue
H. S. Paine and J. Hamilton	US 1,437,816	1922	Invertase	Process for preparing fondant or chocolate soft cream centers
J. Takamine	US 1,460,736	1923	Amylases, protease, lipase	Enzymic substance and process of making same
A. Boidin and J. Effront	US 1,505,534	1924	Amylases	Treatment of textile fabrics or fibers
Wallerstein Co.	UK 355,306	1931	Amylases, protease, lipase	Improvements in process of depilating hides
M. Wallerstein	US 1,854,353-5	1932	Amylases or papain	Method of making chocolate syrups
R. Douglas	US 1,858,820	1932	Amylases	Process of preparing pectin
L. Wallerstein	US 1,919,675	1933	Invertase	Invertase preparation and method of making the same
L. Wallerstein	US 2,077,447-9	1937	Proteases	Process of chillproofing and stabilizing beers and ales
L. Wallerstein	US 2,097,481	1937	Proteases	Rubber
L. Wallerstein	US 2,116,089	1938	Proteases	Deproteinization of rubber latex

cussion concerning the application of biocatalytic methods to commercial processes. There was a major involvement of the brewing industry, but the tanning, laundry, textile, candy, food, and sweetener industries were also active.

Takamine was clearly a dominant force in the genesis of industrial enzymology. One of his favored microorganisms for enzyme production was *Aspergillus oryzae*. In 1914, he had these adoring comments for the fungus[1]: "Curiously enough this tiny and important hustler scarcely attracted attention in the Occident, and this fact made we determine to work for its introduction to industrial use in the United States."

Despite his determination, Takamine's attempts to replace malt by his fungal diastase in the alcohol fermentation industry was not a success. Underkofler *et al.*[2] stated: "Use of mold preparations to replace malt in the fermentation industry was suggested by Takamine, and large-scale tests at the plant of Hiram Walker and Sons, Inc., in Canada in 1913, proved entirely successful, yields of alcohol being better than with malt. However, a slight off-flavor or odor was produced in the alcohol, and since the flavor is of paramount importance in beverage alcohol, Takamine's preparation has not found favor in the alcohol industry. Now, however, with the increasing interest in power alcohol, it would seem that a procedure similar to Takamine's should hold much promise for production of industrial alcohol."

The reason for the development of an off-flavor may be related, at least in part, to the existence of other enzymatic activities in the diastase preparation. The three recognized diastatic enzymes were identified as α-amylase, β-amylase, and amyloglucosidase.[3] However, Harada[4] reported that the diastase preparation obtained by Takamine's method was a mixture of the following enzymes: alcohol oxidase, amidase, amylase, catalase, dextranase, ereptase, esterase, glycerophosphatase, inulase, invertase, lactase, lipase, maltase, peptase, phenolase, rennet, sulfatase, and trypsin. Clearly, unless there was an intestinal or pancreatic contamination, rennet and trypsin were actually fungal enzymes with activities similar to those of the mammalian enzymes. Such an enzymatic polyglot would be likely to give an off-flavor. One such commercial product was appropriately called polyzime.[5]

An area of enzyme application with an atmosphere peculiar to itself was the tanning industry, in which animal hides were treated with the objective of producing acceptable leather for commercial use. Part of the process involved bating the hides in excrement tubs. Two types of tubs were in vogue: one redolent with dog dung, the other with a similar pigeon product. The first variety was the most prevalent. One aspect of the preparation and application of the doggy tub was described by Effront[6]: "Dog dung, generally imported from Asia Minor, is put in barrels and is sprinkled with a certain quantity of water in such a way as to moisten it and bring about a fermentation. After about a month, a certain quantity of this paste is taken, diluted in water, and soon afterward the clear liquid is decanted. It is in this maceration that the hides coming from barking and fleshing are plunged."

Eventually the patents of Röhm (US 886411) and Wallerstein (UK 355306), which employed pancreatic extracts and bacterial liquors, respectively, helped to replace this technology.

2.3. ENZYME SOURCES[7–9]

Enzymes for industrial applications were obtained from plant, animal, and microbial sources as illustrated in Table 2. Beckhorn[9] noted in 1960

Table 2. Major Sources of Pre-1960 Commercial Enzymes

Source	Enzyme
Microbial	
Yeast: *Saccharomyces cerevisiae*	Invertase
Saccharomyces fragilis	Lactase
Bacterial: *Bacillus subtilis*	Amylases, proteases
Micrococcus lysodeikticus	Catalase
Fungal: *Aspergillus flavus*	Amylases
Aspergillus niger	Glucosidases
Aspergillus oryzae	Proteases
Aspergillus niger	Catalase
Aspergillus notatum	Glucose oxidase
Aspergillus niger	Cellulase, lipase, pectinase
Plant	
Barley malt *(Hordeum vulgare)*	Amylases (diastase)
Papaya *(Carica papaya)*	Protease (papain)
Pineapple *(Ananas commosus)*	Protease (bromelin)
Animal	
Beef pancreas (*Bos* sp.)	Trypsin, erepsin, amylase
Cattle and pig gastric extracts (*Bos* sp. and *Sus scrofa*)	Pepsin, lipase
Calf stomach (*Bos* sp.)	Rennin

that the volume of enzymes produced from plant and animal sources exceeded that from microorganisms, but, he predicted that, for both technical and economic reasons, the role of microbial enzymes would become increasingly important. While this observation has been confirmed, another intriguing point is relevant here. Would the aficionados of biocatalysis have believed that, within 20–30 years, even plant and animal enzymes might be produced by microorganisms altered by the technology of molecular biology and genetic engineering?

In the 1950s, there was a technology revolution as well. Plants as enzyme sources were grown in the field; animals as enzyme sources made their contribution as by-products of the meat industry; but enzymes from microorganisms were now produced in submerged fermentations rather than surface culture, which had been prevalent until the institution of this process improvement[8].

What then were the major commercial applications for enzymes up to the early 1950s?

2.4. ENZYME APPLICATIONS[6–8, 10–16]

The use of enzymes in industrial processes of the early 1950s was already broad in scope, as indicated in Table 3. In fact, in scanning the information in the table, one might conclude that little has changed in 30–40 years. One obvious and major addition has been glucose (xylose) isomerase and its role in the production of high-fructose corn syrup. Much of the research in the past 30–40 years has concentrated on process improvement either by discovering and utilizing improved ezymes or by modifying the conditions of the enzymatic reaction and the subsequent isolation, purification, or presentation of products.

The introduction of enzymes into commercial processes has and often still does run the gauntlet of conservatism and tradition that is in the nature of many of the industries involved. Effront[6], over 70 years ago, considered this issue in the cheese and beer domains. On cheese, he commented: "The manufacture of cheeses involves a certain number of

Table 3. Industrial Applications of Enzymes

Industry	Enzyme	Source[a]	Application
Baking	Amylase	M, P	Bread and cracker baking
	Protease	M	Bread and cracker baking
Beer	Amylase	M, P	Mashing
	Protease	A, M, P	Chillproofing
	Glucose oxidase	M	Oxygen removal
Carbonated Drinks	Glucose oxidase	M	Oxygen removal
Cereals	Amylase	M, P	Precooked baby food
	Amylase	M, P	Breakfast food
Chocolate	Amylase	M	Syrup
Coffee	Pectinase	M	Bean fermentation
	Pectinase, hemicellulase	M	Concentrate
Condiments	Protease	A, M, P	Flavor ingredients
Confectionery	Invertase	M	Soft center, fondants
Dairy	Rennin	A	Cheese production
	Lipase	A	Cheese flavor
	Catalase	A, M	Milk sterilization with hydrogen peroxide
	Protease	A	Milk: off-flavor prevention
	Protease	A, M, P	Milk: protein hydrolysates
	Lactase	M	Whey concentrates, ice cream, frozen desserts, milk concentrates
	Glucose oxidase	M	Dried milk: oxygen removal

[a] A = animal, M = microbiological, P = plant.

Table continued

Table 3. (continued)

Industry	Enzyme	Source[a]	Application
Distilled beverages	Amylase	M, P	Mashing
Egg	Glucose oxidase	M	Glucose removal
Fruit juice	Pectinase	M	Clarification, filtration, concentration
Laundry, dry cleaning	Protease, lipase, amylase	A, M	Spot removal
Leather	Protease	A, M	Bating
Meat	Protease	M, P	Meat tenderizing, casing tenderizing, condensed fish solubles
Molasses	Invertase	M	High-test molasses
Paper	Amylase	M, P	Starch modification for paper coating
Pharmaceutical	Amylase	M	Digestive aids
	Protease	A, M, P	
	Lipase	A	
	Cellulase	M	
	Protease (pepsin)	A	Nutritive peptones
	Proteases	A, M, P	Wound debridement
	Proteases	A, M, P	Injection for bruises, inflammation
	Urease	P	Clinical test for urea
	Glucose oxidase plus peroxidase	M, P	Clinical test for diabetes
Photographic	Protease	M	Silver recovery
Starch	Amylase, amyloglucosidase	M	Corn syrup
	Amylase	M	Cold swelling laundry starch
Textile	Amylase	A, M, P	Fabric desizing
	Protease	A, M	Fabric desizing
Wallpaper	Amylase	M	Wallpaper removal
Wine	Pectinase	M	Pressing, clarification, filtration

practices imposed by ancient custom, which we formerly did not dare to dispense with for fear of coming to grief."

Apparently, the cheese industry did incorporate new technology and, therefore, escaped the incinerating ire of Effront. Not so the beer brewing industry.

Effront's concept of beer was as follows: "Beer must be considered primarily as a liquid food. In addition to taste and appearance, unquestionably very important factors, it must not only answer an immediate need of drinking, but must also bring to the organism the maximum nutrition. We can even say that the true excuse for its existence lies in the nourishment offered. On the other hand, beer should not be an expensive food, but a cheap one."

Effront's opinion of beer-makers was less kind: "When we study brewing a little more closely, we find an exaggerated respect on the part of the specialists in this branch for all the so-called natural products and practices which must be used in the course of this industry. The malt, the hops, and the yeast are the three fundamental constituents that characterize beer. Without these, no beer is possible. The defenders of this opinion forget that beer is not a natural beverage, like wine, but that it forms, on the contrary, an artificial product created by man, and that nothing requires us to keep to the same formula."

"The brewing academies of Germany, Austria, America, etc., bear the stamp of institutions created for the maintenance of the brewing traditions. Traditions whose exclusive basis is the holy trinity, Hops, Malt, Yeast. The conservative spirit of the present science of brewing condemns it to sterility. If, in fact, we assume that only the three elements cited above can come into play in the manufacture of beer, the door of all investigation is forever closed."

One of the early problems with beer was that it formed a proteinaceous haze upon storage. Wallerstein[17] described the situation thusly: "At the turn of the century, almost all beer in this country was consumed as draught beer and little attention was paid to its stability and chillproof qualities, since it was intended to be consumed within a short period of time after it had left the brewery. With the developments in bottling, however, beer began to be shipped in bottles instead of kegs over longer distances. As a result, the beer had to keep in good condition for considerably longer periods. While pasteurization, to which American brewers soon resorted, prevented biological breakdown of their products, the pasteurized, bottled beer, after a short time, however, developed a haze which greatly lessened its appeal to the eye. This problem became especially acute in view of the American custom of keeping beer for prolonged periods in the icebox and of serving it extremely cold."

So urgent was the demand throughout the brewing industry in the United States for a practical solution to this problem that in 1909 and again in 1910 the then U.S. Brewmasters Association offered two cash awards in a contest for the best papers on the causes of instability in finished, bottled beer. The submitted papers contained much valuable matter, with, however, a number of contradictory views. In any case, no satisfactory method of achieving a chillproof and stable beer was presented in any of these papers."

Subsequently, as indicated in Table 1, Wallerstein showed that the use of various proteases eliminated the haze problem. Effront,[6] whom we have experienced in his critical mode, was most positive about this advance made by Wallerstein: "In the course of the year 1913, the writer received from a brewery of New York two lots of beer of the same brew, one having undergone "wallerization," the other not having been treated. The analysis of these two lots, which were allowed to stand at the temperature of the laboratory, was made four months after the date of shipping. All the bottles treated had remained clear. On the other hand, the controls all contained a considerable deposit, and the liquid became turbid upon shaking and did not become clear again until after a prolonged standing. Further, the beers which had been treated were of a fresh, perfect taste; the others, on the contrary, had a flavor denoting a marked deterioration."

Early enzyme-based products were plagued by other characteristics that still skulk about to this day and, predictably, will continue to do so into the future: poor production control and unabashed adulteration.

Effront[6] notes the following with regard to the use of pepsin as a digestive aid: "In 1855, Corvisart first recommended the use of pepsin in therapeutics. Boudault, in 1857, prepared a powder with pepsin base which was much in fashion at that period. The success caused a great many limitations to appear, which soon led to a complete discrediting of this remedy. From suits entered against manufactures of pepsin in France at this time, it appears that there had been then delivered to the public products completely inactive, unclean, and dangerous."

Another protease, papain, had similar irregularities. In 1884, Martin[18] reported the following philosophical and practical considerations: "*Papain* was the name given by Wurtz to the proteolytic ferment found in the juice of the unripe fruits, of the stem and leaves of *Carica papaya*, a plant indigenous to the East and West Indies. The occurrence of such a ferment in the vegetable kingdom is in itself remarkable, and may give rise to important considerations on the researches of Darwin and others on the assimilation of animal food by canivorous plants. The leaves and juice have long been known in the native countries of the plant to have an action on uncooked meat, which is rendered tender and subsequently rotten."

Of further historical interest is that the chemist Vanquelin,[19] in 1802, described papaya juice as blood without pigment due to its considerable digestive activity.

From this colorless observation it was only a short step to produce a commercial meat tenderizer using papain. However, there were some tough issues to face, as discussed by Mansfield[20]: "People looking about for ways and means for reducing the high cost of living have overlooked papain. A round steak treated with a solution for an hour or so before cooking becomes as tender and palatable as the best tenderloin. Papain is too valuable a drug to be discredited and forced into disuse merely because of an insatiable desire for profits. During the past few years it has been a common practice to adulterate papain."

One of the common ways to adulterate papain was to incorporate it into unleavened bread or to completely eliminate the papain, selling only the bread. Mansfield[20] described the economic advantages of such skulduggery: "Rice bread costs about eight cents per pound. (No yeast or salt is used in its preparation.) The same eight-cent bread when sold as papain brings over two dollars per pound, thus netting a profit to the adulterator of about 2400 percent. Schemes for extracting gold from sea water or selling gold bricks are Christian acts compared to the getting-rich-quick, and the safe (to date) practice of selling rice bread for papain."

It is worth noting that papain had many other applications in the United States, including use in the treatment of peritoneal and other adhesions, chronic dyspepsia, gastritis, diphtheria, and a recommendation for use in eczema.[21]

It was mentioned above that not all was well with pepsin as a digestive aid. In another example concerning a mixture of enzymes in solution used as a digestive aid, Sollman[22] made the following incisive points: "If the solution is acid (as in the Elixir Digestivum Compositum of the National Formulary and most of the proprietary digestant mixtures), the trypsin and diastase will be destroyed; if it is alkaline, the pepsin and diastase will disappear, and if, as a last resort, it is made neutral, the pepsin will destroy the diastase, and the pepsin, in its turn, will be digested by the trypsin. At room temperature the process will be somewhat slower than in the thermostat, but the final result, in a very short time, will be the same."

On the other hand, it is only fair to balance the case by mentioning an enzyme product including pepsin that was both remarkable in its technical aspects for the time and apparently and mercifully efficacious. In this regard, it was reported by Nemetz[23] that favorable clinical results had been obtained with a new enzyme product marketed by "Norgine A. G. Prague" under the trade name "Enzypan." This product contained an outer shell that dissolved in the stomach, releasing gastric enzymes, while the pancreatic enzymes, trypsin, lipase, and amylase, were incorporated in the inner shell of the tablet, which released its contents into the duodenum. This represents an impressive example of any early timed-release formulation.

A few remarks about teeth and urine will serve to close this section on enzyme applications. The enzyme urease converts urea to ammonia and carbon dioxide and is a means of producing an alkaline medium. This controlled release of ammonia was the basis for a toothpaste containing urease.[13] Further, Smythe[13] pointed out that another toothpaste, as well as a cleaning solution for false teeth, contained a protease and, in principle, these are analogous to the desizing applications noted in Table 3 for the textile industry. These preparations were, then, desizers for incisors.

Smythe in discussing urease, made a suggestion that had technical merit (not really) and no aesthetic appeal (really): "There are other potential applications for this enzyme such as the interesting one of recovering water from urine under conditions of severe water deprivation, as might be imposed by desert warfare."

Even without a major market position for this last potential application, it is clear that the industrial use of enzymes in the first half of the 20th century was varied and intimately associated with many profitable processes and products.

2.5. CONCLUSIONS

In 1926, Waksman and Davison[7] wrote: "Life is just one enzyme reaction after another." On the other hand, it is equally true that industrial catalysis is not just one enzyme reaction after another. Many commercial reactions are chemically catalyzed, not enzymatically catalyzed. However, this historical chapter and the state-of-the-art chapters to follow illustrate that industrial biocatalysis has been around for a long time and will become more entrenched in the future as it gains further respect as an economically acceptable solution to many industrial problems.

This chapter might have been expanded by including examples wherein the biocatalytic agent is a live, intact cell rather than a quasi- or completely purified enzyme. The discussion would, then, have included such items of commercial interest as citric acid,[24,25] gluconic acid,[26] acetone and butanol,[27] and vinegar.[28] But it did not.

It seems appropriate that this historical chapter should give the final words to those reviewing biocatalysis in the period 1939–1960 rather than to a contemporary voice. What were their thoughts about the future?

1. "As our knowledge regarding enzymes advances, it is not too much to anticipate that in time more and more of the enzymes are likely to be put to work industrially."[11]
2. "Through the ceaseless efforts of scientists and technologists, enzymes have been successfully harnessed to the needs of industry. Progress has been furthered, not only by the knowledge of the behavior of enzymes, which research has expanded, but by the development of enzyme-containing preparations designed to meet the requirements of specific processes. The increasing complexity of modern industry, it is believed, will bring about a corresponding increase in new and interesting uses for this fascinating group of organic catalysts."[12]
3. "Many industrial problems have been solved by the proper application of enzyme chemistry; others await improvements or solution. Indeed the fields for exploration offered by catalytic microorganisms and practical enzymology are almost unlimited."[10]

4. "Finally, it must be pointed out that the synthesis of our principal foods—sugars (starches) and amino acids (proteins)—is carried out enzymatically by plants and animals. The simple building stones for many of these processes are sometimes available, and a few of the enzymes involved in these syntheses have been studied.

   "It is not impossible to visualize an era in which the enzymatic synthesis of our common foods could be undertaken."[29]
5. "Currently much enzyme research is underway by various industries including enzyme manufacturers. Such research is devoted to finding new and improved methods for using enzymes, to improving yield of industrial microbial enzymes, and to finding new enzymes for industrial purposes. Continually increasing usage of old and new enzymes will result from such research."[8]
6. "It is generally conceded that we have only begun to tap the full potential of enzymes. The extensive research now in progress will bear fruit, making new systems and new purified forms available. Cooperative research with industrial users of enzymes should continuously uncover new areas in which enzymes can serve a useful role."[9]
7. "The recent growth in the industrial and medical applications of enzymes has resulted from greatly improved methods for their production, especially from plant and microbial sources. In no small degree it has also evolved as a result of our deeper understanding of the chemical and physical properties of enzymes and of the kinetics of the total enzyme–substrate system, with special reference to the role of activators and inhibitors. This fundamental knowledge of enzyme systems will inevitably lead to major growth in the industrial and medical applications of enzymes in the near future."[15]

These statements have an unmistakable similarity to contemporary pronouncements.

REFERENCES

1. J. Takamine, *J. Ind. Eng. Chem. 6*, 824 (1914).
2. L. A. Underkofler, E. I. Fulmer, and L. Schoene, *Ind. Eng. Chem. 31*, 734 (1939).
3. D. P. Langlois, *Food Technol. 7*, 303 (1953).
4. T. Harada, *Ind. Eng. Chem. 23*, 1424 (1931).
5. H. F. Smyth and W. L. Obold, *Industrial Microbiology: The Utilization of Bacteria, Yeasts and Molds in Industrial Processes*, Ch. 32, p. 256, Williams & Wilkins Co., Baltimore (1930).

6. J. Effront, *Biochemical Catalysis in Life and Industry*, Trans. by S. C. Prescott, John Wiley and Sons, Inc., New York (1917).
7. S. A. Waksman and W. C. Davison, *Enzymes*, Williams and Wilkins Co., Baltimore (1926).
8. L. A. Underkofler, R. R. Barton, and S. S. Fennert, *Appl. Microbiol. 6*, 212 (1958).
9. E. J. Beckhorn, *Wallerstein Lab. Commun. 23*, 201 (1960).
10. H. Tauber, *Enzyme Technology*, John Wiley and Sons, New York (1943).
11. L. Wallerstein, *Ind. Eng. Chem. 31*, 1218 (1939).
12. R. A. Gale, *Wallerstein Lab. Commun. 4*, 112 (1941).
13. C. V. Smythe, *Econ. Bot. 5*, 126 (1951).
14. E. J. Beckhorn, M. D. Labbel, and L. A. Underkofler, *J. Agr. Food Chem. 13*, 30 (1965).
15. I. W. Sizer, *Adv. Appl. Microbiol. 6*, 207 (1964).
16. L. A. Underkofler, *Bioeng. Food Process. 62*, 11 (1966).
17. L. Wallerstein, *Wallerstein Lab. Commun. 19*, (1956).
18. S. H. C. Martin, *J. Physiol. (London) 5*, 213 (1884).
19. L. N. Vanquelin, *Ann. Chim. Phys. 18*, 267 (1802).
20. W. Mansfield, *J. Am. Pharm. Assoc. 3*, 169 (1914).
21. K. Kwang and A. C. Ivy, *Ann. N. Y. Acad. Sci. 54*, 161 (1951).
22. T. Sollman, *J. Am. Med. Assoc. 48*, 415 (1907).
23. C. Nemetz, *Dtsch. Med. Wochenschr.*, 1047 (June 22, 1928).
24. C. Wehmer, US Patent 515,033 (1894).
25. J. N. Currie, *J. Biol. Chem. 31*, 15 (1917).
26. P. A. Wells, A. J. Moyer, J. J. Stubbs, H. T. Herrick, and O. E. May, *Ind. Eng. Chem. 29*, 653 (1937).
27. C. Weizmann, US Patent 1,315,585 (1919).
28. R. J. Allgeier and F. M. Hildebrandt, *Adv. Appl. Microbiol. 2*, 163 (1960).
29. G. Reed, *Food Eng. 24*, 105 (1952).

II

BIOCATALYSTS WITHOUT CHEMICAL COMPETITION

3

Enzymatic Depolymerization of Starch

ZIVKO L. NIKOLOV and PETER J. REILLY

3.1. INTRODUCTION

The amylases, a large family consisting of α-amylase, β-amylase, glycoamylase, and a number of lesser-known enzymes, are united in their preferential hydrolysis of the α-(1,4) glucosidic bond in amylose, amylopectin, glycogen, and maltooligosaccharides. They have been used and studied for most of this century, and are among the most commercially important of all enzyme families. They act in two different modes. α-Amylase is an endo-hydrolase, and has more affinity for interior bonds than those located near the ends of substrates composed primarily of α-(1,4) glucosidic bonds. Because the location of the bond it cleaves is not fixed, a mixture of products results. All other amylases are exo-hydrolases, in that they attack bonds at a specific distance from the nonreducing end of the substrate. Glucoamylase, for instance, attacks the bond immediately adjacent to the nonreducing end, and therefore produces glucose along with a molecule one glucosyl residue shorter than the original substrate. β-Amylase attacks the second bond from the end to produce maltose. The single small products of other exo-amylases are likewise determined by the specific bonds they break.

ZIVKO L. NIKOLOV • Department of Food Technology, Iowa State University, Ames, Iowa 50011. PETER J. REILLY • Department of Chemical Engineering, Iowa State University, Ames, Iowa 50011.

These differing specificities in products formed are caused by differences in subsite structure. The active sites of all amylases are composed of multiple subsites, each of which holds a glucosyl residue. With α-amylase and other endo-hydrolases, the substrate can position itself in many different configurations in the active site, and therefore cleavage of different bonds can occur. With exo-hydrolases such as β-amylase and glucoamylase, on the other hand, the substrate is bound in only one productive configuration because of the physical and chemical characteristics of the active site, and therefore only one product is possible.

Despite important differences in their action patterns, endo- and exo-amylases are often considered together, because they attack the same substrates and are sometimes used in the same industrial processes. We will cover both in this chapter, dealing first with α-amylase, then with glucoamylase and β-amylase, and finally with other exo-amylases that are now becoming better known. There have been many reviews of the various amylases, either together or separately. However, the recent rapid increase in the knowledge of enzyme structure and function, obtained by traditional as well as by new methods entailing use of recombinant DNA techniques, makes a new treatment of this very important family appropriate.

3.2. α-AMYLASE

α-Amylase [α-(1,4)-glucan glucanohydrolase, EC 3.2.1.1] is widely distributed in nature, being produced by members of many animal, plant, and protist genera. It is a typical endo-hydrolase that produces oligosaccharides of α-configuration by hydrolysis of α-(1,4) glucosidic bonds in amylose, amylopectin, and glycogen. Although α-(1,6) glucosidic linkages are not attacked, α-amylase can bypass them to yield smaller branched fragments. Besides maintaining the configuration of the bond broken in the new anomeric hydroxyl, α-amylase hydrolyzes interior glucosidic linkages of oligo- and polysaccharides in preference to those at the end of the molecule and is more active on long chains than on short ones.[1]

3.2.1. Physicochemical Properties

Physicochemical properties and mechanism of action of α-amylases depend to some extent on the source of the enzyme. Although mammalian and grain α-amylases have been studied extensively, it is the bacterial α-amylases that have gained commercial significance. The genus *Bacillus* is the single most important bacterial source of this enzyme. Two types of α-amylases from *Bacillus* species, liquefying and saccharifying amylases, are distinguished based on their action pattern on starch. This chapter will focus on liquefying *Bacillus* α-amylases and will review recently described microbial amylases that possess useful or unusual properties from an industrial viewpoint. For further information on other microbial α-amylases the reader is referred to an extensive review by Fogarty.[2]

Most α-amylases have molecular weights of about 50,000,[2] although values of 22,500 for the *Bacillus licheniformis* enzyme[3,4] and 68,000 for that from *Bacillus acidocaldarius*[5] have been reported. Variations in molecular weights are in part due to different methods of determination as well as to chemical pretreatment. More precise values of molecular weights were recently obtained from the nucleotide sequence of the genes encoding α-amylases from *Bacillus amyloliquefaciens*,[6] *Bacillus stearothermophilus*,[7] and *B. licheniformis*[8]; these molecular weight estimates were 54,778, 58,779, and 55,200, respectively.

α-Amylases are calcium metallo-enzymes having at least one atom of calcium per molecule of enzyme.[9] The binding strength of the metal to the protein depends on the source of the enzyme. The presence of calcium is required for both enzyme activity and stability.[10] For example, the removal of the calcium ion from *B. amyloliquefaciens* α-amylase with ethylenediaminetetraacetate (EDTA) is accompanied by a loss of activity, which can be substantially restored by addition of the metal.[9] Approximately four calcium ions per mole of enzyme are required for highest activity and stability.[11,12] As the *B. amyloliquefaciens* α-amylase lacks intramolecular disulfide bridges, unlike other α-amylases, it has been suggested that calcium ions form intramolecular linkages instead as a means of stabilization.[11]

The pH optima of α-amylases are usually between 4.8 and 6.5, but some extreme values such as pH 3.5 for *B. acidocaldarius* α-amylase[5] and pH 9.2 for *Bacillus* sp. NRRL B-3881 α-amylase[13] have been reported. The exact pH optima and shape of the pH-activity profiles vary with the source of the enzyme.[2] α-Amylases are generally stable in a wider pH range (5.5–9.0) in the presence of calcium ion and substrate. The temperature optima of α-amylases also differ with enzyme source.[2] Most purified enzymes lose activity rapidly above 50 °C, but this inactivation can be reduced by calcium and substrate. The most thermostable of all *Bacillus* α-amylases is that from *B. licheniformis*.[3] This enzyme has a temperature optimum of 76 °C at low substrate concentration and a very broad spectrum of activity between pH 5 and 9. With high substrate concentrations (30–35% starch) it has an optimum of 90 °C and can be used at temperatures over 100 °C.[14] Both *B. licheniformis* and *B. amyloliquefaciens* α-amylases are used commercially in the starch liquefaction process, with the former requiring much lower levels of calcium for stability.[4,14] Recently an α-amylase from *B. stearothermophilus* was introduced commercially for use in starch liquefaction.[15]

3.2.2. Action Pattern

The α-amylase from *B. amyloliquefaciens* is the most studied α-amylase from any *Bacillus* strain. This enzyme degrades soluble starch, amylose, amylopectin, and glycogen.[16,17] Native corn starch granules are slowly attacked, while those of potato starch are not touched at all.[18] Amylose and amylopectin hydrolysis initially gives mostly maltose, maltotriose, maltohexaose, and maltoheptaose, with the latter being hydrolyzed slowly to glucose and maltohexaose. With glycogen, the major initial products are maltohexaose and maltoheptaose; later, glucose, maltose, and maltohexaose are found in highest amounts.[16,17] The major end product of prolonged hydrolysis of starch is maltohexaose. Although endo-hydrolases do not have extensive transfer and synthetic capabilities, this enzyme is not completely devoid of them. Greenwood *et al.*[19] found that the enzyme can convert labeled glucose or maltose plus maltohexaose or amylose to labeled products larger than the starting material. Hehre *et al.*[20] observed the polymerization of α-D-glucopyranosyl fluoride to oligosaccharides as large as maltopentaose.

The hydrolysis of starch with α-amylase from *B. licheniformis* initially gives higher-molecular-weight oligosaccharides (G_5–G_9), followed by later production of maltopentaose, maltotriose, maltose, glucose, and maltotetraose, in descending order.[3,21] Maltopentaose, maltotriose, and maltose are the main products following prolonged action on starch.[22]

The *B. stearothermophilus* α-amylase produces equal amounts of maltose, maltotriose, maltotetraose, and maltopentaose from starch,[23] while *Bacillus coagulans* α-amylase gives unidentified oligosaccharides, maltotetraose, maltotriose, maltose, and glucose.[24]

B. acidocaldarius α-amylase is able to hydrolyze starch, amylose, amylopectin, and glycogen, but not maltose.[5] The enzyme did not bind insoluble starch and glycogen under the experimental conditions tested. Starch hydrolysis initially gives maltotetraose as the major product, with smaller amounts of maltopentaose, maltose, maltotriose, and glucose. Maltotriose, maltose, and glucose accumulated after one hour of hydrolysis.

Bacillus subtilis var. *amylosaccharicus*, *B. subtilis* Marburg, and *Bacillus natto* are producers of saccharifying α-amylases.[25,26] These give much higher levels of reducing sugars from starch than the liquefying enzymes. For example, the *B. subtilis* saccharifying α-amylase can produce syrups of DE (dextrose equivalent) 70 with roughly 50% (w/w) glucose on a dry basis, while the *B. amyloliquefaciens* and *B. licheniformis* liquefying α-amylase achieve DE 30–35 with 10% (w/w) glucose.[1] In general, saccharifying α-amylases from *Bacillus* are less stable than liquefying α-amylases and therefore they are not widely used commercially. More on the action pattern and the specificity of these two types of α-amylases can be found in reviews by Reilly[1] and Fogarty.[2]

Another group of microbial saccharifying α-amylases that have commercial significance are fungal α-amylases. Although many fungi produce α-amylase, it is the *Aspergillus oryzae* α-amylase that has been studied the most and is currently used commercially. The enzyme is an endo-hydrolase that attacks α-(1,4) but not α-(1,6) linkages. This enzyme produces up to 60% maltose on a dry basis from starch and is used commercially in the production of maltose and high-conversion syrups. Properties, products, and mode of action of fungal amylases have been a subject of recent reviews.[1,2]

The previously discussed differences between the saccharifying α-amylase produced by *B. subtilis* and liquefying amylases from *B. amyloliquefaciens*, *B. stearothermophilus*, and *B. licheniformis* can be explained by considering the homology of their amino acid sequences. The amino acid sequence of *B. stearothermophilus* α-amylase shows about 64% homology with *B. licheniformis* α-amylase[8,27] and 61% with *B. amyloliquefaciens* α-amylase.[7] The α-amylases from *B. licheniformis* and *B. amyloliquefaciens* closely resemble each other, with 80% total homology.[8] In contrast, the sequence of the liquefying α-amylase from *B. stearothermophilus* is not homologous with that of the saccharifying α-amylase from *B. subtilis.*[7] Therefore, the liquefying amylases belong to the same enzyme family, which probably includes *B. coagulans* α-amylase but not the saccharifying α-amylases of *B. subtilis.*[27]

3.2.3. Novel α-Amylases

A large number of microbial α-amylases have been extensively reviewed by Fogarty.[2] The discussion to follow will focus on endo-acting α-amylases with different properties, substrate specificity and mode of action than those already described. Exo-amylases will be discussed in a separate section of this chapter.

Among novel α-amylases reported since 1980 are maltose- and maltotriose-producing amylases that are referred to as maltogenic or maltotriogenic α-amylases. Hidaka and Adachi[28] reported an endo-acting α-amylase from *Streptomyces hygroscopicus* that produces unusually high levels of maltose (>75% on a dry basis) from potato and corn starch. When incubated with 20% potato starch at pH 5.5 and 60 °C, this enzyme initially produced maltotetraose and maltotriose and small amounts of maltose. In later stages of hydrolysis, maltotetraose was hydrolyzed directly to maltose, and maltotriose was converted to maltose via transglycosylation and/or condensation. The *S. hygroscopicus* α-amylase does not hydrolyze α-(1,6) linkages and it is not active on pullulan. The enzyme has a pH optimum between 5 and 6, a temperature optimum of 55–60 °C, and a molecular weight of 48,000. Calcium ions increase enzyme stability. The enzyme produced 75% maltose, 11% glucose, 6% maltotriose, and 8.5% higher oligosaccharides from corn starch upon prolonged hydrolysis.

A pullulan hydrolase from *B. stearothermophilus* was purified to homogeneity and characterized as a unique type of maltogenic α-amylase.[29] This α-amylase is a dimer with a molecular weight of 115,000 and maximal activity at 55 °C and pH 5.8. The enzyme hydrolyzes pullulan, amylopectin, α- and β-limit dextrins, α- and β-cyclodextrins, phenyl-α-D-maltoside, and maltooligosaccharides. The main product from amylose and pullulan is maltose, with trace amounts of glucose and maltotriose from amylose and panose from pullulan. Amylose is a far better substrate than pullulan and amylopectin; glycogen is not attacked by the enzyme. Maltotriose is hydrolyzed to glucose and maltose; maltotriose and maltose are the main products from a maltopentaose digest. The *B. stearothermophilus* α-amylase does not have an antigenic group in common, and therefore does not cross-react, with either *Aerobacter aerogenes* pullulanase or *B. subtilis* α-amylase.

A new *B. subtilis* strain isolated from soil produces an extracellular α-amylase capable of hydrolyzing starch to maltotriose.[30] The enzyme was purified to almost complete homogeneity and had a molecular weight of about 25,000. The pH and temperature optima are between 6 and 7 and around 50 °C, respectively. This maltotriogenic α-amylase produced approximately 56% maltotriose from soluble starch, 56% from amylopectin, 69% from DP 17 amylose, and 40% from glycogen, but it did not attack pullulan or α- and β-cyclodextrins. Hydrolysis of starch led to maltotriose production at an early stage of the reaction, indicating that maltotriose was the main product. The typical end-product distribution with liquefied starch as a substrate was 2.8% glucose, 12.1% maltose, 55.3% maltotriose, 4.4% maltotetraose, and 25.4% other products. Because there are appreciable amounts of glucose and maltose in the final product and because the enzyme does not hydrolyze maltotriose, it appears that the *B. subtilis* maltotriogenic α-amylase is an endo-hydrolase.

Suzuki *et al.*[31] have recently characterized a similar endo-acting maltotriogenic α-amylase from *Bacillus thermoamyloliquefaciens*. The molecular weight for this enzyme was estimated as 78,000 by SDS gel electrophoresis. The enzyme has highest activity at 70 °C and pH 5.6 and requires calcium for stability. When incubated for 24 h with maltose, isomaltose, maltotriitol, and panose, no product formation was observed by paper chromatography. The *B. thermoamyloliquefaciens* α-amylase continuously produced maltotriose as its major product from glycogen, soluble starch, amylopectin, α- and β-limit dextrins, and amylose through an incubation of 3 h. Maltohexaose accumulated during the first 10 min of reaction, followed by maltopentaose and maltotetraose after 40 min. Trace amounts of glucose and maltose were detected after 2 h of incubation. The authors concluded that their α-amylase and the previously described maltotriogenic α-amylase from *B. subtilis*[30] have very similar modes of action, the important difference being the higher thermostability of their enzyme.

A novel pullulanase–amylase enzyme complex, which hydrolyzes pullulan to maltotriose and soluble starch to maltose and maltotriose, was found in the culture filtrate of a *B. subtilis* strain.[32] Because the purified enzyme has an extremely large molecular weight (450,000), higher than those of pullulanases and amylases, and because the two activities could not be separated, Takasaki[32] considered this enzyme as a pullulanase–amylase complex. The pH optimum of the pullulanase activity is pH 7.0–7.5, while the amylase activity is maximal between pH 6 and 7. The optimal temperatures of pullulanase and amylase activities were observed at about 60 °C and 50 °C, respectively. Calcium ions protected both enzyme activities. The enzyme complex produces maltose from maltotriose without appreciable formation of glucose. During the saccharification of soluble potato starch this enzyme initially produced much more maltotriose than maltose. Sugar syrups obtained from DE 4.2 liquefied starch contained 50–55 % maltotriose, 20–40 % maltose, and 2–5 % glucose, depending on the enzyme concentration.

A raw starch-binding α-amylase from *Clostridium butyricum* had maximal activity at pH 5.0 and 60 °C and stability up to 40 °C on native starch granules in the presence of calcium.[33] The molecular weight of the enzyme estimated by SDS disc-gel electrophoresis was 89,000. This α-amylase produces various maltooligosaccharides from maltose through maltopentaose using 1 % soluble starch as substrate. Maltotriose is hydrolyzed to glucose and maltose in equimolar amounts. The *C. butyricum* α-amylase could easily digest cereal starch granules in 48 h, but only 15 % of potato starch was hydrolyzed during 72 h of incubation.

3.2.4. Molecular Cloning and Expression of the α-Amylase Gene

Cloning and expression of α-amylase have been intensively studied in the last ten years. Interest has been focused on the mode of secretion, regulation of synthesis, protein structure, and expression in new hosts. The latter, to increase production and/or recovery as well as to provide the host with an amylolytic enzyme system, is of highest importance for the industry. The amylase genes of *B. amyloliquefaciens*,[6,34] *B. licheniformis*,[8,35–37] *B. stearothermophilus*,[7,38,39] *B. coagulans*,[40] and *B. subtilis*[41,42] have been cloned and expressed in *B. subtilis* and/or *Escherichia coli.* α-Amylases excreted by transformed *B. subtilis* have biochemical properties indistinguishable from or very similar to those of the native enzymes. Although α-amylase genes from *B. stearothermophilus*[39,43,44] and *B. licheniformis*[45] have been efficiently produced in other bacterial hosts, *B. subtilis* is still the preferred choice for the production of industrially important enzymes.

The α-amylase gene from *S. hygroscopicus* has been recently cloned in *E. coli* and characterized.[46] The amino acid sequence of the structural gene revealed several conserved regions found in other α-amylases. The cloned α-amylase gene coded for a protein of 478 amino acids.

One of the important achievements with rDNA from the industrial perspective is the cloning and successful expression of an α-amylase gene in *Saccharomyces cerevisiae.*[47–49] Manipulating *S. cerevisiae* to synthesize and secrete a functional α-amylase represents a significant step toward producing yeast that can utilize starchy materials and convert them directly to ethanol or single-cell protein.

3.3. GLUCOAMYLASE

Glucoamylase [α-(1,4)-glucan glucohydrolase, EC 3.2.1.3], despite the similarity of both its trivial and its systematic names to those of α-amylase, is a very different enzyme. Glucoamylase is an exo-hydrolase that cleaves primarily α-(1,4) bonds at the nonreducing ends of amylose and amylopectin molecules to produce β-D-glucose. Unlike α-amylase, this enzyme inverts the configuration of the broken bond and can attack α-(1,6), α-(1,3), α-(1,2), and α,β-(1,1) glucosidic linkages, although at much lower rates than that on α-(1,4) bonds.[50–53] Glucoamylase, which was discovered in the early 1950s, has gained widespread use in the starch bioprocessing and brewing industries because of its specificity and very high levels of glucose production.

Glucoamylase is produced by a large number of molds and yeasts,[54] but only those from *Aspergillus* and *Rhizopus* are of commercial importance. Therefore, the enzymes from these two sources will be discussed here and compared to recently discovered glucoamylases from other sources that could be of commercial significance.

3.3.1. Physicochemical Properties

The molecular weights for glucoamylases from *Aspergillus* and *Rhizopus* species range from 48,000 to 112,000,[55] although Flor and

Hayashida[56] reported a value of 250,000 for a glucoamylase from a mutant of *A. awamori* var. *kawachi.* Again, as in the case of α-amylases, great variations within species are probably due to different production and purification procedures as well as to methods of molecular weight determination. The fact that glucoamylases are produced in multiple forms also contributes to the wide range of the observed molecular weight values. Most strains of *A. niger* produce two major glucoamylase forms,[57–61] although some strains produce only one form of the enzyme[62, 63] and others more than two.[64, 65] Among other aspergilli two glucoamylase forms have been isolated from *A. awamori* var. *kawachi,*[66] *A. saitoi,*[67] and *A. foetidus*[68] and three[69] and four forms[70] from *A. oryzae.* Various numbers of forms are also produced by different *Rhizopus* strains.[71–74] Isolation of different numbers of glucoamylase forms from the same strain is probably a result of different production and purification conditions. Hayashida and co-workers[75–78] suggested that multiple forms were obtained by stepwise degradation of the native glucoamylase molecule with proteases and glucosidases during the course of the fermentation and recovery of the enzyme. This hypothesis was supported in the case of glucoamylase from *A. awamori* var. *kawachi,* as Yoshino and Hayashida[77] produced *in vitro* a smaller glucoamylase II from a larger glucoamylase I by exposing the latter to protease and glucosidase activities. A protease-negative, glucosidase-negative mutant of *A. awamori* var. *kawachi* produced only one form.[78] A similar observation was made by Takahashi *et al.*[79] with *Rhizopus* sp. glucoamylase. Recent cloning of glucoamylase genes from *A. niger,*[80] *A. awamori,*[81] and *Rhizopus* sp.[82] demonstrated that there is a single glucoamylase gene. Post-translational modification and differential mRNA splicing were suggested as possible mechanisms for generation of the multiple forms.[80, 81]

The molecular weight of *A. niger* glucoamylase has been estimated from sequencing[83] and cloning[80] studies. Mature glucoamylase I consists of 616 amino acid residues corresponding to a molecular weight for the polypeptide moiety of 65,424. By including the carbohydrate content Svensson *et al.*[83] calculated the molecular weight of the glycosylated form to be about 82,000. In the case of a *Rhizopus oryzae* glucoamylase, the amino acid sequence was predicted from the nucleotide sequence of the *Rhizopus* glucoamylase gene.[82] The glucoamylase consisted of 604 amino acids, including a putative signal peptide, and its molecular weight was calculated as 65,000. Therefore, glucoamylases from these two sources are similar in size based on the polypeptide chain length. Irrespective of the number and origin of glucoamylase forms, their properties such as pH and temperature optima, isoelectric point, and extent of glycosylation are very similar. Most commercial preparations of glucoamylase are mixtures of different forms.

Fungal glucoamylases have a pH optimum for activity between 4 and

5, and are generally stable between pH 3 and 8.[2, 55] The wide pH range for stability comes from the different criteria used in the determination. Temperature optima range from 40 to 60 °C. *Rhizopus* enzymes generally require a lower temperature (40 °C) for optimal activity,[69, 84] while *Aspergillus* enzymes show optimal activity at temperatures between 50 and 60 °C. Glucoamylases from both sources are quite stable up to 60 °C but are rapidly inactivated at higher temperatures.[57, 58] There are conflicting reports on the stability of *Aspergillus* glucoamylase forms. Lineback *et al.*[58] reported that the two enzyme forms had the same heat stability, while others[57, 62, 64, 85] found differences, especially above 40 °C.

Tanaka *et al.*[86] compared the amino acid sequences of *R. oryzae* and *A. niger* glucoamylases and found an overall homology of 44 %. There were four highly homologous regions where the degree of conservation ranged from 45 to 74 %. Immunochemical analysis of glucoamylases from three *A. niger* strains and *A. foetidus* showed that the enzyme forms are immunologically indistinguishable.[58, 87, 88] On the other hand, *Rhizopus niveus* glucoamylase did not cross-react with antibodies for *A. niger* glucoamylases.[88]

Glucoamylases are generally glycoproteins with a carbohydrate content varying between 5 and 20 %, depending on the source.[2] The carbohydrate moieties of *Aspergillus* enzymes contain mainly D-glucose, D-mannose, and D-galactose and in some cases D-glucosamine and D-xylose, while the enzymes from *Rhizopus* have only D-mannose and D-glucosamine.[55] The carbohydrate units of *A. niger* glucoamylase are single or oligosaccharide chains, comprising 15–20 % of the total molecule, attached *O*-glycosidically to L-serine and L-threonine residues of the protein.[87, 89–91] Pazur and co-workers[91, 92] presented evidence for a random distribution of the carbohydrate chains along the polypeptide chain that could explain the unusual thermal and storage stability of the enzyme and its high resistance to proteolysis. This finding was contradicted by the reports by Svensson *et al.*,[93, 94] who claimed that more than 80 % of the carbohydrate chains were located in a polypeptide segment composed of 70 amino acids. The most comprehensive study on carbohydrate side-chain structure in a *Rhizopus* glucoamylase was performed by Watanabe and Fukimbara[95, 96] and by Watanabe.[97] The carbohydrate sequence of the *R. javanicus* glucoamylase revealed large branched structures of D-mannose residues linked to each other by different α-glycosidic bonds. The branched aggregates were attached through D-acetylglucosamine to L-asparagine, L-glycine, or L-threonine. In addition, Watanabe[97] found 3-*O*-α-D-mannosyl-D-mannose fragments linked to L-threonine and L-serine residues of the same enzyme. The function of carbohydrate residues of glycoproteins at the molecular level is still a rather intriguing area. The suggestion by Pazur and co-workers[91, 92] that the carbohydrate chains stabilize the enzyme structure and therefore its activity seems to be the most plausible explanation of their role.

3.3.2. Action Pattern

The action pattern of glucoamylase with regard to the end-product distribution is straightforward: glucose is released as the only product from glucoamylase action on various substrates ranging from α-glucosyl disaccharides to starch and glycogen. As its systematic name suggests, glucoamylase primarily attacks nonreducing-end α-(1,4) glucosidic bonds of starch and maltooligosaccharides. This enzyme is also capable of hydrolyzing α-(1,6) glucosidic bonds in starch dextrins, although at a much slower rate. Nevertheless, that property of glucoamylases is an important factor in the industrial production of glucose from starch.

The rate of enzymatic hydrolysis is affected by the chain length of the substrate; both maximum rate and Michaelis constant (K_M) are affected.[52, 98, 99] Hiromi and co-workers[98, 99] were the first to examine and explain this phenomenon. They postulated that the glucoamylase active site contained a number of subsites capable of accommodating and holding the α-(1,4)-linked glucosyl residues of maltooligosaccharides. Longer substrates are bound by more subsites, and consequently they have lower K_M's. In addition, longer substrates have higher probabilities of formation of productive complexes between the enzyme and substrate. The subsite theory proposed by Hiromi and co-workers[98] has been recently confirmed and applied to α-(1,6)-linked glucosyl residues of isomaltooligosaccharides.[100] The rate constant for maltooligosaccharide hydrolysis increases through maltotetraose and then stays essentially constant, while K_M decreases monotonically with increase of substrate chain length up to maltoheptaose. With isomaltooligosaccharides a similar trend was observed: the rate constant increases through isomaltopentaose and K_M decreases through isomaltohexaose. It should be noted that the binding and subsequent hydrolysis of isomaltooligosaccharides is less preferred than that of maltooligosaccharides, as evidenced by the consistently much lower values of the rate constant and much higher K_M values for substrates of the same chain length. The calculated subsite binding energies suggest that α-(1,6)-linked glucosyl residues have a different orientation in the active site that prevents the firm binding of isomaltooligosaccharides. Both forms of *A. niger* glucoamylase have the same subsite map.[100]

An important property of glucoamylases relevant to industrial starch processing is their ability to synthesize by condensation various α-linked di- and trisaccharides, commonly called reversion products. The formation of these products during the starch conversion process reduces glucose yield to approximately 95% of the theoretical. It has been known for years that glucoamylase has relatively low specificity and that it can hydrolyze and synthesize α-(1,4), α-(1,6), and α-(1,3) glucosidic bonds.[74, 101–104] Recently, Nikolov *et al.*[105] identified two new condensation products, α,β-trehalose (α-D-glucopyranosyl-β-D-glucopyranoside) and kojibiose (2-*O*-α-D-glucopyranosyl-D-glucose). Therefore, glucoamylase can hydrolyze and form all naturally occurring α-D-glucosidic bonds between adjacent D-glucosyl residues except the D-glucosidic bond found in α, α-trehalose (α-D-glucopyranosyl-α-D-glucopyranoside).[53, 105] However, α-(1,6) linkages are formed to much greater extent than other bonds, and therefore the α-(1,6)-linked di- and trisaccharides isomaltose and isomaltotriose, respectively, are the main reversion products that occur in high-glucose syrup production. A novel approach to increase the specificity and reduce the synthesis of α-(1,6) glucosidic bonds has been described recently by Sierks.[106] He changed Trp178 to Arg178 in the *A. awamori* enzyme by cassette mutagenesis. The value of K_M on isomaltose doubled over the wild-type enzyme, while those on maltose decreased slightly. Rates on both substrates decreased six- to eightfold with the first mutation, doubling the relative selectivity of maltose hydrolysis to isomaltose hydrolysis over that of the wild-type enzyme. This has potential industrial interest, in that a change in hydrolysis selectivity must be matched by an equal change in synthesis selectivity because of the invariability of the equilibria between glucose, maltose, and isomaltose, but clearly to be commercially feasible an enzyme of higher reflectivity must have activity similar to that of the wild-type enzyme.

3.3.3. Novel Glucoamylases

The impetus to screen for new glucoamylases comes from their lower thermostability compared to α-amylase. A thermostable and thermoactive glucoamylase that can be used at the same temperatures as α-amylase would result in a more energy-efficient process and would reduce the risk of retrogradation (aggregation) of starch dextrins. There are several new glucoamylases described recently that seem to have a higher thermostability than the current industrial glucoamylases. Glucoamylase from a thermophilic fungus *Thermomyces lanuginosus* was stable at 50 °C in 0.1 *M* phosphate buffer (pH 6) during a 6-h incubation and lost 50% of its initial activity at 60 °C.[107] The enzyme was highly specific for α-(1,4) glucosidic linkages and did not hydrolyze isomaltose even after prolonged incubation at high enzyme and substrate concentrations. Fogarty and Benson[108] purified a glucoamylase from a strain of *A. niger* that was stable at 50 °C and lost 10% of its activity at 60 °C during a 30-min incubation at pH 4.5 with 1% maltose or soluble starch. Katkocin *et al.*[109] isolated a glucoamylase from *Clostridium thermoamylolyticum* that had a half-life of 3 h at pH 6.0 and 70 °C.

3.3.4. Molecular Cloning and Expression of the Glucoamylase Gene

A driving force to isolate and clone glucoamylase genes is the possibility of producing alcohol directly from starch. This probably explains why glucoamylase genes have been cloned and expressed almost exclusively in yeast. Glucoamylase genes from *A. niger*[80] and *A. awamori*[81] were cloned in 1984. From the nucleotide sequence of the respective complementary DNAs (cDNA) it was concluded that the two glucoamylase genes were the same. Subsequently, Innis *et al.*[110] cloned and expressed *A. awamori* glucoamylase gene in *S. cerevisiae* using the promoter and termination region of the yeast enolase gene. Glucoamylase I was fully glycosylated and more than 90% of its activity was secreted into the culture medium. The glucoamylase expression level was increased fivefold by additional modification of the yeast promoter and the sequence upstream of the start of translation.[111] Ashikari *et al.*[82] have also cloned and expressed a *Rhizopus* glucoamylase gene in *S. cerevisiae.* The enzyme was excreted into the medium but the expression level was too low for practical production of glucoamylase and ethanol from starch. The recombinant *Rhizopus* glucoamylase was similar to native glucoamylase I, but it degraded raw starch more efficiently than native glucoamylase.[112]

In the current alcohol fermentation process raw starch is gelatinized by cooking, liquefied with α-amylase, and then saccharified to glucose by either acid or glucoamylase. To eliminate the separate saccharification step a laboratory strain of *S. cerevisiae* was engineered to secrete glucoamylase derived from *A. awamori.*[111] This novel amylolytic strain of *S. cerevisiae* was able to ferment soluble starch to ethanol.

3.4. β-AMYLASE

β-Amylase [α-(1,4)-D-glucan maltohydrolase, EC 3.2.1.2] is an exohydrolase that cleaves alternate α-(1,4) linkages from the nonreducing ends of amylose and amylopectin to produce β-maltose. Since it is unable to bypass α-(1,6) linkages in amylopectin, soluble starch hydrolysis by β-amylase results in the production of 50–69% maltose and β-limit dextrins. The hydrolytic action of β-amylase on starch dextrins proceeds from the nonreducing end by multiple attack to produce β-maltose molecules. The multiple attack mechanism decreases with chain length and change of pH and temperature.[113] As with other amylases, the rate of hydrolysis increases with chain length of the substrate; K_M decreases and maximum rate increases with chain length.[114–116] Since the enzyme hydrolyzes alternate α-(1,4)-D-glucosidic bonds, maltose is the only product of hydrolysis if the substrate is a straight-chain molecule with an even number of D-glucosyl residues. When the enzyme attacks a molecule made of an uneven number of D-glucosyl residues, some glucose and maltotriose are also found among the endproducts.[117] Amylopectin is attacked by β-amylase in two steps. The first and faster step leaves stubs of three or four D-glucosyl units beyond the branch. The second and much slower step gives the final β-limit dextrin.[116] β-Amylase has also synthetic ability similar to other amylases. Maltose of either anomeric form can be added to β-maltose to form maltotetraose. The equilibrium constant for α-(1,4)-D-glucosidic bond formation is similar to that of the glucoamylase-catalyzed reaction,[101, 105] as it theoretically must be.

β-Amylase is present in many higher plants, and the majority of information on the properties and mode of action of the enzyme comes from β-amylases from sweet potato,[118] soybean,[119] wheat,[120, 121] and barley.[122–124] Plant β-amylases, mainly from barley and soybean, have been for years the only commercial enzyme sources used to convert starch to maltose. In the last fifteen years a number of microbial β-amylases have been discovered, and wider use of these enzymes is anticipated in the future. In this section we will summarize the available information on plant β-amylases and discuss the extracellular microbial enzymes of commercial potential.

3.4.1. Plant β-Amylases

Plant β-amylases are sulfhydryl enzymes with molecular weights ranging from 53,000 to 64,000.[118–124] These enzymes also occur in multiple forms, such as dimers and tetramers, which may be linked by disulfide bonds. Some of the sulfhydryl groups,[125] maybe as few as one,[126, 127] seem to be essential for β-amylase activity. Apparently no metal ions are required for β-amylase activity. The enzyme is inactivated by sulfhydryl-complexing agents as well as by oxidation. Schardinger dextrins (cyclodextrins) and maltose are competitive inhibitors.[128] Plant β-amylases are most active at pH values between 4.5 and 6.5, but their temperature optima and stabilities depend on their source. Temperature optima for soybean and barley β-amylase are around 60 °C. They lose 50% of their initial activity when incubated in 1% substrate solution at 65 °C for 30 min. The soybean and barley β-amylases are used for saccharification at 55 °C and pH 5.5.

3.4.2. Microbial β-Amylases

Most of the β-amylases discovered are extracellular enzymes secreted by several species of *Bacillus*, including *B. cereus*,[129, 130] *B. circulans*,[2, 131] *B. megaterium*,[2, 132] and *B. polymyxa*[133, 134] as well as *Clostridium thermosulfurogenes*[135] and *Psudomonas* sp.[136] Their molecular weights range from 32,000 to 63,000 and their pH optima are between 6.5 and 7.5. The *Bacillus* enzymes have higher pH optima than plant β-amylases but similar temperature optima and thermostabilities. Most of these enzymes are stable only below 50 °C, which does not make them commercially attractive.

B. cereus var. *mycoides* β-amylase[129] purified almost to homogeneity produced only β-maltose from amylose, amylopectin, and glycogen. The β-amylase displayed low activity on maltotriose and was strongly inhibited by maltose. Its temperature and pH optima are 50 °C and pH 7.0, respectively. The enzyme is stable in the pH range from 6.0 to 9.6 at temperatures below 50 °C for 10 min. At 65 °C β-amylase activity is completely lost in 10 min. The enzyme is inhibited by *p*-chloromercuribenzoate (PCMB), but the activity is restored by addition of L-cysteine. The advantage of using this strain for β-amylase production is that it produces pullulanase, a debranching enzyme. These two enzymes effectively convert starch dextrins into high-maltose syrup.

B. circulans β-amylase is distinguished from other *Bacillus* β-amylases by its higher temperature optimum (60 °C) and lower susceptibility to PCMB.[2] This enzyme does not degrade maltotriose.

The most prominent microbial β-amylase to hold a commercial future is *C. thermosulfurogenes* β-amylase.[135] Fortunately, the gene coding for this thermophilic β-amylase has been cloned and expressed in *B. subtilis*, so the regulatory barriers for using this enzyme have been removed. The *C. thermosulfurogenes* β-amylase is a tetramer composed of 51,000-Da monomers. The optimal temperature for activity on soluble starch is 75 °C at pH 6.0. Ca^{2+} enhances enzyme stability but is not required for activity. The β-amylase is stable at 80 °C in the presence of either Ca^{2+} (5 m*M*) or 1% substrate solution, or between pH 3.5 and pH 7.0 at 70 °C with 5 m*M* $CaCl_2$. The purified β-amylase shows high hydrolytic activity with amylopectin, glycogen, and soluble starch, producing only maltose and β-limit dextrins as end products. As with other β-amylases, *C. thermosulfurogenes* β-amylase is inhibited by PCMB, as well as by α- and β-cyclodextrin.

3.4.3. Molecular Cloning and Expression of the β-Amylase Gene

There are only five reports in the literature that deal with the isolation and cloning of genes coding for β-amylase. The primary structures of mature barley[137] and soybean[138] β-amylases were determined from their nucleotide sequences. The barley β-amylase has 535 amino acids while soybean β-amylase has 495, giving molecular weights of about 60,000.

As for microbial β-amylases, the gene encoding *B. polymyxa* β-amylase was cloned and expressed in *B. subtilis*[139, 140] and *E. coli.*[140] Friedberg and Rhodes[139] demonstrated by restriction endonuclease mapping that the inserted DNA fragment containing the gene is approximately 3 kb in length. *B. subtilis* excretes a β-amylase of about 68,000 Da that hydrolyzes starch to maltose but does not attack either maltotriose or cyclodextrins. Kawazu *et al.*[140] cloned the gene coding for β-amylase from *B. polymyxa* into *E. coli.* Nucleotide sequence analysis of the cloned 3.1-kb DNA revealed that it contains one open reading frame of 2808 nucleotides; this translates into 936 amino acids including a signal peptide of 33 or 35 residues. Therefore, the mature β-amylase consists of more than 900 amino acid residues, giving a molecular weight greater than 100,000. When expressed in both *E. coli* and *B. subtilis,* multiple forms of β-amylase of 70,000, 58,000–56,000, and 42,000 Da were detected by SDS-gel electrophoresis and Western blot analysis. The authors confirmed that the same forms were produced by *B. polymyxa* in the absence of protease inhibitor. In the presence of high concentration of chymostatin, a protease inhibitor, 70,000- and 100,000-Da forms were detected in the *B. polymyxa* culture. Murao *et al.*,[134] who first isolated and characterized the *B. polymyxa* β-amylase, found only two forms of 44,000 Da each. This may suggest that the smallest β-amylase form is a proteolytic product. It should be noted that all of the forms were enzymatically active, producing maltose from starch but not from cyclodextrins.

The gene coding for thermophilic β-amylase from *C. thermosulfurogenes* was cloned into *B. subtilis.*[141] The nucleotide sequence analysis suggested one open reading frame encoding a polypeptide of 551 amino acid residues. The deduced amino acid sequence of the mature β-amylase contained 519 residues with a molecular weight of 57,167, which is slightly higher than the estimate reported for *C. thermosulfurogenes* enzyme.[135] *B. subtilis* excreted a thermophilic β-amylase that had thermostability comparable to the *C. thermosulfurogenes* enzyme. The amino acid sequence of the *C. thermosulfurogenes* β-amylase is 54% homologous to the *B. polymyxa* enzyme, 32% to that from soybean, and 32% to barley β-amylase. Twelve well-conserved regions were found among the amino acid sequences of the four β-amylases. Comparison of the hydropathy profiles of the *C. thermosulfurogenes* and *B. polymyxa* β-amylases revealed that the former has a higher content of hydrophobic amino acid residues. Kitamoto *et al.*[141] suggest that the lower content of hydrophilic amino acid residues increases the internal hydrophobicity of the enzyme molecule; this forces it to fold into a heat-stable form with stronger internal packing in an aqueous environment.

3.5. OTHER EXO-AMYLASES

The term α-amylase has been used for years to define an endo-acting hydrolase that produces α-maltooligosaccharides from starch by hydrolysis of α-(1,4) glucosidic bonds. There is another group of amylases that produces α-maltooligosaccharides, but by exo hydrolysis. They differ from the exo-hydrolases just described that yield products of β-configuration from α-linked substrates. Because of their exo mode of hydrolysis, they are not considered typical α-amylases and are most often classified as maltooligosaccharide-forming amylases. None has yet reached commercial application, but they are of interest for specialized applications and for what they can tell us of exo-acting hydrolases as a family.

3.5.1. Maltose-Forming Amylase

In 1984 Outtrup and Norman[142] discovered a thermostable maltose-forming amylase from *B. stearothermophilus.* The *B. stearothermophilus* amylase has a molecular weight of about 70,000 estimated by SDS-gel electrophoresis. This enzyme is thermostable and thermoactive with a temperature optimum around 60–65 °C and a pH optimum between 5.0 and 5.5. The enzyme was stable for 60 min at 60 °C in acetate buffer, pH 5.5, and lost 25% of its activity at 70 °C. By using low-molecular-weight substrates (methyl-β-D-maltotrioside, maltotriose, and maltotetraose), Outtrup and Norman established that the *B. stearothermophilus* amylases acted in an exo fashion, producing α-maltose rather than β-maltose, the product of β-amylase, from the nonreducing end. In the initial stage of hydrolysis of starch this maltose-forming amylase produces mainly maltotetraose, maltotriose, and maltose. As the reaction progresses maltotetraose disappears and maltotriose is converted into equimolar amounts of maltose and glucose. In contrast to β-amylase, the *B. stearothermophilus* amylase is not inhibited by sulfhydryl reagents, can hydrolyze cyclodextrins, and degrades amylopectin more efficiently. The latter property suggests that this enzyme can hydrolyze α-(1,4) linkages closer to the branch point than can plant β-amylases.

Because the *B. stearothermophilus* strain gave poor enzyme yields, the gene encoding the maltose-producing amylase was isolated and cloned into *B. subtilis.*[142] Enzyme production by *B. subtilis* was significantly improved, and the expressed protein had the same properties as the amylase produced by *B. stearothermophilus.*

An unusual maltose-forming amylase, capable of cleaving not only α-(1,4) but also α-(1,6) glucosidic bonds in amylopectin, was described recently by Suzuki *et al.*[31] This enzyme is produced extracellularly by *B. thermoamyloliquefaciens* and has a molecular weight of 67,000. The highest enzymatic activity was measured at 63 °C and pH 6.8; the pH optimum was around pH 6.0. The maltogenic amylase form *B. thermoamyloliquefaciens* degraded amylopectin and *β*-limit dextrins in exo fashion by preferentially cleaving α-maltose units from the nonreducing ends and producing small amounts of glucose. Maltotriose was hydrolyzed to glucose and maltose but panose was not degraded. When incubated for 6 h with 1 % pullulan, the enzyme formed only glucose, maltose, and panose. Its activity on pullulan was confirmed by immunotitration tests. The evidence suggests that this is a unique type of maltose-forming amylase with debranching activity on amylopectin only; the enzyme did not attack glycogen.

3.5.2. Maltotriose-Forming Amylase

Wako and co-workers[143] discovered a new exo-amylase from *Streptomyces griseus* that produces maltotriose from the nonreducing end of starch and related glucans. The purified enzyme is stable up to 40 °C and has maximal activity at 45 °C and pH 5.6–6.0. This amylase hydrolyzes soluble starch, waxy corn starch, and glycogen, producing maltotriose in yields of 55 %, 51 %, and 40 %, respectively. Amylose is degraded completely to maltotriose and small amounts of maltose and glucose in 3 h. The *S. griseus* amylase does not degrade pullulan, *β*-cyclo-dextrin, and *β*-limit dextrins from waxy corn starch.

3.5.3. Maltotetraose-Forming Amylase

A maltotetraose-forming amylase (Exo-maltotetraohydrolase, EC 3.2.1.60) was first discovered and described by Robyt and Ackerman.[144] This enzyme is produced extracellularly by *Pseudomonas stutzeri*, and it was the first enzyme since the discovery of *β*-amylase to be classified as a specific maltooligosaccharide-forming amylase. The *P. stutzeri* amylase has high activity between pH 6.5 and pH 10.5, with an optimum at pH 8.0. The optimal temperature for activity is 47 °C, but the enzyme rapidly loses activity above 40 °C. A number of different enzyme forms ranging from 12,500 to 58,000 Da were identified. This enzyme specifically attacks the fourth glucosidic bond from the nonreducing ends of maltohexaose and maltoheptaose. When incubated with amylopectin and glycogen, it produces maltotetraose and high-molecular-weight limit dextrins but does not attack cyclodextrins. The properties of the *P. stutzeri* exo-amylase have been studied and confirmed by others.[145–147] Sakano *et al.*[146] purified a *P. stutzeri* exo-amylase of approximately 57,000 Da and determined that it produced α-maltotetraose from starch. The enzyme activity was inhibited by Cu^{2+}, Hg^{2+}, *N*-bromosuccinimide, and 2,3-butanedione. The *P. stutzeri* exo-amylase was recently immobilized for continuous production of maltotetraose.[148] Tao *et al.*[149] demonstrated an interesting use of the enzyme when they converted a ^{13}C-enriched glycogen made from α-D-[^{13}C]glucopyranosyl fluoride with amylosucrase to [1-^{13}C]maltotetraose for the determination of the mechanism of α-amylase activity.

3.5.4. Maltohexaose-Forming Amylase

Aerobacter aerogenes produces a maltohexaose-forming amylase (Exo-maltohexahydrolase, EC 3.2.1.98) that cleaves α-maltohexaosyl units from the nonreducing ends of starch, amylose, and amylopectin but does not hydrolyze α- and β-cyclodextrin, pullulan, and maltohexitol.[150–152] The optimal pH and temperature are 6.8 and 50 °C, respectively. An interesting property of the enzyme is its degradation of β-limit dextrins to yield branched oligosaccharides, suggesting that the enzyme could bypass α-(1,6)-branch points on the substrate. The *Aerobacter* amylase slowly hydrolyzes maltohexaose to maltotriose and maltose. Recently, Takasaki[153] and Taniguchi *et al.*[154] found in separate studies that *Bacillus circulans* also produces exo-maltohexahydrolase with molecular weights of 76,000 and 93,000, respectively. Takasaki reported an optimal temperature of 60 °C and an optimal pH between 6.0 and 6.5, while Taniguchi and co-workers found the same optimal temperature and an optimal pH of 8.0. The purified enzyme in the early stages of starch hydrolysis produces maltohexaose, which is degraded later to maltotetraose and maltose, so it has an action pattern similar to the *A. aerogenes* amylase. It is unclear whether the two exo-amylases from *B. circulans* are different enzymes or different forms of the same enzyme. The significant difference between the two amylases is that the amylase described by Taniguchi *et al.*[154] is larger in size and can digest starch granules.

3.6. APPLICATIONS AND FUTURE TRENDS

3.6.1. α-Amylase

The most important application of thermostable α-amylases is in the first step of the starch depolymerization process, that of liquefaction. In the liquefaction process starch slurry (30–40% dry solids) is gelatinized by heat and partially hydrolyzed by α-amylases. Two different liquefaction schedules are used commercially, depending on the enzyme source (Table 1).[14, 155, 156]

The *B. licheniformis* α-amylase is injected into the starch slurry before steam addition in a jet cooker. Because of its high thermostability, only about 20% of its activity is lost during jet-cooking. After flash-cooling of

Table 1. Typical Liquefaction Schedules

	B. *licheniformis* α-amylase	*B. amyloliquefaciens* α-amylase
Thinning	Jet cooking 103–107 °C, pH 6–6.5 2–5 min	Heating 85–90 °C, pH 6–6.5 15–30 min Jet cooking 140–150 °C 5 min
Flash cooling	90–100 °C	80–90 °C
Additional enzyme	No	Yes
Dextrinization	95 °C 1–2 h	85 °C 1–2 h

the thinned slurry, usually no additional enzyme is required for the completion of the dextrinization process. In case of the *B. amyloliquefaciens* enzyme a jet-cooking is required to complete the starch gelatinization and dissociation of the amylose–lipid complex after the initial thinning of the starch slurry at temperatures near 90 °C. After the dextrinization step the liquefied starch has a DE of 10–15 and is used as a substrate for subsequent saccharification.

The liquefying α-amylases are also employed in the manufacturing of maltodextrins and ethanol. In the production of ethanol and maltodextrins of DE > 12, glucoamylase addition is required to increase the amount of lower-molecular-weight dextrins and glucose, respectively.

It is difficult to predict what new developments will occur in the coming years, since the quest for higher thermostability has been resolved by the discovery of the *B. licheniformis* and more recently the *B. stearothermophilus* α-amylases. Amylase thermostability is further discussed in chapter 12 in this book. Obtaining greater stability at the present operating conditions is not vital, but finding or engineering a new thermostable α-amylase that would operate at pH 4.0–5.0 and would not require Ca^{2+} ions for stability would be of immediate significance. Such an enzyme would eliminate the need for pH adjustment before the saccharification step and would reduce the required ion exchange capacity for the removal of metal ions before the isomerization step. That would also enable dextrinization and saccharification of starch to be carried out in the same reactor.

3.6.2. Glucoamylase

Regardless of the manner in which starch is liquefied, by acid or α-amylase, saccharification to produce high-glucose and other syrups for human consumption is conducted with glucoamylase. More than 80% of the high-glucose syrup is subsequently converted to high-fructose corn syrup, while about 10% is processed into crystalline dextrose. A typical saccharification process employs liquefied starch solutions of 30–35% (w/w) and DE 10–15 that are cooled to 60–65 °C and adjusted to pH 4.0–4.5. The saccharification is performed in stirred reactors with sufficient glucoamylase to achieve a maximum DE level in 48–72 h.[1,156] A typical high-glucose syrup contains 95–96% glucose on a dry basis. Glucose loss occurs due to maltulose formation in the liquefaction step and accumulation of reversion products during the saccharification. The yield of glucose in high-glucose syrups can be increased by using debranching enzymes, such as pullulanase and isoamylase, which also reduces the amount of glucoamylase added and the total time of saccharification.[156]

Glucoamylase in combination with fungal α-amylase is used to produce high conversion syrups of DE 60–70. Different glucose-to-maltose ratios can be obtained by altering the proportions of the two enzymes in the saccharification.[157]

Another important application of glucoamylase is in industrial and fuel ethanol production. Glucoamylase in combination with α-amylase converts primarily corn starch into fermentable sugars. When wet-milled starch is used as a feedstock, the liquefaction and saccharification steps are carried out as in glucose syrup production except that saccharification is stopped at DE 85. Hydrolyzed starch solution is cooled to 30 °C, and then yeast is added to initiate the fermentation process. The omission of the liquefaction step is a possible alternative, but increased enzyme consumption is required—a not economically viable concept at present prices.[158] In this process starch slurry is mixed with α-amylase, glucoamylase, and yeast, and the fermentation is carried out at about 30 °C. Hydrolysis of the granular starch is the limiting step.

When ground whole grain is used as a starch source, starch is liquefied, presaccharified, and then transferred to the fermentor. In this procedure fermentation and the saccharification occur simultaneously at about 30 °C.

Attractive future developments would include improved glucoamylase thermostability and specificity. While thermostability can be increased by either genetic engineering or by further screening for new thermophilic producers of glucoamylase, specificity will probably be increased at a molecular level by site-directed mutagenesis. Increased thermostability would allow the use of glucoamylase at higher temperatures and eliminate the flash-cooling step, reducing at the same time the risk of microbial contamination and dextrin retrogradation. With the development and application of debranching enzymes in the saccharification process, the α-(1,6) cleaving and synthetic activity of glucoamylase is no longer beneficial. On the contrary, it decreases yields through formation of isomaltose and eventually isomaltotriose from product glucose.

3.6.3. β-Amylase

β-Amylase has its main application in the production of various maltose syrups. Normal maltose syrups containing 50–55% maltose on a dry basis are produced from either acid- or enzyme-liquefied starch using plant β-amylase or fungal α-amylase.[156] Due to the difference in action patterns, the syrups produced by β-amylases are higher in β-limit dextrins and lower in maltotriose content than those produced by fungal α-amylases. To produce high-maltose syrups that contain more than 80% maltose, addition of debranching enzymes such as isoamylase or pullulanase is necessary during the saccharification.[155] The initial DE of the liquefied starch needs to be as low as possible to achieve high maltose concentrations in the final product. For that reason, β-amylases of high thermostability and thermoactivity should be developed and utilized in high-maltose syrup production to prevent retrogradation of the low-DE hydrolyzates. The *C. thermohydrosulfuricum* β-amylase has the necessary properties, and its use in the production of maltose and high-maltose syrups has recently been demonstrated.[159]

3.6.4. Other Exo-Amylases

Exo-amylases other than glucoamylase and β-amylase are not used commercially; their future application would be to produce specialty syrups with well-defined saccharide distributions and properties. These would include syrups high in maltotriose, maltotetraose, and maltohexaose. They will probably take some of the maltodextrin market but certainly will generate their own market and new applications. On the enzyme side, the trend will be toward increasing the thermostability of these exo-enzymes and reducing their cost to make them competitive with the other amylolytic enzymes used in syrup production. Unless large markets for specialty syrups are developed, enzyme cost will probably limit their use.

REFERENCES

1. P. J. Reilly, in: *Starch Conversion Technology* (G. M. A. van Beynum and J. A. Roels, eds.), p. 102, Marcel Dekker, New York (1985).
2. W. M. Fogarty, in: *Microbial Enzymes and Biotechnology* (W. M. Fogarty, ed.), p. 1, Applied Science, London (1983).
3. N. Saito, *Arch. Biochem. Biophys. 155*, 290 (1973).
4. J. P. Chiang, J. E. Alter, and M. Sternberg, *Starch/Staerke 31*, 86 (1979).

5. V. Buonocore, C. Caporale, M. de Rosa, and A. Gambacorta, *J. Bacteriol. 128*, 515 (1976).
6. K. Takkinen, R. F. Pettersson, N. Kalkkinen, I. Palva, H. Soderlund, and L. Kaariainen, *J. Biol. Chem. 258*, 1007 (1983).
7. R. Nakajima, T. Imanaka, and S. Aiba, *J. Bacteriol. 163*, 401 (1985).
8. T. Yuuki, T. Nomura, H. Tezuka, A. Tsuboi, H. Yamagata, N. Tsukagoshi, and S. Udaka, *J. Biochem. 98*, 1147 (1985).
9. E. H. Fisher and E. A. Stein, in: *The Enzymes* (P. D. Boyer, H. Lardy, and K. Myrback, eds.), Vol. 4, p. 313, Academic Press, New York (1960).
10. E. A. Stein and E. H. Fisher, *J. Biol. Chem. 232*, 867 (1958).
11. J. Hsiu, E. H. Fisher, and E. A. Stein, *Biochemistry 3*, 61 (1964).
12. A. Imanishi, *J. Biochem. 60*, 381 (1966).
13. E. W. Boyer and M. B. Ingle, *J. Bacteriol. 110*, 992 (1972).
14. G. B. Madsen, B. H. Norman, and S. Slott, *Staerke 25*, 304 (1973).
15. P. J. Brumm, R. E. Hebeda, and W. M. Teague, *Food Biotechnol. 2*, 67 (1988).
16. J. F. Robyt and D. French, *Arch. Biochem. Biophys. 100*, 451 (1963).
17. N. E. Welker and L. L. Campbell, *Biochemistry 6*, 3681 (1967).
18. G. J. Walker and P. M. Hope, *Biochem. J. 86*, 52 (1963).
19. C. T. Greenwood, E. A. Milne, and G. R. Ross, *Arch. Biochem. Biophys. 126*, 244 (1968).
20. E. J. Hehre, D. S. Genghof, and G. Okada, *Arch. Biochem. Biophys. 142*, 382 (1971).
21. F. J. Morgan and F. G. Priest, *J. Appl. Bacteriol. 50*, 107 (1981).
22. G. E. Inglett, *J. Food Biochem. 11*, 249 (1988).
23. S. L. Pfeuller and W. H. Elliott, *J. Biol. Chem. 244*, 48 (1969).
24. S. Medda and A. K. Chandra, *J. Appl. Bacteriol. 48*, 47 (1980).
25. H. Matsuzaki, K. Yamane, K. Yamaguchi, Y. Nagata, and B. Maruo, *Biochim. Biophys. Acta 365*, 235 (1974).
26. Y. Yoneda, K. Yamane, K. Yamaguchi, Y. Nagata, and B. Maruo, *J. Bacteriol. 120*, 1144 (1974).
27. G. L. Gray, S. E. Mainzer, M. W. Rey, M. H. Lamsa, K. L. Kindle, C. Carmonia, and C. Requadt, *J. Bacteriol. 166*, 635 (1986).
28. H. Hidaka and T. Adachi, in: *Mechanisms of Saccharide Polymerization and Depolymerization* (J. J. Marshall, ed.), p. 101, Academic Press, New York (1980).
29. Y. Suzuki and T. Imai, *Appl. Microbiol. Biotechnol. 21*, 20 (1985).
30. Y. Takasaki, *Agric. Biol. Chem. 49*, 1091 (1985).
31. Y. Suzuki, T. Nagayama, H. Nakano, and K. Oishi, *Starch/Staerke 39*, 246 (1987).
32. Y. Takasaki, *Agric. Biol. Chem. 51*, 9 (1987).
33. T. Tanaka, E. Ishimoto, Y. Shimomura, M. Taniguchi, and S. Oi, *Agric. Biol. Chem. 51*, 399 (1987).
34. I. Palva, *Gene 19*, 81 (1982).
35. S. A. Ortlepp, J. F. Ollington, and D. J. McConnell, *Gene 23*, 267 (1983).
36. R. P. Piggott, A. Rossiter, S. A. Ortlepp, J. T. Pembroke, and J. F. Ollington, *Biochem. Biophys. Res. Commun. 122*, 175 (1984).
37. M. Sibakov, *Eur. J. Biochem. 155*, 577 (1986).
38. J. R. Mielenz, *Proc. Natl. Acad. Sci. USA 80*, 5975 (1983).
39. N. Tsukagoshi, S. Iritani, T. Sasaki, T. Takemura, H. Ihara, Y. Idota, H. Yamagata, and S. Udaka, *J. Bacteriol. 164*, 1182 (1985).
40. P. Cornelis, C. Digneffe, and K. Willemot, *Mol. Gen. Genet. 186*, 507 (1982).
41. M. Yang, A. Galizzi, and D. Henner, *Nucleic Acids Res. 11*, 237 (1983).
42. H. Yamazaki, K. Ohmura, A. Nakayama, Y. Takeichi, K. Otozai, M. Yamasaki, G. Tamura, and K. Yamane, *J. Bacteriol. 156*, 327 (1983).
43. S. Aiba, K. Kitai, and T. Imanaka, *Appl. Environ. Microbiol. 46*, 1059 (1983).

44. K. Thudt, K. H. Schleifer, and F. Goetz, *Gene 37*, 163 (1985).
45. A. Filloux, P. Joyet, M. Murgier, and A. Lazdunski, *FEMS Microbiol. Lett. 30*, 203 (1985).
46. S. Hoshiko, O. Makabe, C. Nojiri, K. Katsumata, E. Satoh, and K. Nagaoka, *J. Bacteriol. 169*, 1029 (1987).
47. T. F. Gorozhankina, M. Yu. Beburov, A. V. Sorokin, and A. I. Stepanov, *Dokl. Akad. Nauk SSSR 285*, 717 (1985).
48. L. Ruohonen, P. Hackman, P. Lehtovaara, J. K. C. Knowles, and S. Keranen, *Gene 59*, 161 (1987).
49. I. S. Pretorius, E. Laing, G. H. J. Pretorius, and J. Marmur, *Curr. Genet. 14*, 1 (1988).
50. J. H. Pazur and T. Ando, *J. Biol. Chem. 235*, 297 (1960).
51. J. H. Pazur and K. Kleppe, *J. Biol. Chem. 237*, 1002 (1962).
52. M. Abdullah, I. D. Fleming, P. M. Taylor, and W. J. Whelan, *Biochem. J. 89*, 35 p (1963).
53. M. M. Meagher and P. J. Reilly, *Biotechnol. Bioeng. 34*, 689 (1989).
54. B. C. Saha and J. G. Zeikus, *Starch/Staerke 41*, 57 (1989).
55. P. Manjunath, B. C. Shenoy, and M. R. R. Rao, *J. Appl. Biochem. 5*, 235 (1983).
56. P. Q. Flor and S. Hayashida, *Appl. Environ. Microbiol. 45*, 905 (1983).
57. J. H. Pazur and T. Ando, *J. Biol. Chem. 234*, 1966 (1959).
58. D. R. Lineback, I. J. Russell, and C. Rasmussen, *Arch. Biochem. Biophys. 134*, 539 (1969).
59. I. M. Freeberg, Y. Levin, C. M. Kay, W. D. McCubbin, and E. Katchalski-Katzir, *Biochim. Biophys. Acta 391*, 361 (1975).
60. B. T. Svensson, T. G. Pedersen, I. Svensson, T. Sakai, and M. Ottesen, *Carlberg Res. Commun. 47*, 55 (1982).
61. D. Alazard and J. F. Baldenspenger, *Carbohydr. Res. 107*, 231 (1982).
62. P. Manjunath and M. R. R. Rao, *J. Biosci. 1*, 409 (1979).
63. J. Abe, Y. Takeda, and S. Hitzukuri, *Biochim. Biophys. Acta 703*, 26 (1982).
64. I. D. Fleming and B. A. Stone, *Biochem. J. 97*, 13 (1965).
65. K. Venkataramu, P. Manjunath, and M. R. R. Rao, *Ind. J. Biochem. Biophys. 12*, 107 (1975).
66. S. Hayashida, *Agric. Biol. Chem. 39*, 2093 (1975).
67. T. Takahashi, N. Inokuchi, and M. Irie, *J. Biochem. 89*, 1055 (1981).
68. D. R. Lineback and W. E. Baumann, *Carbohydr. Res. 14*, 341 (1970).
69. Y. Morita, K. Shimizu, M. Ohga, and T. Korenaga, *Agric. Biol. Chem. 30*, 114 (1966).
70. M. N. N. Miah and S. Ueda, *Starch/Staerke 29*, 191 (1977).
71. S. Ueda and S. Kano, *Starch/Staerke 27*, 123 (1975).
72. T. Takahashi, Y. Tsuchida, and M. Irie, *J. Biochem. 84*, 1183 (1978).
73. B. C. Saha and S. Ueda, *J. Ferment. Technol. 61*, 67 (1983).
74. J. H. Pazur and S. Okada, *Carbohydr. Res. 4*, 371 (1967).
75. S. Hayashida, T. Nomura, E. Yoshino, and M. Hongo, *Agric. Biol. Chem. 40*, 141 (1976).
76. S. Hayashida and E. Yoshino, *Agric. Biol. Chem. 42*, 927 (1978).
77. E. Yoshino and S. Hayashida, *J. Ferment. Technol. 56*, 289 (1978).
78. S. Hayashida and P. Q. Flor, *Agric. Biol. Chem. 45*, 2675 (1981).
79. T. Takahashi, Y. Tsuchida, and M. Irie, *J. Biochem. 92*, 1623 (1982).
80. E. Boel, I. Hjort, B. Svensson, F. Norris, K. E. Norris, and N. P. Fiil, *EMBO J. 3*, 1097 (1984).
81. J. H. Nunberg, J. H. Meade, G. Cole, F. C. Lawyer, P. McCabe, V. Schweickart, R. Tal, V. P. Wittman, J. E. Flatgaard, and M. A. Innis, *Mol. Cell Biol. 4*, 2306 (1984).
82. T. Ashikari, N. Nakamura, Y. Tanaka, N. Kiuchi, Y. Shibano, T. Tanaka, T. Amachi, and H. Yoshizumi, *Agric. Biol. Chem. 50*, 957 (1986).
83. B. Svensson, K. Larsen, I. Svendsen, and E. Boel, *Carlsberg Res. Commun. 48*, 529 (1983).

84. C. T. Greenwood and E. A. Milne, *Advan. Carbohydr. Chem. 23*, 281 (1968).
85. G. K. Lee, Ph. D. Dissertation, Iowa State University, Ames, Iowa (1978).
86. Y. Tanaka, T. Ashikari, N. Nakamura, N. Kiuchi, Y. Shibano, T. Amachi, and H. Yoshizumi, *Agric. Biol. Chem. 50*, 965 (1986).
87. J. H. Pazur, H. R. Knull, and A. Cepure, *Carbohydr. Res. 20*, 83 (1971).
88. K. Venkataramu, P. Manjunath, and M. R. R. Rao, *Ind. J. Biochem. Biophys. 17*, 388 (1980).
89. J. H. Pazur, H. R. Knull, and D. L. Simpson, *Biochem. Biophys. Res. Commun. 40*, 110 (1970).
90. D. R. Lineback, L. A. Aira, and R. L. Horner, *Cereal Chem. 49*, 283 (1972).
91. J. H. Pazur, Y. Tominaga, L. S. Forsberg, and D. L. Simpson, *Carbohydr. Res. 84*, 103 (1980).
92. J. H. Pazur, B. Liu, S. Pyke, and C. R. Baumrucker, *J. Protein Chem. 6*, 517 (1987).
93. B. Svensson, K. Larsen, and I. Svendsen, *Carlsberg Res. Commun. 48*, 517 (1983).
94. B. Svensson, K. Larsen, and A. Gunnarsson, *Eur. J. Biochem. 154*, 497 (1986).
95. K. Watanabe and T. Fukimbara, *Agric. Biol. Chem. 38*, 1973 (1974).
96. K. Watanabe and T. Fukimbara, *Agric. Biol. Chem. 39*, 1711 (1975).
97. K. Watanabe, *J. Biochem. 80*, 379 (1976).
98. K. Hiromi, *Biochem. Biophys. Res. Commun. 40*, 1 (1970).
99. K. Hiromi, Y. Nitta, C. Numata, and S. Ono, *Biochim. Biophys. Acta 302*, 362 (1973).
100. M. M. Meagher, Z. L. Nikolov, and P. J. Reilly, *Biotechnol. Bioeng. 34*, 681 (1989).
101. E. J. Hehre, G. Okada, and D. S. Genghof, *Arch. Biochem. Biophys. 135*, 75 (1969).
102. T. Watanabe, S. Kawamura, S. Sasaki, and K. Matsuda, *Staerke 21*, 18 (1969).
103. T. Watanabe, S. Kawamura, S. Sasaki, and K. Matsuda, *Staerke 21*, 44 (1969).
104. J. H. Pazur, A. Cepure, S. Okada, and L. S. Forsberg, *Carbohydr. Res. 58*, 193 (1977).
105. Z. L. Nikolov, M. M. Meagher, and P. J. Reilly, *Biotechnol. Bioeng. 34*, 694 (1989).
106. M. R. Sierks, Ph. D. Dissertation, Iowa State University, Ames, Iowa (1988).
107. V. B. Rao, N. V. S. Sastri, and P. V. S. Rao, *Biochem. J. 193*, 379 (1981).
108. W. M. Fogarty and C. P. Benson, *Eur. J. Appl. Microbiol. Biotechnol. 18*, 271 (1983).
109. M. Katkocin, N. S. Word, and S.-S. Yang, US Patent 4,536,477 (1985).
110. M. A. Innis, M. J. Holland, P. C. McCabe, G. E. Cole, V. P. Wittman, R. Tal, K. W. K. Watt, D. H. Gelfand, J. P. Holland, and J. H. Meade, *Science 228*, 21 (1985).
111. G. E. Cole, P. C. McCabe, D. Inlow, D. H. Gelfand, A. Ben-Bassat, and M. A. Innis, *Bio/Technology 6*, 417 (1988).
112. Y. Tanaka, T. Ashikari, N. Nakamura, N. Kiuchi, Y. Nakao, T. Amachi, and H. Yoshizumi, *Agric. Biol. Chem. 50*, 1727 (1986).
113. T. Saganuma, M. Ohnishi, K. Hiromi, and Y. Morita, *Agric. Biol. Chem. 44*, 1111 (1980).
114. D. French and R. W. Youngquist, *Staerke 15*, 425 (1963).
115. E. Y. C. Lee, *Arch. Biochem. Biophys. 146*, 488 (1971).
116. M. Kato, K. Hiromi, and Y. Morita, *J. Biochem. 75*, 563 (1974).
117. J. F. Robyt and W. J. Whelan, in: *Starch and Its Derivatives* (J. R. Radley, ed.), 4th ed., p. 477, Chapman & Hall, London (1969).
118. A. K. Balls, R. P. Thompson, and M. K. Walden, *J. Biol. Chem. 163*, 571 (1946).
119. A. Gertler and Y. Birk, *Biochem. J. 95*, 621 (1965).
120. K. H. Meyer, P. F. Spahr, and E. H. Fisher, *Helv. Chim. Acta 36*, 1924 (1953).
121. R. Tkachuk and K. H. Tipples, *Cereal Chem. 43*, 62 (1966).
122. R. Shinke and N. Mugibayashi, *Agric. Biol. Chem. 35*, 1381 (1971).
123. K. Visuri and M. Nummi, *Eur. J. Biochem. 28*, 555 (1972).
124. R. Lundgard and B. Svensson, *Carlsberg Res. Commun. 52*, 313 (1987).
125. J. A. Thoma, J. E. Spradlin, and S. Dygert, in: *The Enzymes* (P. D. Boyer, ed.), p. 115, Academic Press, New York (1971).

126. B. Mikami, S. Aibara, and Y. Morita, *J. Biochem. 88*, 103 (1980).
127. Y. Morita, B. Mikami, and S. Aibara, *J. Jpn. Soc. Starch Sci. 28*, 59 (1981).
128. J. A. Thoma and D. E. Koshland, Jr., *J. Biol. Chem. 235*, 2511 (1960).
129. Y. Takasaki, *Agric. Biol. Chem. 40*, 1523 (1976).
130. T. Nanmori, R. Shinke, K. Aoki, and H. Nishira, *Agric. Biol. Chem. 47*, 941 (1983).
131. E. J. Napier, US Patent 4,011,136 (1977).
132. M. Higashihara and S. Okada, *Agric. Biol. Chem. 38*, 1023 (1974).
133. W. M. Fogarty and P. J. Griffin, *J. Appl. Chem. Biotechnol. 25*, 229 (1975).
134. S. Murao, K. Ohyama, and M. Arai, *Agric. Biol. Chem. 43*, 719 (1979).
135. G. J. Shen, B. C. Saha, Y. E. Lee, L. Bhatnagar, and J. G. Zeikus, *Biochem. J. 254*, 835 (1988).
136. R. Shinke, Y. Kunimi, H. Nishira, *J. Ferment. Technol. 53*, 698 (1975).
137. M. Kreis, M. Williamson, B. Buxton, J. Pywell, J. Hejgaard, and I. Svendsen, *Eur. J. Biochem. 169*, 517 (1987).
138. B. Mikami, Y. Morita, and C. Fukazawa, *Seikagaku 60*, 211 (1988).
139. F. Friedberg and C. Rhodes, *J. Bacteriol. 165*, 819 (1986).
140. T. Kawazu, Y. Nakanishi, N. Uozumi, T. Sasaki, H. Yamagata, H. Tsukagoshi, and S. Udaka, *J. Bacteriol. 169*, 1564 (1987).
141. N. Kitamoto, H. Yamagata, T. Kato, N. Tsukagoshi, and S. Udaka, *J. Bacteriol. 170*, 5848 (1988).
142. H. Outtrup and B. E. Norman, *Starch/Staerke 36*, 405 (1984).
143. K. Wako, C. Takahashi, S. Hashimoto, and J. Kanaeda, *Denpun Kagaku 26*, 175 (1979).
144. J. F. Robyt and R. J. Ackerman, *Arch. Biochem. Biophys. 145*, 105 (1971).
145. J. Schmidt and M. John, *Biochim. Biophys. Acta 566*, 88 (1979).
146. Y. Sakano, Y. Kashiwagi, and T. Kobayashi, *Agric. Biol. Chem. 46*, 639 (1982).
147. M. J. Lee and M. Jae, *Han'guk Nonghawa Hakhoechi 28*, 1 (1985).
148. T. Kimura, M. Ogata, M. Yoshida, and T. Nakakuki, *Biotechnol. Bioeng. 32*, 669 (1988).
149. B. Y. Tao, P. J. Reilly, and J. F. Robyt, *Biochim. Biophys. Acta 995*, 214 (1989).
150. K. Kainuma, S. Kobayashi, T. Ito, and S. Suzuki, *FEBS Lett. 26*, 281 (1972).
151. K. Kainuma, K. Wako, S. Kobayashi, A. Nogami, and S. Suzuki, *Biochim. Biophys. Acta 410*, 333 (1975).
152. T. Nakakuki, M. Monma, K. Azuma, S. Kobayashi, and K. Kainuma, *Denpun Kagaku 29*, 179 (1982).
153. Y. Takasaki, *Denpun Kagaku 29*, 145 (1982).
154. H. Taniguchi, C. M. Jae, N. Yoshigi, and Y. Maruyama, *Agric. Biol. Chem. 47*, 511 (1983).
155. B. E. Norman, in: *Enzymes and Food Processing* (G. G. Birch, N. Blakebrough, and K. J. Parker, eds.), p. 15, Applied Science, London (1981).
156. R. E. Hebeda, in: *Corn: Chemistry and Technology* (S. A. Watson and P. E. Ramstad, eds.), p. 501, American Association of Cereal Chemists, St. Paul (1987).
157. C. G. Bohnenkamp and P. J. Reilly, *Biotechnol. Bioeng. 22*, 1753 (1980).
158. N. W. Lutzen, *Adv. Biotechnol. 2*, 161 (1981).
159. B. C. Saha and J. G. Zeikus, *Biotechnol. Bioeng.* 34, 299 (1989).

4

Biocatalysis in the Production of Carbohydrates for Food Uses

ANDREW J. HACKING

4.1. INTRODUCTION

Carbohydrates have many functional applications in foods. Some are well known, such as the role of sucrose and glucose as nutritive sweeteners, starch as a thickener in soups and gravies, pectin in jams, and many others. Others are known in outline, but are often more complex than they appear at first sight. An example of this category is the use of starch in bread and cake making, which will be described shortly. Similarly, sucrose in chocolate and candy manufacture is not just used for its sweetness, but also because it contributes toward the product's texture and mouthfeel. Other carbohydrates can also have a variety of functions. Some polysaccharides can be used as emulsifiers and stabilizers in products such as salad dressings to prevent the separation of oil and water phases, starch can impart "body" and smoothness in drinks such as beer, and sucrose gives chocolate its snap.

In many cases the functionality of carbohydrates is due to their interaction with water. This is especially true with polysaccharides when used as gelling agents, thickeners, and emulsifiers,[1] but it can also apply to mono- or disaccharides. Sucrose can act as a preservative at high concentrations

ANDREW J. HACKING • Dextra Laboratories Limited, Philip Lyle Building, PO Box 68, Reading RG6 2BX, United Kingdom.

because it binds water, reduces the water activity of a solution, and prevents the water being available to permit the growth of spoilage microorganisms. Other sugars, notably trehalose, can prevent or reduce the denaturation of proteins during freezing and thawing by replacing protein–water interactions.

4.1.1. Structure–Function Relationships in Foods

In some cases, the properties of carbohydrates in foods are determined by quite subtle structural properties. An example which may be cited is the degree of branching of starch. When starch solutions are heated, they gelatinize, and on cooling will form gels. On standing, intermolecular association leading to precipitation often occurs, a process termed retrogradation. It is generally associated with the amylose fraction; high amylopectin starches do not undergo this process as rapidly or as extensively. Their gels, however, are soft rather than stiff.[2]

A second process associated with the degree of branching is syneresis—the loss of water from a gel structure—which can be seen when gels are cut. Gels that have been thawed after freezing also show this phenomenon. Ice crystals seem to disrupt the gel structure. This has been overcome by using waxy maize starch (pure amylopectin) in freeze–thaw foods.[3] The gels will resume their consistency on thawing, but they are often too viscoelastic for many food uses. Pastes made with waxy maize starch are thick and clear with little tendency to gel. They can be cross-linked to improve their gelling properties, but this impairs their freeze–thaw stability.

The process of staling of bread and cakes is also associated with water loss. Although all the changes which occur during staling cannot be explained simply by retrogradation of starch, it does appear to be the most important single factor.[3] Staling is a rapid process in bread because the moisture content is optimal; it is much slower in dried starchy foods such as crackers. In cakes, lipids or surfactants are added, both as texture modifiers and because they delay and reduce staling. Highly branched starches cannot be used to replace the lipids directly because the cakes will not rise (cakes will also not rise if high-amylose starch is used).[4] This is because the granular structure of starch is important and this is conferred by both amylose and amylopectin components; their ratio is critical. The nature of the intermolecular forces is the predominant factor in determining the subsequent behavior of starch on cooking.

4.1.2. Trends in the Food Industry

In recent years there have been many significant changes and new developments in the processing, packaging, storage, and distribution of foods. In overall terms this can be regarded as a form of continuous development from the early processes such as drying, through canning, to freezing, vacuum packing, and others. The principal objective has been to combine the preservation or cooking of food with the retention of maximum flavor or "naturalness" quality. Foods that have successfully achieved this have earned the highest margins for their manufacturers. Recently, it has become even more important because of increased demand for preserved foods that retain flavor as close as possible to fresh products without the use of chemical agents. Examples include retaining maximum flavor and texture in chill distribution foods and during freeze–thawing processes, or in microwave cooking. Such developments may permit beneficial advances such as achieving preservation and flavor retention without refrigeration, needed in hot climates. Discovery or manufacture of new carbohydrates or the modification of existing structures can play a significant role in new developments in the food industry. In this chapter, an attempt will be made to show how this might be achieved using enzyme technology.

4.1.3. Raw Materials

Starch and sugars derived from starch, such as glucose, maltose, maltodextrins, and high-fructose syrups, are the most commonly encountered food carbohydrates along with sucrose (from sugar cane and sugar beet) and invert syrups derived from sucrose. In addition to starch, the plants processed for food carbohydrates contain cellulose, glucans with α and β links, and hemicelluloses which are polymers of pentoses, chiefly xylose and arabinose, but with other lesser components. The full list of monomeric sugars found in hemicelluloses is given in Table 1. Celluloses and hemicelluloses could prove to be very useful starting materials to manufacture a whole range of carbohydrates with different properties.

Other polysaccharides commonly used in foods are shown in Table 2. Most are obtained from plant sources, but agar, alginate, and carrageenan are algal (seaweed) in origin, and dextran and xanthan are obtained from bacteria. A huge range of extracellular polysaccharides with different physical properties are made by microorganisms, and much research has been done on their applications. So far only the two listed have widespread

Table 1. Principal Components of Hemicellulose

D-Xylose	D-Glucuronic acid
D-Glucose	D-Galacturonic acid
D-Galactose	4-*O*-Methyl-D-Glucuronic acid
L-Rhamnose	
L-Arabinose	L-Fucose

Table 2. Polysaccharides Used in Foods

Guar gum	Agar
Gum arabic	Alginate
Pectins	Carrageenans
Locust bean gum	Xanthan
Carboxymethylcellulose	Dextran
Uses	
Thickening	Texture
Gelling	Viscosity/rheology modification
Stability	

approval for food uses, although several companies are working on new products.

As will be shown, one approach to producing a new polysaccharide with novel properties is to modify an existing product rather than isolate and grow another organism. One drawback to manufacturing a food-grade polysaccharide from a new organism is that methods to separate the product from cells and cellular debris must be developed. In particular, with gram-negative organisms, the lipopolysaccharide or endotoxin must be removed in processing. The cost of obtaining regulatory approval for a new food product can be very high, and the time needed to reach the marketplace can be very lengthy. There are advantages to developing new products from existing food-approved sources.

4.1.4. Alternative Manufacturing Technologies

Most sugars are obtained by extraction from natural sources, such as cereal grains or other plant tissue. These can be used as the starting materials, but for the production of higher added-value products with more specialist functions a number of technologies are available.

Sugars are a difficult class of compounds for chemical processing because they possess many reactive hydroxyl groups. Achieving specificity is difficult and often necessitates a sequence of protecting nonreacting groups followed by reaction of the desired group(s) and deprotection. The number of steps required can be considerable, and the expense precludes most food uses, except perhaps some flavor compounds. As will be described later, the solubility of sugars in nonaqueous solvents is restricted, and this also presents a barrier to chemical synthesis.

Fermentation can be used as a route to many sugars, both polysaccharides and monomers such as isomaltulose,[5] aldonic and uronic acids,[6] or sugar esters.[7] It can, however, be an expensive technology because it involves the use of dilute solutions with correspondingly high downstream processing costs and frequently high capital costs. Other factors, such as effluent treatment, the maintenance of asepsis, and energy usage also contribute to costs. It can be tolerated as a method of manufacture for specialty products such as xanthan, but is often too expensive for bulk products.

Biotransformations using enzyme technology have had substantial success in the starch industry for the manufacture of glucose, maltose, and high-fructose syrups, but their use elsewhere is still relatively limited. In scientific terms, they offer promise because of their selectivity and fewer toxicological problems. Many enzyme reactions in nature, however, involve sugar phosphates or nucleoside sugars, and they often occur at specific subcellular locations or need membrane-bound enzymes. This makes them too expensive for most food uses, and new technology will be a great advantage to extend the range of commercial products.

4.2. ENZYME TECHNOLOGY

4.2.1. Biotransformations Based on Starch

Enzyme technology for the manufacture of a range of products from cereal starches is well developed and is the basis of large-scale manufacturing plants in several countries, particularly, in the United States. So far it has been most successful in the wet-milling process for maize, but transfer to wheat has presented problems, although most have now been overcome. Developing a new range of products from other sugars is likely to be even more difficult, but there does seem to be a large number of potential opportunities. If the production of glucose, maltose, fructose syrups, and maltodextrins is used as a model, it may be possible to start with cellulose, xylans, or arabinans and make monomers, dimers, short-chain soluble oligomers, branched structures, and possibly even cyclic oligosaccharides (Figure 1). There are many hurdles to be overcome before such a scheme can be realized. The xylans and arabinans are rarely found in a pure form, but more often as mixtures comprising the hemicelluloses. Therefore, often a first fractionation step is necessary, and this may be difficult and expensive. Celluloses are insoluble, and even the short-chain oligomers are poorly soluble, so their potential uses are limited. Even so a number of possibilities are present to be tested. New enzyme systems, such as xylanases and arabinases, could be screened for and their properties evaluated.

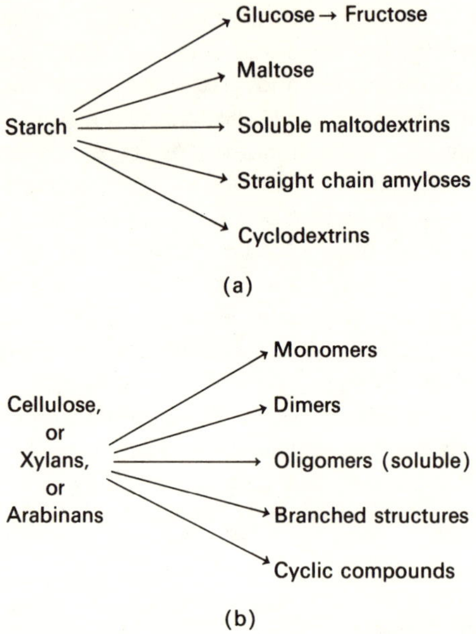

Figure 1. Potential biotransformations using cellulose and hemi-cellulose as raw materials modeled on products from starch. (a) Existing transformations based on starch; (b) Possible transformations based on cellulose and hemicellulose.

4.2.2. Pentose Metabolism

The pathways involving metabolism of the pentoses in bacteria have been known for many years following the pioneering work of Wood, Mortlock and colleagues (Figure 2). Many enzymes have been isolated, and studied and work has been done on reactions with an extensive range of sugars.[8] Unfortunately, with many reactions the enzymes are specific for sugar phosphates, and although there are some ways of overcoming this, they frequently involve toxic or non-food-approved compounds. There are however some enzymes which work on unsubstituted sugars, such as glucose isomerase (in nature a xylose isomerase) which has activity toward a number of sugars (Table 3). There are though ways to greatly extend the range of products by exploiting the properties of other classes of enzyme.

4.2.3. Sugar Transfer Reactions

One interesting property of enzyme catalysis that can be exploited commercially is that many enzymes that hydrolyze carbohydrates can also catalyze the transfer of sugar residue(s) from a glycoside donor to another,

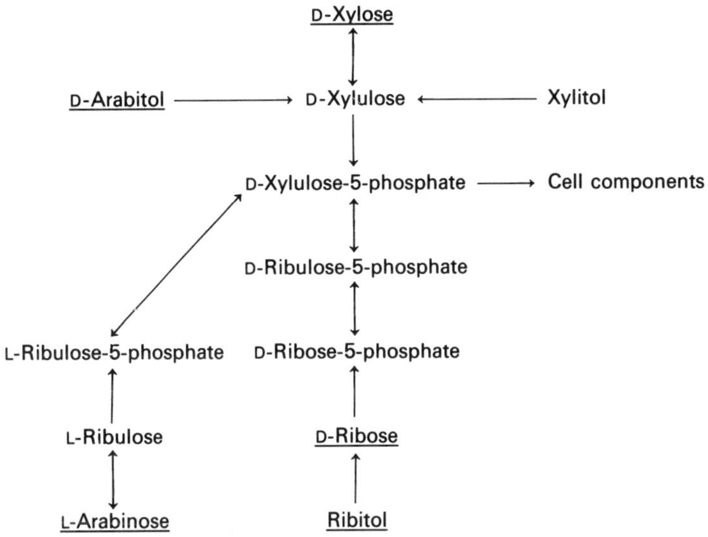

Figure 2. Metabolism of pentoses found in nature. Compounds underlined are found in nature. (Adapted from the work of Mortlock, Wood, and colleagues.[8])

acceptor, sugar. This class of reactions was termed transglycosylation by Hehre.[9] Under most reaction conditions the concentration of water is very high, and it acts as an acceptor, so that a mono- or disaccharide unit is split off a chain and released as a free sugar. However, if the reaction is conducted in the presence of high concentrations of a second sugar, this molecule can then act as an acceptor, and often a number of new di-, tri-, and oligosaccharides can be made.

There are now a number of cases where this property has been reported, and some where it has been exploited commercially. For example, fructosyl transferases have been used to make new di- and oligosaccharides of fructose plus glucose, galactose, and xylose (Table 4). Galactosyl transfer can be achieved by β-galactosidases, giving oligosaccharides with both α-1–4 and α-1–6 links depending on the enzyme and source,[13–15] and by

Table 3. Substrates of Xylose Isomerase

Xylose ⇌ Xylulose
Glucose ⇌ Fructose
Ribose ⇌ Ribulose
Rhamnose ⇌ Rhamnulose
L-Arabinose ⇌ L-Ribulose

Table 4. Fructosyl Transferases

Levansucrase		
G–F	→ G–F–F–F–F	(α-2-6)[10]
G–F + X	→ X–F	Xylsucrose[11,19]
G–F + Gal	→ Gal–F	Galsucrose[11,12,19]
Erwinia and some other organisms		
G–F (1-2)	→ G–F (1-6)	Isomaltulose[5]

Key G = Glucose; F = Fructose; X = Xylose; Gal = Galactose.

α-galactosidases.[16] Another widely studied example of this phenomenon is the enzyme which makes cyclodextrins from starch, cyclodextringlycosyltransferase, which will transfer glucose on to quite a large range of sugars, forming an α-1–4 glycosidic link[17] (Table 5). Similarly, the sugar phosphorylase enzymes can often use different acceptor molecules and make new compounds. In this case, instead of making a phosphate ester, they can make compounds such as alkyl glucosides. The most widely studied example has been the sucrose phosphorylase from bacteria such as *Pseudomonas saccharophila*[18] (Table 6), but this property could be shared by similar enzymes such as maltose phosphorylase or trehalose phosphorylase.

The potential of transfer reactions to make new sugars is extensive, and it is well worth undertaking new enzyme screens for this class of activity. For example, in a recent report, a new fructosyl transferase isolated from a *Bacillus* strain was used to synthesize a number of hitherto unrecorded sugars.[19] The enzymes which possess sugar transferase activity seem to have a number of different roles in nature such as polysaccharide synthesis,[19] carbohydrate degradation,[13] and cyclodextrin synthesis.[17] It is

Table 5. Glucosyl Transfer

Cyclodextringlycosyltransferase[17]	
G–G–G–G + X →	G–X
	G–G–X
	G–G–G–X
G–G–G–G + S →	G–S
	G–G–S
	G–G–G–S

Key G = Glucose; X = Xylose; S = Sorbose.

Table 6. Transfer Activity of Sucrose Phosphorylase

$G-F+P_i$	$\rightarrow F+G\text{-}1\text{-}P$
$+CH_3OH$	$\rightarrow F+\alpha$-methylglucoside
+ ethylene glycol	$\rightarrow$ F + hydroxyethylglucoside
+ *n*-butanol	$\rightarrow$ F + *n*-butylglucoside

perhaps reasonable to assume that other classes of carbohydrate activity such as transport enzymes might also catalyze similar reactions. These reactions are particularly valuable in the synthesis of new oligosaccharides, where protection and deprotection reactions are often long and complicated.

4.2.4. Carbon–Carbon Bond Formation

Unlike the transferases which will make new glycosidic links, enzymes such as transketolases and aldolases can be used to make new carbon–carbon links in sugars and thus new monosaccharides.

Some of the transketolases which are found in the series of sugar rearrangements in the pentose phosphate pathway or the assimilation of one-carbon compounds in certain microorganisms have a broad substrate specificity. They are, for example, able to use a range of aliphatic and aromatic aldehydes as acceptors to make a number of new sugar derivatives. The reaction may be represented as

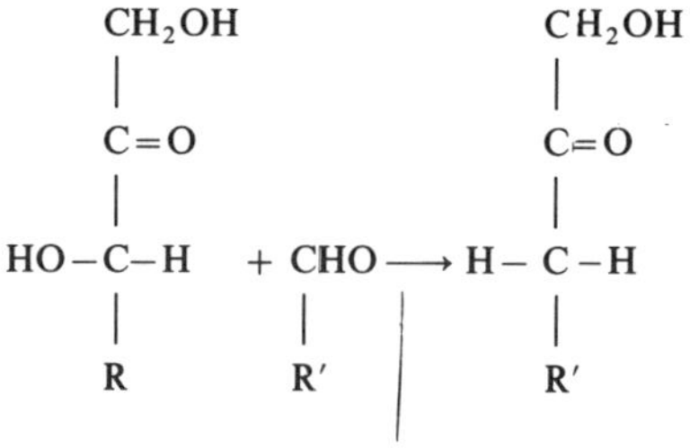

where R′ is the acceptor. The normal donor is D-xylulose-5-phosphate, although fructose-6-phosphate can be used. Enzymes that can use a range of acceptors include those isolated from yeasts[20] and mammalian liver.

Aldolases may also be used to synthesize new carbohydrate compounds.[21] These enzymes normally cleave a ketose phosphate to produce an aldehyde and a ketone as in the case of fructose diphosphate aldolase (EC 4.1.2.13):

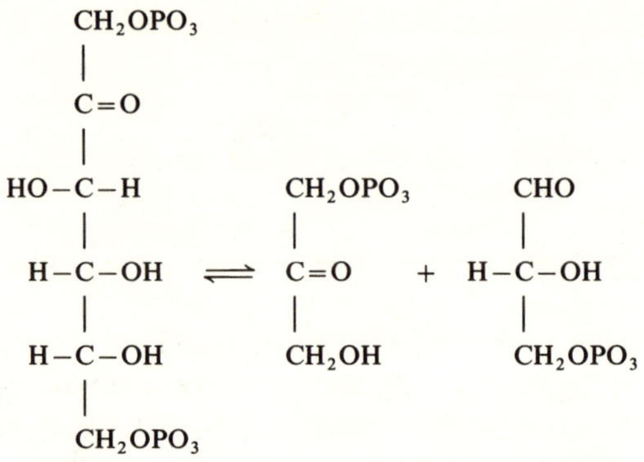

Fructose 1,6-diphosphate — Dihydroxyacetone phosphate — Glyceraldehyde 3-phosphate

The reverse reaction, an aldol condensation—the nucleophilic addition of the enolate anion of an aldehyde or ketone to the carbonyl group of an aldehyde—can also be catalyzed by this class of enzymes. The reaction is difficult to achieve using synthetic organic chemistry because the intermediate enolate complex is unstable and dissociates in aqueous solution. Over 20 enzymes have now been found which will catalyze aldol addition reactions, each using different substrates with different specificity in carbon–carbon bond formation.[21]

One of the most widely used enzymes has been the fructose diphosphate aldolase from rabbit muscle, which has been used to synthesize deoxysugars[22] and isotopically labeled sugars.[21] It has been proposed that it could also be used for other uncommon sugars such as halogenated sugars and hexoses with a different heteroatom in the ring.[21] The enzyme is specific for dihydroxyacetone phosphate, but it will accept a wide variety of unphosphorylated aldehydes.[23] In terms of small scale synthesis of relatively expensive compounds this is a useful enzyme; it is commercially available, has a high specific activity, and is stable. For larger-scale uses, cheaper bacterial or fungal enzymes will probably be required. Other aldolases that have been used to synthesize new sugars in the laboratory include fuculose-1-phosphate,[24] rhamnulose-1-phosphate,[25] and sialic acid.[26]

At present the applications that these enzymes have been put to are small-volume, high-added-value compounds which might be used in research or in the production of pharmaceutical intermediates. Nevertheless, one or two food products have been made on a test basis such as fructose and furaneol,[21] and there may be possibilities at first in production of flavor or aroma compounds. Again sugar phosphate substrates can be a drawback, although dihydroxyacetone in the presence of inorganic arsenate has been found to be an effective substrate for fructose biphosphate aldolase.[22] If nontoxic alternatives could be found, then many new applications could be opened up.

4.2.5. Solvents

The range and potential for using enzyme reactions can be extended by running the reactions in nonaqueous systems or by conditions of low water activity. Many enzymes are active in organic solvents, supercritical fluids, and gaseous solvents, and this permits the reversal of hydrolytic reactions, the use of nonpolar substrates, improved product recovery, and other advantages.[27,28] In general, the chemical industry has replaced water with more suitable organic solvents, but carbohydrates present problems because their solubility in solvents is restricted and some of the most commonly used solvents, such as dimethylformamide, dimethylsulfoxide, and pyridine, reduce or eliminate the activity of many enzymes. Thus as yet there are fewer examples of enzyme catalysis using carbohydrates in solvents than, say, aromatic or alphatic compounds, but the list is increasing as more knowledge about enzyme activity in solvents is accrued.

One group of reactions where solvents are needed is the esterification of sugars principally using esterase or lipases. Since in these cases a hydrolytic reaction is being reversed, they must be carried out in lower-water-activity systems. The use of enzymes here permits regioselectivity. For example, several unrelated commercially available lipases have been shown to catalyze transesterification reactions between trichloroethyl butyrate and monosaccharides with blocked C-6 hydroxyl groups in dry organic solvents.[29] In common with other systems described here, the first applications are in complex organic synthetic schemes, but as will be described, the principles can be extended to cheaper food ingredients.

4.3. POTENTIAL PRODUCTS

4.3.1. Sugar Esters

By forming ester links between polar sugars and nonpolar aromatic or aliphatic groups, it is possible to make a range of surface-active compounds with valuable emulsification, interfacial-tension-lowering or micelle-forming properties. A generalized structure for this class of compounds is shown in Figure 3. It is possible to use a number of sugars and a range of fatty acids in these compounds to create a wide variety of combinations. Sucrose esters have been used as emulsifiers in foods for some time, but using

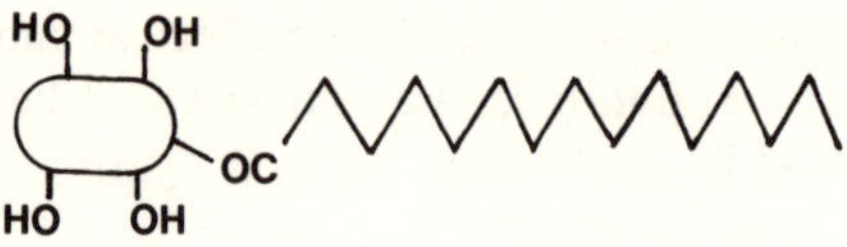

Sugar	Fatty Acid
glucose	oleate
rhamnose	palmitate
sorbitol	stearate
sucrose	

Figure 3. Carbohydrate-based surfactants.

chemical techniques it is difficult to control the degree or specificity of substitution.

The recent progress made using lipases and esterases in nonaqueous solvents to esterify sugars at specific positions[30] opens up many opportunities to make this class of compounds. Specific mono- or diesters may be synthesized and compounds with defined emulsification, surface-tension-lowering or interfacial-tension-lowering properties, obtained. Another possibility is in fat substitution, where it may be possible to utilize the enhanced surface activity of a molecule to replace a greater amount of lipid in a food without any loss of functional properties.

The main obstacle to the more widespread use of these enzymes is economics. The production of sucrose esters by chemical methods is much cheaper than enzyme technology at the present time, so any advantages such as specificity would have to command a premium price. Alternatively, the cost of the enzyme treatment must be reduced, which in turn calls for greater enzyme productivity in solvent systems and perhaps recovery and re-use of the enzyme.

4.3.2. Trehalose

Trehalose (α-D-glucopyranosyl-α-D-glucopyranoside) (Figure 4) is a naturally occurring sugar which has recently been found to have some

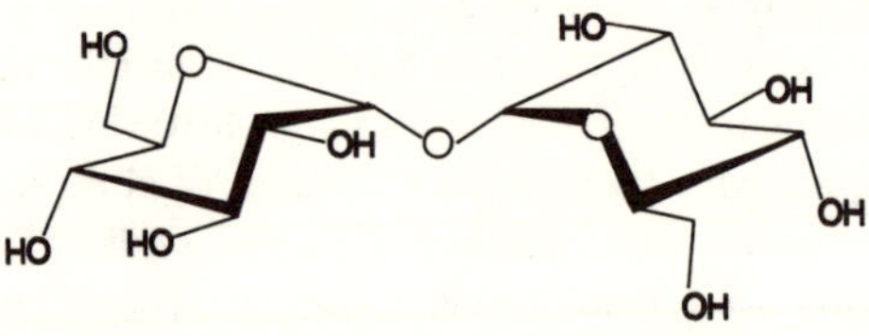

Figure 4. The structure of $\alpha\alpha$-trehalose.

unique and interesting properties. When proteins are dried in the presence of trehalose, it replaces the protein-interacting water molecules; the 3D structure of the protein is maintained, and on addition of water, trehalose is displaced and the original structure is regained. Many potentially high-value applications for trehalose as a stabilizer for proteins are being considered in the health-care field, especially in diagnostics or therapeutics. In the food industry too there are many opportunities for retaining the flavor and mouthfeel of dried food products and perhaps in some situations of replacing refrigeration as a method of preservation. For example, much of the flavor loss that occurs in drying milk or eggs is due to protein denaturation. Trehalose which is nonsweet and odorless can be added in very small quantities to prevent this.

Trehalose is widespread in nature. It has been found in bacteria, yeasts, mushrooms, algae, some plants, and invertebrates such as shrimp. It can be extracted from plants such as trehala manna, but the most favored route at present is extraction from dried baker's yeast, which can be up to 15% by weight trehalose.[31] Even so it is still too expensive for large-scale food uses. The current price is about $400 per kg, and although improved processing from yeast and larger-scale operations could perhaps reduce this considerably, it is thought that new methods are needed to achieve the price reductions demanded by large-scale food applications.

One route which has been proposed is a two-step enzymatic synthesis from maltose.[32] This uses maltose phosphorylase (EC 2.4.1.64) from *Lactobacillus brevis* and trehalose phosphorylase (EC 2.4.1.8) from *Euglena gracilis*:

$$\text{maltose} + \text{Pi} \xrightarrow{\text{maltose phosphorylase}} \text{glucose} + \text{glucose-1-phosphate}$$

$$\text{glucose} + \text{glucose-1-phosphate} \xrightarrow{\text{trehalose phosphorylase}} \text{trehalose} + \text{Pi}$$

This process will require optimization on a large scale and probably a cheap source of enzyme, but it could offer long-term savings over extraction and possibly improved fermentation methods. Another related route could be to use starch as a raw material and produce glucose-1-phosphate using starch phosphorylase plus glucose via α-amylase and amyloglucosidase. Trehalose phosphorylase could again be used as the second step.

4.3.3. Cyclodextrins

The cyclodextrins are a group of homologous oligosaccharides comprising 6, 7, or 8 (α, β, or γ) glucose units linked α-1-4 in a cyclic structure.

They are synthesized from starch by certain microorganisms, notably *Bacillus* species with cyclodextrin glycosyltransferase enzymes.[33] They have a doughnut-like shape, and the inside of the hollow space in the molecule is a relatively hydrophobic region as opposed to the hydrophilic surface. They have unique properties to form inclusion complexes in the ring (which has a size of 6, 8, and 10 Å in the α-, β-, and γ-oligomers, respectively) with compounds which can fit in to this structure. This property of molecular encapsulation has found applications in food as well as pharmaceuticals, cosmetics, agricultural chemicals, and others, but until recently it has been restricted by price. Now, as improvements in manufacture are taking place, there do seem to be commercially feasible applications.[34]

In foods cyclodextrins can be utilized mainly for molecular encapsulation of flavors and fragrances, which in turn can protect the active molecule against oxidation, light-induced decomposition, heat or loss by evaporation.[35] They enable manufacturers to make stable preparations that retain flavors through cooking, packing, storage, and distribution processes. They can protect flavors through manufacturing processes such as baking, extrusion, deep frying, and powder coating, and can be used for flavors such as vanillin, jasmine, cinnamon, and basil. For example, one particularly useful application is in onion and garlic oils for snack foods such as crisps and nuts.[34] There have also been reports of cyclodextrins being used to reduce unpleasant tastes and odors and to improve water retention and emulsion stability.[35]

Improvements in enzyme technology have reduced the costs of cyclodextrins and, as described earlier, created opportunities for the enzyme to be used with other sugar acceptors to make new oligosaccharides for other food applications. There are also opportunities to make new and improved cyclodextrins. One example might be to put acyl or other nonpolar groups on the inside of the molecule to increase its hydrophobicity. Branched cyclodextrins have also been made and reported to have new properties.[36] This could be extended by attaching other sugar residues. The prospects are very wide-ranging and have attracted the attention of a number of research laboratories. Also it would be very interesting to find new enzyme systems that will make other cyclic oligosaccharides, either with different carbohydrate residues or with different links, such as the cyclic (1-2)-β-D-glucan made by *Agrobacterium*.[37]

4.3.4. Polysaccharides

Polysaccharides are used widely in the food industry for thickening, gelling, stabilization of emulsions, and texture modification. (Table 2). Most have been obtained from plants or seaweed for many years, and their

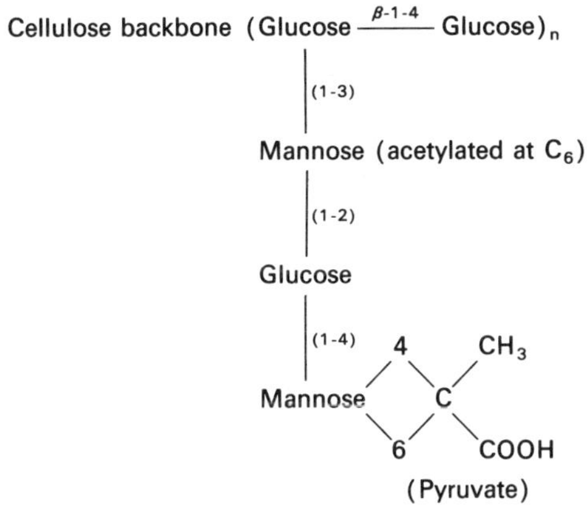

Figure 5. Chemical structure of xanthan.

use has mainly grown from a base in traditional foods and subsequent empirical developments. This in turn has led to the development of a scientific methodology and technology to measure and quantify their effects. In recent years much basic and applied research has been directed toward the way in which complex heteropolysaccharides modify the viscosity or flow characteristics of solutions. There is a need for new polysaccharides with new functionalities, such as emulsification, and they can command higher prices on the basis of their usefulness.

This work has been strongly influenced by the success of xanthan, a complex polysaccharide with a cellulose backbone and sugar/pyruvate side chains (Figure 5) produced extracellularly by the bacterium *Xanthomonas campestris*. Xanthan is an exceptionally good stabilizer and thickener that has achieved widespread use in the food industry. Its success has led many companies conducting extensive screens of microbial polysaccharides to find new compounds with desirable physical characteristics. This work is often empirical, and the measurements are complex. If a product is identified and successfully tested, it subsequently requires the development of an economic fermentation process and also downstream processing to separate it from other products of the organism, which might have adverse toxicological effects, as referred to earlier.

One interesting and relevant property of polysaccharides is that even relatively small changes in their sugar composition such as the substituents along the main chain[38] can radically alter their physical properties. For example, gellan gum produced by *Pseudomonas elodea* is comprised of a repeating tetrasaccharide unit with the structure

-3)-β-D-Glc-(1-4)-β-D-GlcA-(1-4)-β-D-Glc-(1-4)-α-L-Rha-(1-

It forms firm, brittle gels.[39]

Welan gum produced by *Alcaligenes* sp. does not gel, but forms viscous solutions. It differs structurally from gellan only in that the tetrasaccharide has an additional mannose or rhamnose moiety[40]:

-3)-β-D-GlcA-(1-4)-β-D-Glc-(1-4)-β-D- Glc -(1-4)-α-L-Rha-(1-
| (1-3)
Man

This suggests an alternative way of making new polysaccharides with different properties by enzymically modifying existing structures. The enzyme might be used to add groups on to a structure, or remove them to epimerize one sugar to another within a polysaccharide structure, to break down the main structure of the polysaccharide or to cross-link chains. This approach also has the attraction of avoiding the development of a new fermentation process.

An example of the enzyme modification approach being put into practice is the action of α-galactosidase on galactomannans.[41] Galactomannans are storage polymers in seeds. They are comprised of a backbone of β-1-4 linked mannose residues with single galactose side chains linked by an α-1-4 bond. The degree of substitution by galactose is variable and confers different functional properties on the polymer. Guar gum, produced from the plant *Cyamopsis tetragonolopus*, is comprised of 62% mannose with 38% galactose side chains, while locust bean gum obtained from *Ceratonia siliqua* is 77% mannose and 23% galactose. As a consequence of this difference in structure, locust bean gum forms more stable gels with polysaccharides such as xanthan and carageenan, and it commands a higher price than guar gum. Treatment of guar gum with α-galactosidase under controlled conditions enables the production of a product with properties equivalent to locust bean gum. Reaction conditions have been developed which allow the enzyme to function at 50% solids.

This method can lead to a great number of potentially valuable commercial opportunities. Many different sugar backbones are obtainable from cheap polymers such as cellulose, starch, xylans, and arabinans. New developments in enzyme technology are constantly increasing the possibilities for adding different groups to these backbones and for adding them in a controlled manner. For example, it may be possible to add acyl groups using lipases, esterases, or even proteases. Sugar transferase reactions may be possible using polysaccharides as acceptors rather than the mono-, di-, or short oligosaccharides described so far. If side chains can be added systematically and in a controlled way, it may also be possible to develop a series of predictive rules linking structure and function and thereby leading to "designer polysaccharides." Cross-linking of polymer chains could also lead to significant commercial potential in the formation of different types of gels with varying degrees of flexibility, heat stability, shear stability, etc.

This technology is still in its infancy; polysaccharides can be difficult to work with, some are poorly soluble in water, and even fewer are soluble in many nonaqueous solvents, so much of the pioneering work on sugar substitution in solvents will be much more difficult to achieve with larger molecules. As a consequence, most initial work has been done with hydrolytic systems, but this has also been the case in other areas of enzyme technology, and it has led to synthetic applications later. A particularly encouraging development is the improvement of analytical methods for polysaccharide structures, notably NMR. This technique has been used to determine accurately the degree of branching in starch[42] and has been used to determine the mannuronic–guluronic acid ratio in alginates.[43] Such developments will simplify the task of enzyme technologists in assaying the reaction products.

4.4. CONCLUSION

4.4.1. The Pressures of the Marketplace

Inevitably, the applications of carbohydrate enzyme technology will be shaped by market forces. There are a number of pressures on the food industry which can be addressed by the type of technology described here. The first is in food preservation, where there is increasing demand to reduce or eliminate preservatives or in some cases to replace refrigeration. Carbohydrates can have significant impact because of their water binding or water retention properties, but these need to be enhanced so that a greater effect can be achieved with a small amount of material. Some polysaccharides can act as hydrogels—they can absorb a large fraction of their own weight of water. If this property can be attributed to structural properties, then great advances can be made in synthesizing new and more effective compounds.

A second very topical area is the search for fat replacers. There is increased demand from individuals to reduce their caloric intake for medical or health reasons. Fats contribute nine calories per gram as opposed to the four calories per gram of carbohydrates or proteins, but are often retained in foods for reasons such as flavor, mouthfeel, preservation, or prevention of staling. It follows that if carbohydrates with some or all of these properties could be found, their commercial applications would be considerable. There are some products already in use in this area.[44] For example, some salad dressing formulations have used a polysaccharide system to suspend flavorings and stabilize emulsions.[45]

Short chain-maltodextrins can be used as partial replacements for fats and oils in some applications such as spreads, desserts, and salad dressings because their helical structure has a hydrophobic core which has a somewhat oily or slippery mouthfeel. As with cyclodextrins, it may be possible to enhance this property by adding substituents. Polysaccharides such as xanthan too can replace some of the functions of oils in reduced-calorie salad dressings in which the oil content has been reduced. Nondigestible sugar esters such as Procter&Gamble's "Olestra" may also find applications as fat replacers.

Another area where carbohydrates may find uses is cooking. Some fat replacers can be used for frying, and there is a market for food ingredients which can facilitate microwave cooking. For instance, one problem encountered in microwave cooking is that some particles in food can be heated in the center causing steam formation while the outside of the particle is dried and sealed, giving a form of case-hardening effect with a hot center. It may be possible to overcome this by altering the nature and degree of branching or cross-linking of food polysaccharides.

4.4.2. Economic Considerations

The principal determinant of the extent of use of new carbohydrates will be their price. If we look at the present prices of sugars, we find that a few, namely glucose, sucrose, and sorbitol, are cheap; the price of fructose has been falling recently, and maltose and galactose are relatively low-priced. The vast majority of sugars, however, are very expensive. The sugars which are cheap have a wide range of applications, and this illustrates a more general economic principle that as the price of a commodity falls, more applications become possible and the volume of consumption rises. The corollary is that frequently it is the prospect of larger-scale applications that encourages the development of new technology to reduce manufacturing costs. These relationships can be seen in amino acids, where large-scale production of monosodium glutamate and aspartate (flavor compounds) and methionine and lysine (animal feed supplements) has resulted in low prices; phenylalanine has fallen in price because of demand for aspartame manufacture, and if genetic engineering technology can reduce the cost of tryptophan, it can also be used as a feed supplement. Other amino acids remain expensive. If this reasoning is extended to carbohydrates, it can be predicted that a sugar such as trehalose will fall in price and more uses for it will become possible. There may be a cascade of applications starting with diagnostics or specialist enzymes, spreading to high-value food uses, and finally larger-volume markets.

Sugars have been difficult to make chemically. But there might develop a form of synergy between improvements in biotransformations and the growth of carbohydrate usage, first in specialist areas, then in larger markets. In particular, there seems to be a potential for utilizing the carbohydrates which are presently discarded or have low-value applications, such as the celluloses and hemicelluloses. Improvements in enzyme technology could open up new markets and new industrial developments.

REFERENCES

1. D. Blenford, *Food*, December 1986, p. 26.
2. H. F. Zobel, *Starch/Starke 40*, 44 (1988).
3. E. J. Rolfe, in: *Molecular Structure and Function of Food Carbohydrates* (G. G. Birch and L. F. Green, eds.), p. 1, Applied Science, Barking (1973).
4. W. Banks, C. T. Greenwood, and D. D. Muir, in: *Molecular Structure and Function of Food Carbohydrates* (G. G. Birch and L. F. Green, eds.), p. 177.
5. P. S. J. Cheetham, in: *Methods in Enzymology* (K. Mosbach, ed.), Vol. 136, p. 432, Academic Press, London (1987).
6. E. Percival, in: *Comprehensive Biochemistry* (M. Florkin and E. H. Stotz, eds.), Vol. 5, p. 67, Elsevier, New York (1963).
7. R. B. Duff, D. M. Webley, and V. C. Farmer, *Nature 179*, 104 (1957).
8. R. P. Mortlock, in: *Advances in Microbial Physiology* (A. H. Rose and D. W. Tempest, eds.), Vol. 13, p. 1, Academic Press, London (1976).
9. E. J. Hehre, in: *Advances in Enzymology* (F. F. Nord, ed.), Vol. 11, p. 297, Interscience, New York (1951).
10. T. Tanaka, S. Oi, M. Lizuka, and T. Yamamoto, *Agric. Biol. Chem. 42*, 323 (1978).
11. K. Sugimoto, S. Okada, and S. Yoshikawa, UK Patent 2, 046, 757 (1983).
12. D. S. Feingold, G. Avigad, and S. Hestrin, *J. Biol. Chem. 224*, 295 (1957).
13. M. Aronsen, *Arch. Biochem. Biophys. 39*, 370 (1953).
14. J. H. Pazur, *J. Biol. Chem. 208*, 439 (1954).
15. K. Wallenfels and A. Arens, *Biochem. Z. 332*, 247 (1960).
16. W. M. Watkins, *Nature 181*, 117 (1958).
17. S. Kitahata, S. Okada, and T. Fukui, *Agric. Biol. Chem. 42*, 2369 (1978).
18. J. J. Mieyal, M. Simon, and R. H. Abeles, *J. Biol. Chem. 247*, 532 (1972).
19. P. S. J. Cheetham, A. J. Hacking, and M. Vlitos, *Enzyme Microb. Technol. 11*, 212 (1989).
20. J. R. Quayle, personal communication (1983).
21. C.-H. Wong, in: *Enzymes as Catalysts in Organic Synthesis* (M. P. Schneider, ed.), p. 199, D. Reidel, Munich (1986).
22. J. R. Durrwachter, D. G. Druekhammer, K. Nozaki, H. M. Sweers, and C.-H. Wong, *J. Am. Chem. Soc. 108*, 7812 (1986).
23. C.-H. Wong and G. M. Whitesides, *J. Org. Chem. 48*, 3199 (1983).
24. M. A. Ghalambor and E. C. Heath, *J. Biol. Chem. 237*, 2427 (1962).
25. T. H. Chiu and D. S. Feingold, *Biochemistry 8*, 98 (1969).
26. D. G. Comb and S. Roseman, *J. Biol. Chem. 235*, 2529 (1960).
27. A. M. Klibanov, *Chemtech 16*, 354 (1986).
28. J. S. Dordick, *Enzyme Microb. Technol. 11*, 194 (1989).
29. M. Therisod and A. M. Klibanov, *J. Am. Chem. Soc. 109*, 3977 (1987).
30. S. Riva, J. Chopineau, A. P. G. Kieboom, and A. M. Klibanov, *J. Am. Chem. Soc. 110*, 584 (1988).

31. N. K. Richtmeyer, in: *Methods in Carbohydrate Chemistry* (R. L. Whistler and M. L. Wolfrom, eds.), Vol. 1, p. 370, Academic Press, New York (1962).
32. S. Murao, H. Nagano, S. Ogura, and T. Nishino, *Agric. Biol. Chem. 49*, 2113 (1985).
33. K. Horikoshi, *Process Biochem. 14*, 26 (1979).
34. J. S. Pagington, *Food*, September 1985, p. 51.
35. J. Szejtli, *Starch/Starke 34*, 379 (1982).
36. S. Kobayashi and K. Kainuma, UK Patent 2, 165, 549 (1986).
37. M. Hisamatsu, A. Anemura, T. Matsuo, H. Matsuda, and T. Harada, *J. Gen. Microbiol. 128*, 1873 (1982).
38. I. C. M. Dea, in: *Industrial Polysaccharides*: *Genetic Engineering, Structure Property Relations and Applications* (M. Yalpani, ed.), p. 207, Elsevier, Amsterdam (1987).
39. P. E. Jansson, B. Lindberg and P. A. Sandford, *Carbohydr. Res. 124*, 135 (1983).
40. P. E. Jansson, B. Lindberg, G. Widmalm, and P. A. Sandford, *Carbohydr. Res. 139*, 217 (1985).
41. R. Jeffcoat, in: *Industrial Applications of Enzymes* Proceedings of a conference at the University of Reading, p. 10 (1988).
42. M. J. Gidley and S. M. Bociek, *J. Am. Chem. Soc. 107*, 7040 (1985).
43. H. Gradsalen, B. Larsen, and O. Smidsrod, *Carbohydr. Res. 56*, C11 (1977).
44. R. G. LaBarge, *Food Technol.*, January 1988, p. 84.
45. B. Fitch Haumann, *J. Am. Oil Chem. Soc. 63*, 278 (1986).

5

Biocatalysts for the Synthesis and Modification of Biopolymers

ROBERT J. LINHARDT and ALI AL-HAKIM

5.1. INTRODUCTION

Biopolymers are polymers that are biosynthesized from monomeric building blocks. Although biopolymers have been used for many years by the food, pharmaceutical, and specialty chemical industries, they have recently taken on increasing industrial importance often for novel applications.[1–4] The heightened interest in these materials has primarily been the result of new biotechnological methods[5–7] which have given industry access to large quantities of highly-pure biopolymers of various structures representing both natural and nonnatural products.

Three major classes of biopolymers have commanded the most attention from the scientific community; these are proteins, polysaccharides, and nucleic acids. Proteins can be conveniently broken into two classes—those with catalytic activity and those without. Enzymes have catalytic activity and thus are not consumed in the chemical reaction which they facilitate, while many other bioactive proteins (i.e., antibodies, growth factors, polypeptide hormones, etc.) participate stoichiometrically in a reaction. Other proteins, such as collagen, may play structural roles, supporting tissues in higher organisms. Polysaccharides are similar to proteins in their

ROBERT J. LINHARDT and ALI AL-HAKIM • Division of Medicinal and Natural Products Chemistry, College of Pharmacy, The University of Iowa, Iowa City, Iowa 52242.

multiple roles as both bioactive (catalytic and noncatalytic) and structural biopolymers.[8] Nucleic acids, although primarily acting as informational molecules,[9] also have demonstrated catalytic or pseudocatalytic activities.[10]

Biopolymers can be prepared and modified in several ways. The first and oldest method is by extraction from plant and animal tissues followed by fractionation or chemical modification. Although extractive techniques are still widely used to prepare biopolymers, this method has certain shortcomings. Extraction generally only affords the natural product, and chemical modification of a biopolymer resulting from extraction procedures is usually uncontrolled and often undesirable.

Fermentation is an ancient method used to prepare natural products including biopolymers. When it is coupled with modern methods such as recombinant genetics and tissue culture technology, biopolymers having very specific structural characteristics can often be produced. For the preparation of proteins, genetic engineering and site-directed mutagenesis[11,12] have become the methods of choice to prepare both structurally modified proteins and protein-based natural products. Chemical methods have also been widely used to synthesize both natural and nonnatural biopolymers. Chemical synthesis of proteins and nucleic acids has been accomplished using solid-phase supports utilizing automated synthesizers.[13] Polysaccharide synthesis is somewhat less developed, in part due to the high level of chirality present in these molecules. Recent advances in the chemical synthesis of oligosaccharides and polysaccharides suggest that someday automated solid-phase synthesizers will be available even for the complex-carbohydrate-based biopolymers.

Enzymatic synthesis as an additional method for the preparation and modification of biopolymers has recently received a great deal of attention.[14–16] Increased interest in the use of enzymes as tools to synthesize and modify biopolymers has resulted from several recent scientific developments. First, enzymes of high purity have become increasingly available due to recent progress in biotechnology.[17] These enzymes can now be modified using site-directed mutagenesis to impart different specificity or increased stability.[18] Designer enzymes represent a set of precision tools useful in the construction and modification of biopolymers. Second, advances in the use of enzymes as reagents have resulted from the discoveries that enzymes can often catalyze nonnatural reactions, that enzymes can work well in organic solvents,[19–22] and that many enzymes can be immobilized and used efficiently in bioreactors. Third, the scientific community has begun to recognize that enzymes often represent the best way to prepare or modify biopolymers both from the standpoint of their high regio- and stereoselectivity and that of their ability to be used under very mild reaction conditions.

This chapter will focus on the use of enzymes to prepare and modify biopolymers including proteins, nucleic acids, and polysaccharides. The preponderance of biopolymers and modified biopolymers, having high commercial value, are being investigated for their therapeutic potential. Therefore, the major emphasis of this chapter will be on biopolymers which have potential importance to the pharmaceutical industry.

5.2. PROTEINS AND POLYPEPTIDES

5.2.1. Biosynthesis

Proteins are prepared biosynthetically in ribosomes by the addition of amino acids to the carboxyl terminal of the growing peptide chain. The activating enzymes are aminoacyl-tRNA synthetases. These enzymes construct a protein having a sequence directed by a mRNA template. This class of enzymes requires a cofactor such as a nucleoside triphosphate like ATP, to form a peptide linkage between the amino acid residues of the growing peptide chain. Following their template-directed synthesis, proteins are often further processed by cleavage at certain sites catalyzed by specific proteolytic enzymes. The reader is referred to more detailed reviews in this field.[23,24]

The enzymes that are responsible for the biosynthesis of proteins have not been useful for *in vitro* synthesis because of problems associated with cofactor regeneration, enzyme availability, cost, and stability. Therefore, alternative chemical methods for the synthesis of proteins and polypeptides have been developed.

5.2.2. Chemical Synthesis

Various chemical methods have been developed by which single amino acids, dipeptides, or polypeptides are covalently coupled, affording a target polypeptide sequence. Most of the reported methods deal with the production of polypeptides that have properties similar or identical to natural products.

Although chemical synthesis offers different ways for the preparation of desired peptides in reasonable quantities, problems such as racemization and side reactions still exist. The development of solid-phase synthesis and its constant improvement have resulted in convenient methods for obtaining pure peptides, particularly since the introduction of automated synthesis. Solid-phase synthesis is based on the addition of a single amino acid residue at a time to a growing chain by coupling the α-COOH group of one amino acid derivative with the α-NH_2 group of another, as shown in Figure 1. Peptide synthesis utilizes a variety of approaches to active

$$R^1NH\text{-}R^2CH\text{-}COOH + NH_2\text{-}R^3CH\text{-}CO_2\text{-}R^4 \longrightarrow R^1\text{-}NH\text{-}R^2CH\text{-}CO\text{-}NH\text{-}R^3CO_2\text{-}R^4$$

Figure 1. Coupling of the α-COOH of a protected amino acid with the α-NH_2 group of a second protected amino acid. R^1 and R^4 are the amino and carboxyl protecting groups while R^2 and R^3 designate the first and second amino acids.

functional groups including azide coupling,[25] carbodiimide coupling,[26] and mixed anhydride coupling.[27] Enzyme-catalyzed coupling has also been used in the formation of polypeptide bonds[28] and is discussed in detail later.

Amino group protection, of the reactant with the available carboxyl group, and carboxyl group protection, of the reactant with the available amino group, are blocking steps essential for the prevention of side reactions. There are a number of amine protective groups including urethane derivatives[29] and trifluoroacetyl derivatives that can be removed under very mild conditions.[30] Carboxyl groups are generally protected by esterification.[31] Other functional groups, such as the nitrogen of the histidine side chain, are protected as an imidazole derivative[32] and tyrosine and serine hydroxyl groups are protected by etherification.[33]

There are two general strategies of protein synthesis both of which use these protection and coupling procedures. The stepwise chemical approach starts at a terminal group and proceeds by the addition of single amino acids in the desired sequence. The peptide condensation approach requires the assembly of separate portions of the peptide chain often prepared by stepwise chemical synthesis with appropriate blocking and activating groups in place. These peptides are then coupled to form a larger peptide of the desired sequence.

Liquid-phase peptide synthesis, carried out in a homogeneous solution,[34] requires multiple synthetic (blocking and deblocking) steps. Purification by crystallization, extraction, or chromatography is required after each synthetic step. Consequently, the yield is considerably lower than solid-phase synthesis. Thus large-scale peptide synthesis is typically performed using the solid-phase peptide synthesis method introduced by Bruce Merrifield[35] and subsequently modified.[36] Its practical application of insoluble, polymer-bound reagents has enhanced its utility particularly with the advent of automated solid-phase peptide systems. These automated systems can produce large quantities of synthetic peptides in a relatively short time.

The synthetic scheme used in automated, modified Merrifield synthesis begins with the attachment of the C-terminal residue of the peptide to be synthesized to an insoluble solid support. The peptide chain is then synthesized from the amino terminus of the growing peptide chain. When the desired peptide chain has been completed, it is cleaved from the support (Figure 2).

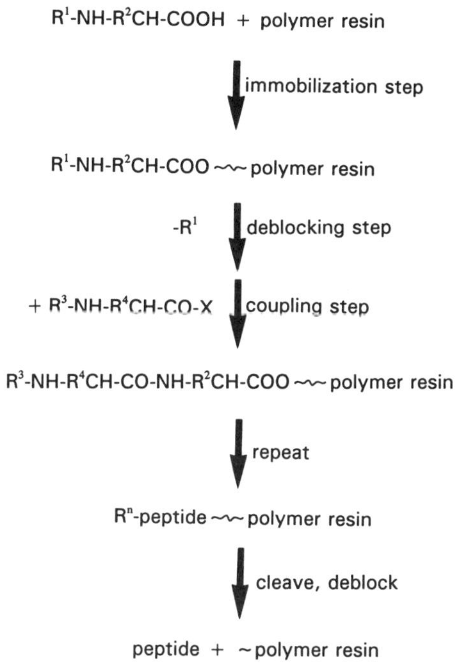

Figure 2. Automated solid-phase synthesis of peptides. The C-terminus of a protected amino acid is coupled to a solid, polymer resin support. A peptide of the desired amino acid sequence is then synthesized on the polymer resin by adding amino-blocked carboxy-activated (X) amino acids. The synthesized peptide is then released from the support by treatment with mild acid. R^1 is the amino-protecting group, while R^2 and R^3 designate the first two amino-acid residues added to the growing peptide chain.

Thousands of synthetic peptides having biological activity have been prepared using chemical methods. One of many applications is the chemical synthesis of peptides resembling specific regions of viral coats as immunogens or for blocking viral infection.[37,38] For example, an 18-residue synthetic peptide corresponding to the region of the envelope glycoprotein of HIV (responsible for acquired immune-deficiency syndrome) has been chemically synthesized and found to elicit antibodies that recognize intact precursor envelope glycoprotein.[39]

5.2.3. Enzymatic Synthesis and Modification

Protease-catalyzed enzymatic synthesis of bioactive proteins and polypeptides is an area of great interest and activity.[40–43] These proteolytic enzymes are used in organic solvents or in the presence of molecular traps to form peptide bonds.[44] Peptide synthesis using reverse enzymic catalysis decreases the requirements for protective blocking groups and often results in enhanced stereo- and regioselectivity, representing the main advantage over chemical methods of peptide synthesis.

Semisynthetic techniques, which combine chemical and enzymatic methods, have been widely used in the preparation and coupling of bioactive peptides.[45] For example, a peptide isolated from naturally occurring protein can be coupled to a synthetic peptide to form the same protein[46] or an entirely new protein. This approach was applied successfully to the preparation of ribonuclease-S[1–20],[46] cytochrome C,[47] human insulin (HUMULIN),[48] and somatotropin[49] and other proteins.[50]

Aspartame is a low-calorie dipeptide sweetener, commercially important as a sugar substitute. Enzymatic synthesis of aspartame is based on the coupling of *N*-carbobenzoxy-L-aspartic acid and D, L-phenylalanine methyl ester. The reaction is catalyzed by thermolysin, a metalloprotease, in either aqueous[51] or organic solvents.[52]

The two main strategies used in peptide synthesis by enzymatic condensation are thermodynamic control and kinetic control.

5.2.3.1. Thermodynamically Controlled Synthesis

Thermodynamically controlled synthesis involves the direct reversal of protease-catalyzed peptide hydrolysis (Figure 3). The equilibrium of reactions (a) and (b) in Figure 3 favors hydrolysis (a) in an isolated or undisturbed system under physiologic conditions. However, if a substrate of reaction (b), R^1CO_2H and/or R^2NH_2, is used at a high concentration, reaction (b) can be driven to produce the condensation product.[53] The formation reaction can also be driven by continuously removing the product R^1CONHR^2 through precipitation[54] or adsorption to a solid phase, or by using a molecular trap (such as an antibody which binds product but not starting materials). The use of a molecular trap couples the peptide formation reaction to a second, thermodynamically highly favorable, ligand-binding reaction. This approach was used for reversible modification of the active site of soybean trypsin inhibitor.[55] The use of an organic cosolvent which can accommodate both the substrates and the enzyme but precipitates the peptide product will drive the reaction towards product formation.[56] The equilibrium of a reversible reaction (Figure 3, reactions (a) and (b)) can be altered by using an organic cosolvent such as glycerol, dimethylformamide, ethylene glycol, or dimethylsulfoxide. Water–organic

a. $R^1CONHR^2 + H_2O \longrightarrow R^1CO_2^- + {}^+H_3NR^2$ (hydrolysis)

b. $R^1CO_2H + H_2NR^2 \longrightarrow R^1CONHR^2 + H_2O$ (formation)

c. $R^1CO_2H + H_2NR^2 \rightleftharpoons R^1CO_2^- + {}^+H_3NR^2$

Figure 3. Reactions important in thermodynamically controlled peptide formation and hydrolysis. These are: (a) peptide hydrolysis; (b) peptide formation; and (c) deprotonation and protonation of carboxyl and amino groups. In peptide hydrolysis (a), the carboxyl and amino groups in the products are charged as is typically favored under physiologic conditions. In peptide formation (b), the carboxyl and amino groups in the reactants are uncharged. R^1 and R^2 designate different amino acids.

solvent systems increase the pKa value of the carboxyl component. This favors the protonated form R^1CO_2H over $R^1CO_2^-$ by shifting the equilibrium (reaction (c) in Figure 3) to the left. The increased concentration of the protonated form of the acid, a substrate in the reaction (b) (Figure 3) results in increased peptide formation. For example, the synthesis of Cbz–Try–Gly–NH_2 from Cbz–Try–OH and Gly–NH_2 was maximum in 60% trimethylene glycol. The increased peptide formation in the presence of an organic cosolvent is attributable to the suppression of the transfer of the acidic proton from the carboxyl group to the amino group. This method has been used to prepare human insulin from porcine insulin[43] by condensation of desoctapeptide insulin with synthetic tripeptide using trypsin in 50% dimethyl formamide.[57] Human insulin is also prepared by tryptic transpeptidation of porcine insulin and threonine derivatives using cosolvents.[58]

5.2.3.2. Kinetically Controlled Synthesis

Enzymatic synthesis of proteins and peptides can also be achieved by taking into account the kinetic considerations governing the reaction shown in Figure 4.

Kinetically controlled synthesis involves the enzymatic coupling of an *N*-blocked amino acid ester (acyl donor) with an amino acid or amino acid ester (acyl acceptor).[59,60] This approach is limited to enzymes forming an acyl-enzyme intermediate such as lipases and pig liver esterases[61,62] and having no amidase activity. Proteases which form acyl-enzyme intermediates can also be used; however, cleavage of peptide bonds in the resulting products can be caused by their amidase activity. Sometimes proteases can be modified to remove their amidase activity while retaining their esterase activity.[61] For example, the active-site serine of subtilisin could be modified, reducing its amidase activity while maintaining its weak esterase activities.[63]

A number of biologically important peptides containing unusual

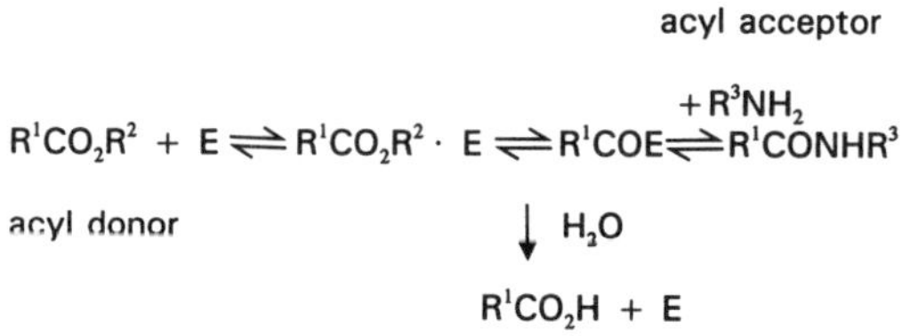

Figure 4. Kinetically controlled synthesis of peptides. $R^1CO_2R^2$ is an *N*-blocked amino acid ester representing an acyl donor and R^3NH_2 is an amino acid or an amino acid ester representing an acyl acceptor. E is an esterase, lipase, or modified protease having no amidase activity capable of forming an acyl-enzyme intermediate R^1COE.

D-amino acids have been prepared. Proteases will often not accept peptide linkages composed of D-amino acids as substrates. Thus protease-catalyzed bond formation involving these D-amino acids is not possible, nor is recombinant technology useful in preparing peptides or proteins from such unusual amino acids. However, esterase-catalyzed, kinetically controlled peptide synthesis has been used successfully to prepare peptides containing D-amino acids.[62,64]

5.3. POLYSACCHARIDES AND OLIGOSACCHARIDES

5.3.1. Structure

Oligosaccharides and polysaccharides are comprised of monosaccharide residues in either the furanose (5-membered ring) or the pyranose (6-membered ring) form, attached through glycosidic linkages.[4] The difference between an oligosaccharide and a polysaccharide is primarily one of size. Typically, oligomers having less than ten sugar residues are called oligosaccharides while those with more than twenty residues are called polysaccharides. Glycosidic linkages occur between the 1-position (anomeric position) of one saccharide unit and a nonanomeric hydroxyl group on the second saccharide unit. A typical disaccharide unit composed of two simple hexopyranose saccharide units can be linked between the $1 \rightarrow 2$, $1 \rightarrow 3$, $1 \rightarrow 4$, or $1 \rightarrow 6$ positions in either of two linkage configurations α or β owing to the chirality of the anomeric center. Thus there are sixteen possible disaccharides comprised of any two simple hexopyranose saccharide residues. If the chirality, ring size, and substitution by nonsugar moieties (i.e., acetate, amino groups, phosphate, sulfate, etc.) are considered, thousands of different disaccharide structures can be drawn. While many oligosaccharides and polysaccharides are linear, like proteins and nucleic acids, branching is often observed, increasing the number of possible structural isomers. Glycoconjugates are carbohydrates attached to small molecules (aglycones) such as steroids, and lipids or larger molecules such as peptides and proteins. Glycoproteins (and glycopeptides) are biopolymers in which an oligosaccharide is attached to a structurally dominant protein component. Alternatively, in proteoglycans (and peptidoglycans) a structurally dominant polysaccharide is attached to a smaller protein or peptide core. In this section pure carbohydrate containing biopolymers as well as biopolymeric glycoconjugates are discussed.

This high level of structural diversity in complex carbohydrates suggests the utility of enzyme-assisted synthesis and structural modification with its expected high level of regio- and stereoselectivity. The large-scale production of recombinant proteins for pharmaceutical applications has demonstrated the need for developing methods for the enzyme-assisted synthesis and enzymatic modification of the carbohydrate portion of glycoproteins. Many recombinant proteins either totally lack attached carbohydrate or are glycosylated differently from the natural products which they are intended to supplant. Carbohydrate "remodeling" has been proposed and uses purified enzymes to modify these glycoproteins to enhance their activities or alter their *in vivo* half-lives.[65] Before such modification of synthesis can be undertaken, the structure of the glycoprotein to be remodeled and the desired carbohydrate structure (often that of the natural product) must first be determined. The determination of the carbohydrate structure of glycoproteins—molecules that are very complex, structurally heterogeneous, and available in very small quantities—often requires an involved, integrated approach using chemical, enzymatic, and spectroscopic methods.[66]

5.3.2. Biosynthesis

Polysaccharide and oligosaccharide biosynthesis differs from nucleic acid and protein biosynthesis in that it is not template directed. However, this does not mean that it is any less complex or that information cannot be stored (coded) within the resulting carbohydrate biopolymer. The biosynthesis of polysaccharides and oligosaccharides is a highly regulated, multienzyme process which often takes place in specialized cellular com-

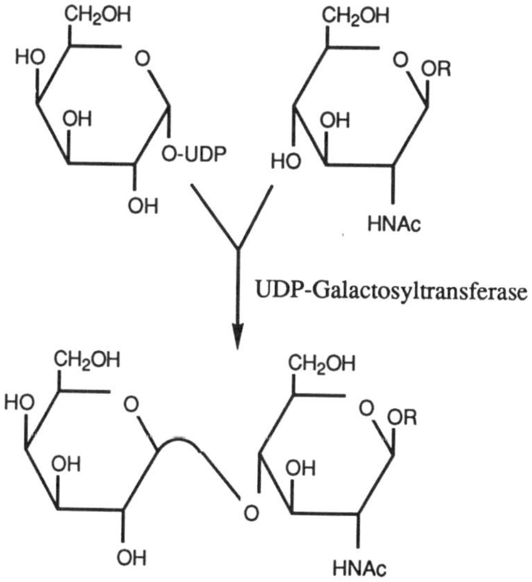

Figure 5. Carbohydrate synthesis by stepwise addition of nucleotide diphosphate sugar. UDP is uridine diphosphate, Ac is acetate, and R represents one or more sugar residues.

partments. Once the glycosidic linkages are formed, further enzymatic modification, such as acetylation, sulfation, phosphorylation, etc., takes place. The final polysaccharide or oligosaccharide component that "decorates" a glycoprotein is often more structurally complex than the protein portion of the molecule.

There are three major types of glycosylation reaction.[67] The first type

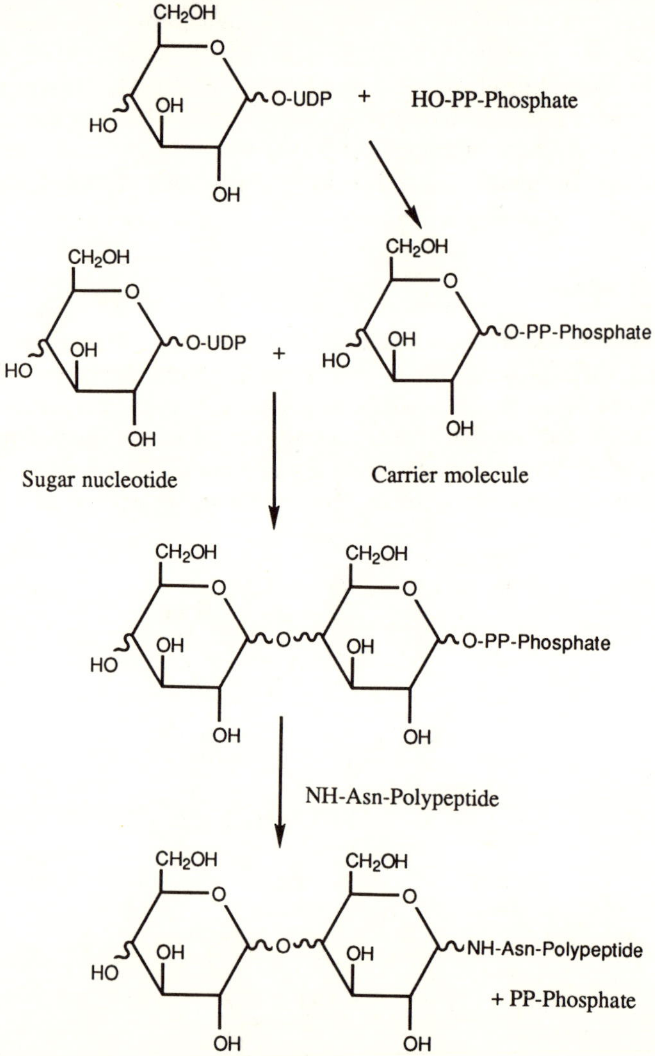

Figure 6. Carbohydrate synthesis by transfer of a glycosyl from a sugar nucleotide to a carrier molecule. UDP is uridine diphosphate, PP-phosphate is polyprenyl phosphate, and Asn is asparagine.

(Figure 5) is the formation of carbohydrate chains usually on core protein by stepwise addition of nucleotide diphosphate sugar to an acceptor with the release of nucleotide diphosphate. This irreversible transfer reaction resulting in glycosylated acceptor is catalyzed by glycosyl transferase. This enzyme is specific for both the acceptor and the donor nucleotide sugar.

The second glycosylation process (Figure 6) involves the transfer of glycosyl residue from a sugar nucleotide to a carrier molecule such as

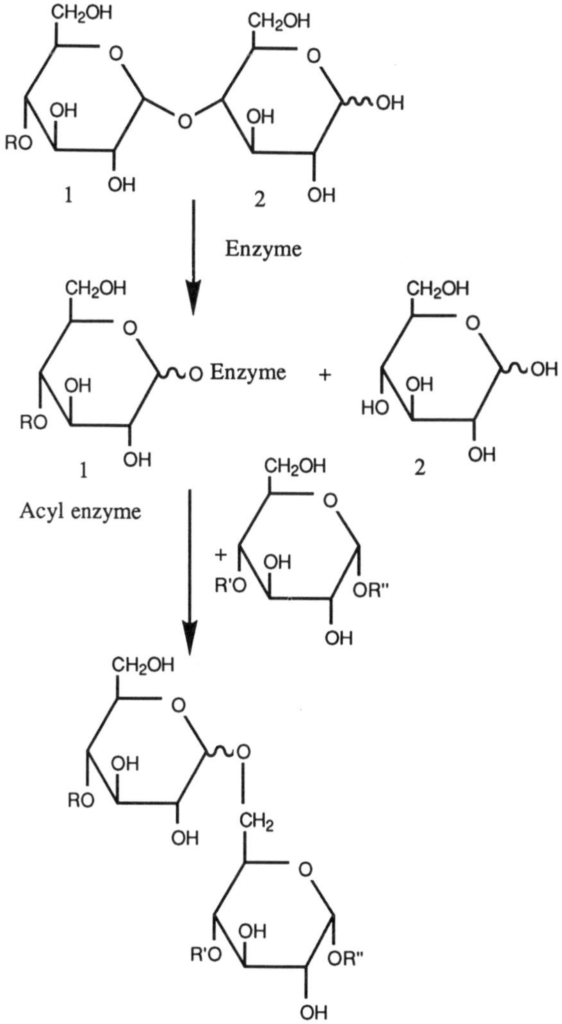

Figure 7. Transglycosylation of preassembled oligosaccharides leading to branching. R, R′, and R″ represent one or more sugar residues.

polyprenyl phosphate or diphosphate with the release of inorganic phosphate. Core oligosaccharides are then assembled on polyprenyl diphosphate carriers. Glycosyl transfer from carrier to protein is followed by glycosidase-catalyzed remodeling of oligosaccharide, eventually leading to the formation of the final carbohydrate chain.

The last type of glycosylation (Figure 7) is the transglycosylation of preassembled oligosaccharide fragments, which often leads to polysaccharide branching. This type of assembly is usually applied to the preparation of storage polysaccharides, resulting in the formation of branched structures from linear structures. This mechanism does not involve phosphate-activated species; rather, the enzyme is glycosylated as an acyl-enzyme-activated intermediate.[67]

5.3.3. Chemical Synthesis

The chemical synthesis of oligosaccharides requires careful planning due to the large number of possible arrangements of monosaccharide residues. The pivotal step in the chemical synthesis of oligosaccharide is the stereospecific and regiospecific formation of glycosidic linkages between polyfunctional sugar components. This coupling process generally involves the reaction of an activated glycosyl donor[68,69] such as a glycosyl halide with a suitably protected saccharide acceptor (Figure 8). The coupling step itself must be controlled to give a product with the required configuration (either α or β). All steps, i.e., activation, formation of glycosidic linkage, protection, deprotection, and other modifications must have high regio- and stereoselectivity to prevent the formation of complex product mixtures.

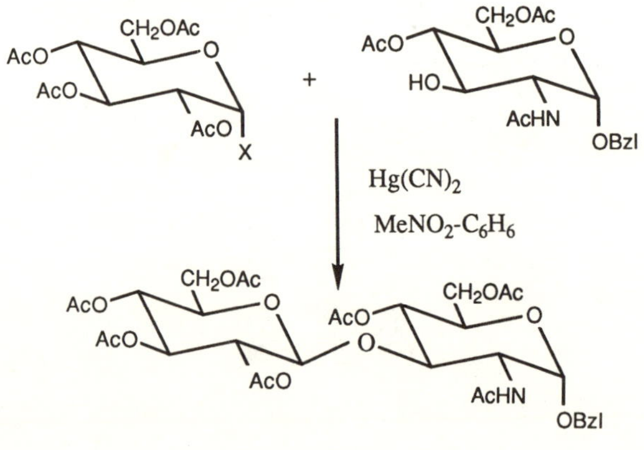

Figure 8. Chemical synthesis of a disaccharide using a glycosyl halide and appropriately protected sugars. X represents a halide, Ac is acetate, and Bzl is benzyl.

Multiple synthetic steps, required for appropriate blocking and deblocking reactions, lower the yields, as do chromatographic and other purification steps.

5.3.4. Enzymatic Synthesis

Enzymatic synthesis of oligo- and polysaccharides is less laborious than chemical synthesis.[70] With the increased availability of enzymes acting on sugars and the development of new approaches for enzymatic synthesis, enzymatic methods are becoming a routine practice in polysaccharide synthesis. The advantages of enzymatic synthesis over chemical synthesis are a decrease in the number of required synthetic steps and higher regio- and stereoselectivity. The chemical synthesis of a simple disaccharide may take as many as six synthetic steps due to required protection and deprotection, while the same enzymatic synthesis can often be performed in a single step. As discussed earlier, although regioselectivity can be controlled in chemical synthesis by the appropriate use of well-placed protective groups, stereoselectivity is often more difficult to control. Enzymes are chiral catalysts and thus have very high levels of stereoselectivity. Thus, by decreasing the number of steps while increasing the regio- and stereoselectivity, higher yields and greater product purity are obtained.

The main two types of enzymes which are used for the preparation of complex oligo- and polysaccharides are the glycosyltransferases (EC 2.4.–.–) and the glycosidases (EC 3.2.–.–).

5.3.4.1. Glycosyltransferase-Catalyzed Synthesis

The glycosyltransferases are a group of enzymes that catalyze the specific glycosidic synthesis by the transfer of a monosaccharide from glycosylnucleotide donor substrate to an acceptor saccharide substrate (Figure 5). This transfer process is both stereo- and regiospecific. Glycosyltransferases are present at very low concentrations on the intracellular membrane and are difficult to purify; thus only a small number are commercially available.[70] In addition, these enzymes require expensive activated nucleotide sugars[71,72] as substrates, and are highly specific enzymes catalyzing only certain reactions, which limits their general utility.[70]

A few glycosyltransferases that are used in oligosaccharide synthesis have been isolated and purified.[73] A pure glycosyltransferase typically catalyzes the synthesis of only one linkage and the product can be purified with relative ease. Some examples include the enzyme-catalyzed synthesis of *N*-acetyllactosamine using galactosyltransferase.[71] UDP galactosyltransferase has also been utilized in the coupling reaction to form di- and trisaccharides (Figure 9).[74]

Solid-phase synthesis of oligosaccharides using glycosyltransferases represents an important advance.[75] An interesting compound, the trisaccharide Neu-5-Ac-$\alpha(2 \rightarrow 3)$Gal-$\beta(1 \rightarrow 3)$Glc-NAc, has been synthesized by the use of the immobilized enzyme, sialytransferase. This sequence is of particular interest because it is frequently bound to the terminus of glycoproteins.[72]

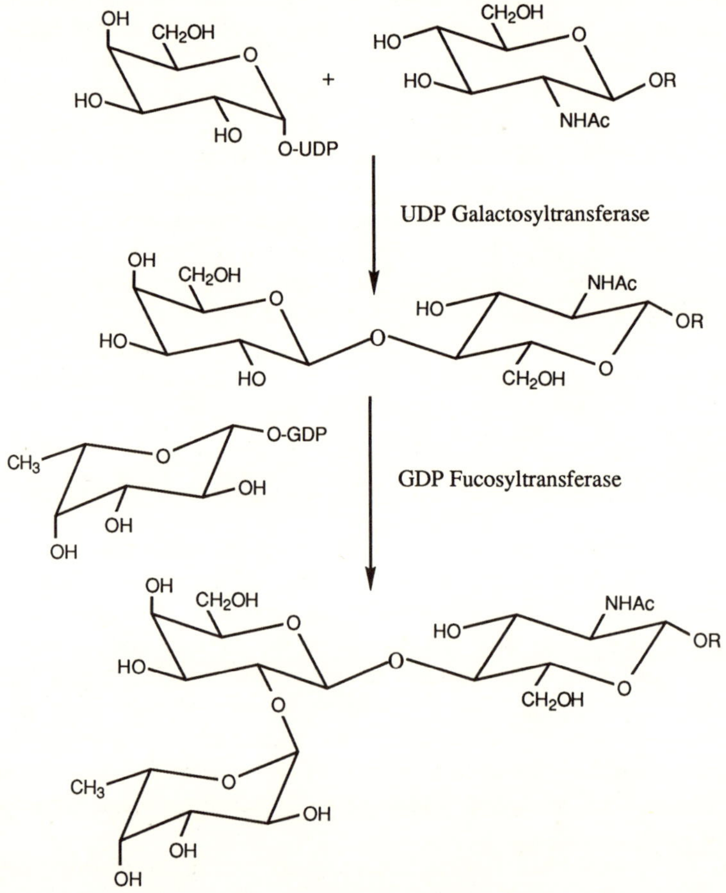

Figure 9. Enzymatic synthesis of a disaccharide and a trisaccharide using glycosyltransferases and nucleotide sugars. UDP is uridine diphosphate, GPD is guanidine diphosphate, and Ac is acetate.

5.3.4.2. Glycosidase-Catalyzed Synthesis

Glycosidases are enzymes that are typically thought to hydrolyze glycosides, but this reaction only represents a specialized case of the transfer of a glycosyl moiety from a substrate to a water molecule acting as an acceptor. This transfer could as easily take place to a sugar acceptor (Figure 10). A wide variety of glycosidases (both endolytic and exolytic enzymes) are commercially available.[66] These enzymes are named for the glycosyl portion of the donor substrate (Figure 10) and are usually quite specific for this sugar residue and its anomeric configuration, but are at the same time quite tolerant of a wide variety of acceptor molecules. This can lead to problems in the regioselectivity of glycosidase-catalyzed transglycosylation reactions resulting in mixtures of (1 → 2)-, (1 → 3)-, (1 → 4)-, and (1 → 6)-linked products.[76,77]

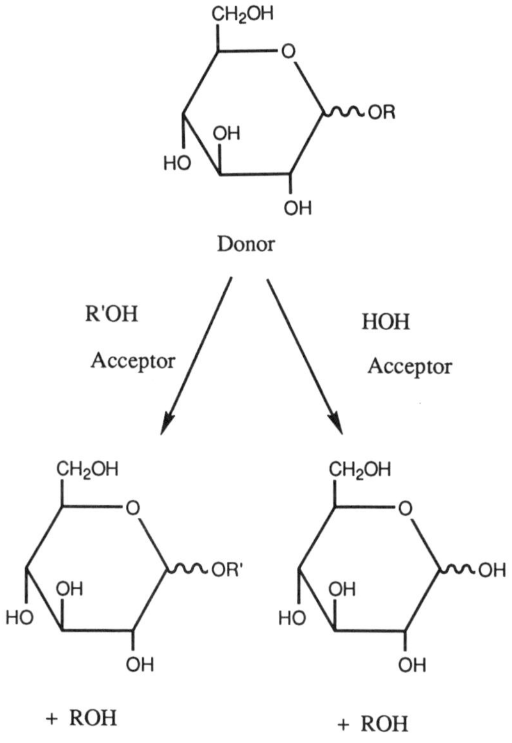

Figure 10. Glycosidase-catalyzed synthesis using a glucoside as the example. Transfer of a glycosyl moiety from donor to acceptor is shown. In the glucoside donor R typically represents a good leaving group such as *p*-nitrophenol. When the acceptor is water, glucose is the product. When a sugar or a simple alcohol (R′OH) is the acceptor, a new glucoside is formed.

As with protease-catalyzed synthesis of peptides, glycosidases can be used under either kinetic or thermodynamic control.[73] Synthesis under equilibrium control uses high monosaccharide substrate concentrations[78] or molecular traps[78] to drive the reaction to glycoside formation (Figure 11).

Kinetically controlled formation of glycosides using glycosidases first requires the synthesis of glycoside substrate. This substrate effectively glycosylates the enzymes which then serve as the glycosyl donors (Figure 7). This uses considerably less enzyme than does thermodynamically controlled synthesis, and it does not require the use of high substrate concentrations or the use of molecular traps. The glycoside substrate is consumed in the reaction, and its hydrolysis often represents a competing side reaction.[70] Generally, aromatic aglycones are used in such synthesis,[76,77] but inexpensive disaccharides can also be used as donors.

Compared to protease-catalyzed peptide synthesis, glycosidase-catalyzed glycoside synthesis has several serious drawbacks. First, hydroxyl groups are less nucleophilic than amino groups, leading to equilibria in glycosidase-catalyzed reactions, lying further toward hydrolysis than formation. Second, the presence of multiple nucleophiles on the acceptor molecule often results in multiple products ((1 → 2)-, (1 → 3)-, (1 → 4)-, (1 → 6)-linked glycosides) with 1 → 6 formation being favored because the primary hydroxyl group at the 6-position is usually the most nucleophilic. This regioselective preference for the 1 → 6 linkage can be countered by using specific glycosidases that favor other linkages[79] or by using aglycones that have demonstrated the ability to alter the regioselectivity of the transfer reaction.[80] Third, organic solvents, which favor protease-catalyzed peptide formation, result primarily in decreased glycosidase activity and thus decreased oligosaccharide formation. This may be due to an increase in the binding affinity of polar sugar substrate (or product) to the enzymes active site, resulting in substrate (or product) inhibition of the enzyme.[19]

Both glycosyl transferases and glycosidases can be used in concert to synthesize complex oligosaccharides. By sequentially using β-galactosidase and β-D-galactoside-3-α-sialyltransferase, α-D-Neup5Ac-(2 → 3)-β-D-Galp-(1 → 3)-α-D-GalpNAc-OR could be synthesized.[76]

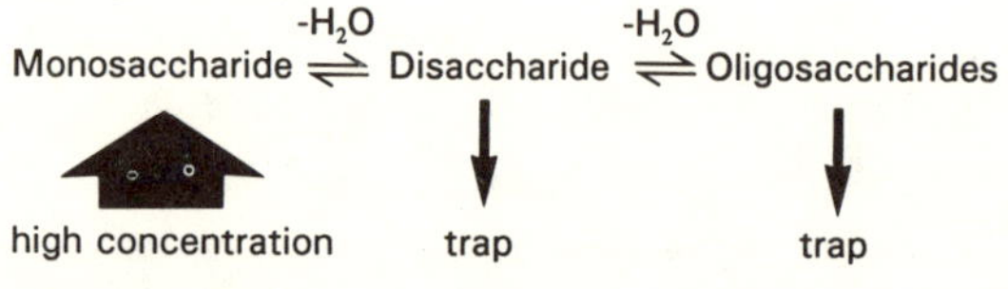

Figure 11. Thermodynamically controlled glycosidase-catalyzed synthesis of disaccharides and oligosaccharides. Synthesis is driven by using high concentrations of monosaccharide or by using molecular traps such as activated carbon.

Future prospects might include combined chemical and enzymatic (glycosidase and glycosyltransferase) synthesis.[70] Genetic engineering might be an appealing approach to preparing such complex molecules as oligosaccharides and polysaccharides. Polysaccharide biosynthesis, however, is highly compartmentalized and requires many enzymes making the near term success of such an approach highly improbable. Some recent success in this area—the preparation of cyclodextrin from genetically engineered *Bacillus subtilis*[81]—gives hope that efforts in this direction will bear fruit.

5.3.5. Enzyme-Assisted Modification or Degradation

Enzymatic hydrolysis of oligosaccharides and polysaccharides is an important tool in structure determination, sequencing and characterization of carbohydrate composition. Precursors used in organic synthesis, representing chiral synthons, are often obtained from degradation of polysaccharides. These range from simple disaccharides such as Gal(α-1 $\rightarrow$ 4)-Gal(β)-OMe, obtained from the glycosidase-catalyzed hydrolysis of polygalacturonic acid,[82] to structurally complex oligosaccharides obtained from glycoproteins.[66]

5.3.5.1. Neutral Polysaccharides

Glycoproteins represent the final product of many recently developed biotechnological processes such as recombinant genetic engineering, monoclonal technology, and cell-tissue culturing. Although these glycoproteins have many applications, the most important so far has been as pharmaceuticals. Recent animal and human testing of these glycoprotein-based drugs has demonstrated that their glycosylation plays an important role in their biological activities.[66] The term "carbohydrate remodeling" has recently been coined when talking about chemical and enzymatic approaches aimed at modifying the carbohydrate portion of glycoproteins. First the carbohydrate composition, the oligosaccharide sequence, and the site(s) of glycosylation in glycoprotein obtained from cell culture or recombinant organisms are determined. Then the glycosylation requirements for appropriate activity, specificity, and pharmacodynamics are established and a structure–activity relationship (SAR) is developed to understand the glycosylation requirements for ideal pharmacological activity. Once a target has been selected, the glycoprotein is remodeled, primarily by using specific glycosidases. These enzymes are tools that can be used with precision to cut and trim the oligosaccharide groups of a glycoprotein to a form which results in an ideal pharmacological agent.

These enzymes are also used for determination of oligosaccharide content and sequence.[66] Selective hydrolysis using an exolytic glycosidase removes one sugar residue at a time from the nonreducing end of an oligosaccharide chain.[66] Alternatively, hydrolysis or partial hydrolysis using a specific endolytic glycosidase results in smaller oligosaccharides that can be used in structural determination.[66,83] Enzymatic processes are also important in the food industry for the modification, mostly through partial hydrolysis, of a variety of polysaccharides, important as food additives and in baking and brewing. The hydrolysis of starch using amylase is an example.[84]

5.3.5.2. Acidic Polysaccharides

Depolymerization of acidic polysaccharides using lyases catalyzes the eliminative cleavage (Figure 12) of a glycosidic linkage (distinctly different from the hydrolysis mechanism associated with glycosidases) and results in one product chain containing an unsaturated sugar residue at its nonreducing end.[85] These enzymes can cleave acidic polysaccharides with either an endolytic or exolytic action pattern resulting in either small or large oligosaccharide products.[85–88] Lyases which act on plant polysaccharides, such as pectin/pectate and alginate are used by the food industry to reduce their molecular weight or to remove these polysaccharides from complex mixtures.[89] Complex acidic polysaccharides derived from animal tissue, including proteoglycans and glycosaminoglycans, have important biological functions.[90] Heparin, for example, has been intensively studied for over

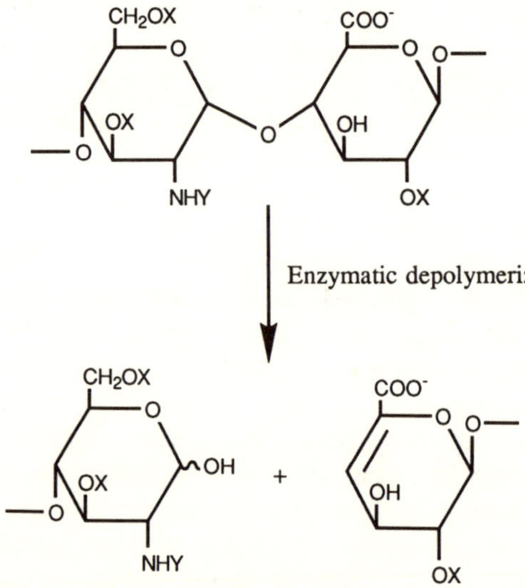

Figure 12. Eliminative cleavage of an acidic polysaccharide by a polysaccharide lyase. X is SO_3 or H, and Y is SO_3 or $COCH_3$.

fifty years by many research groups including ours, in an effort to improve its properties as an anticoagulant drug.[2,3,8] Heparin lyases[90,91] have been used in the controlled depolymerization of the drug heparin to prepare low-molecular-weight heparins with improved anticoagulant properties.[8] Heparin oligosaccharides, prepared enzymatically from heparin, exhibit exhibit other important activities including anti-atherosclerotic activity, the ability to inhibit complement activation, and also anti-angiogenic, antitumor, and antiviral activities.[8] The chemical structures of many of these oligosaccharides have been characterized using spectroscopic methods, and this has led to a better understanding of the precise structure of the complex, microheterogeneous, polydisperse polysaccharide heparin.[92,93] One of these oligosaccharides which has been produced by chemical and enzymatic controlled depolymerization of heparin[94,95] as well as by *de novo* multistep chemical synthesis[96] is heparin's antithrombin III binding site. This unusual pentasaccharide is believed to be primarily responsible for heparin's anticoagulant activity.[8] Although controlled enzymatic depolymerization of heparin can result in low recoveries if intact binding site,[94] exhaustive treatment with heparin lyases destroys this binding site.[8,94] Work is currently underway in our laboratory to control this enzymatic reaction in an effort to prepare a structurally defined oligosaccharide having an intact antithrombin III binding site in high yield in a single enzymatic step in an effort to rival the complicated and time-consuming chemical synthesis currently used.[96] There is no reported reverse enzymic catalysis involving polysaccharide lyases. Such an approach might be extremely useful in reassembling oligosaccharides into acidic polysaccharides of defined structure and with desirable biological properties.

5.4. NUCLEIC ACIDS

5.4.1. Biosynthesis

Nucleic acids, deoxyribonucleic acids, and ribonucleic acids (DNA and RNA) are informational molecules that are biosynthesized by all cells. The template-directed synthesis of DNA takes place in the nucleus of eukaryotes and in the cytoplasm of prokaryotes. The DNA polymer backbone consists of furanose residues linked $3' \rightarrow 5'$ through a phosphate diester linkage. In DNA synthesis, the diester linkage is constructed by coupling a deoxyribonucleotide triphosphate to a short oligonucleotide primer sequence bound to a DNA template with the release of inorganic pyrophosphate. Synthesis is catalyzed by a DNA-directed DNA polymerase exclusively in the $5' \rightarrow 3'$ direction.[97] The specificity of each step (i.e., which nucleoside is added next to the growing chain) is determined by the DNA template.[98]

The biosynthesis of RNA from a DNA template is called transcription. The enzymes involved (DNA-dependent RNA polymerases) are responsible for the catalysis of the transcription reaction. RNA polymerases are also responsible for synthesizing cellular RNA. The activated intermediates are the ribonucleoside triphosphates. Polymerization starts with the formation of a phosphodiester bond between the 5′-α-phosphate of the incoming nucleotide triphosphate and the 3′-OH group of the growing RNA chain. The RNA polymerases, unlike DNA polymerases, do not require a primer and do not have nuclease activities. The overall reaction is irreversible due to the release and subsequent hydrolysis of the pyrophosphate group. As with DNA, the synthesis of RNA proceeds in the $5' \rightarrow 3'$ direction. RNA is not a self-replicating molecule, and thus any mistakes in its sequence made during its synthesis do not become genetically inherited. RNA synthesis is highly regulated, promoters direct the binding of RNA polymerase and the unwinding of the DNA double strand to reveal the template, and the transcription process stops once a termination signal is reached.

5.4.2. Chemical Synthesis

In the last two decades, numerous synthetic methods have been developed and modified for the chemical synthesis of nucleic acids. The rapid advances in nucleic acid chemistry can be ascribed to the great biological potential of and the high demand for synthetic oligonucleotides.

The key step in oligo- and polynucleotide synthesis is the preparation of ultrapure reagents and the use of stable and protected nucleotides. Most of these required nucleotides and reagents are now commercially available.

The classical method for the chemical synthesis of oligonucleotides is the solution-phase phosphotriester approach which was popular prior to the introduction of solid-phase synthesis. This method was complicated, usually requiring an experienced organic chemist to perform the complex synthetic steps. A major achievement of this method was the chemical synthesis of a specific gene for somatostatin, which was expressed in *E. coli* to produce protein product.[99]

The introduction of solid-phase synthesis has resulted in its application throughout synthetic organic chemistry for the preparation of oligomers and polymers of known sequence through the stepwise addition of monomers.[100]

The advantages of solid-phase synthesis are many; the most important is the isolation of final product in high purity by simply washing the resin and releasing the product. High reaction yields can be obtained by using excess reagents, and all of the reaction and addition steps can be carried out in one flask, minimizing product loss. Finally, the introduction of automated synthesizers[101] makes it possible for any scientist, even one without a background in organic synthesis, to prepare the desired products in high yield and purity.

There are two general methods used in solid-phase synthesis; the phosphotriester method and the phosphite triester method (Figure 13). In the phosphotriester method the activated nucleotide, protected at its 5′ position, reacts with the 5′-hydroxyl group of a nucleoside which is already linked to the solid support to form a new 3′ → 5′ phosphotriester bond. The protecting group is then removed, exposing a new reactive 5-hydroxyl group ready for the next coupling step. Details about the coupling reaction, capping, deprotection, support type, and purification procedures can be found in many recent articles.[102–107]

The phosphite triester method is similar to the phosphotriester approach, except that the phosphite product formed in the coupling reaction is subsequently oxidized to the desired 3′ → 5′ phosphotriester linkage (Figure 13). Both of these synthetic methods are reliable, easy to perform, and have been automated. The major application of synthetic oligonucleotides is as primers to prepare recombinant DNA,[108] but they are also valuable as probes for analysis and in site-directed mutagenesis.[109]

5.4.3. Enzymatic Synthesis and Modification

Compared to other biopolymers, the use of enzymes in the synthesis and modification of nucleic acids is most well-developed. This is primarily due to three factors: (1) the importance of nucleic acids as informational molecules; (2) the uncomplicated nature (compared to the linkage chemistry of polysaccharides) of the 3′ → 5′ linkage; and (3) the availability of a variety of inexpensive, pure enzyme tools. This section will only examine the more important aspects of enzymatic synthesis and modification of nucleic acids.

5.4.3.1. Types of Enzymes Used on Nucleic Acids

Biologically-based synthesis of DNA and RNA using recombinant technology is beyond the scope of this review. Two types of enzymes used by molecular biologists—ligases and restriction endonucleases—play an important role in the synthesis of nucleic acids.[110–112] Ligases and endonucleases that act on DNA are readily available and are widely used in genetic engineering.[113,114] Restriction endonucleases recognize specific sequences and act to break phosphodiester bonds in DNA at these sites.[111] Ligases are used to rejoin these restriction-endonuclease-modified DNA

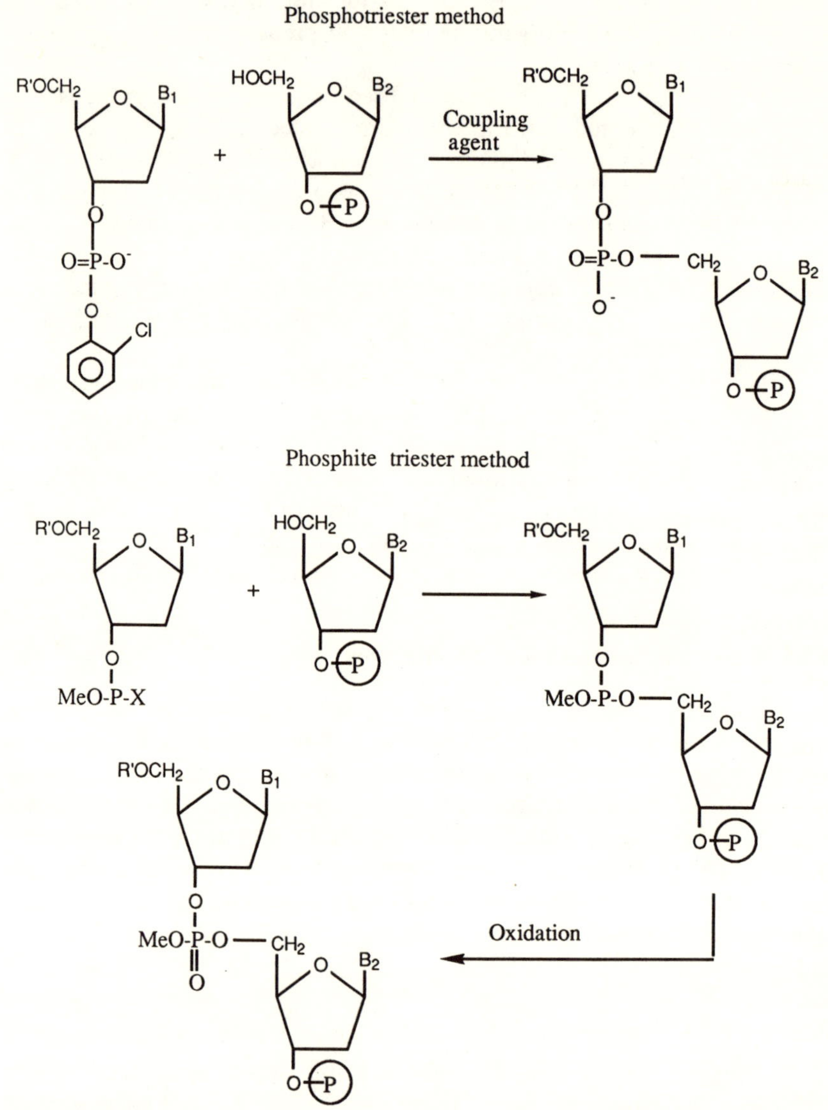

Figure 13. Solid-state chemical synthesis of nucleic acid polymers. The phosphotriester method couples an ortho-chlorophenol-activated nucleotide phosphate protected at the 5′ position (R′) with a nucleotide bound to a polymeric support (P). B_1 and B_2 are two different bases.

oligomers through reformation of the phosphodiester linkage. Small quantities of DNA oligomers are prepared *in vitro* and are then inserted into a microbe which can be grown in a fermentor to high cell densities resulting in the *in vivo* production of large quantities of the desired DNA or gene product. Large-scale enzymatic synthesis of DNA oligomers using these enzymes will become practical only when suitable quantities of stable ligases and endonucleases become available.[20]

5.4.3.2. Enzymatic Synthesis

For biological studies, oligonucleotides longer than the ones obtained using chemical methods and having well-defined sequences, are required. This has led to the frequently used approach which combines the chemical synthesis of small oligomer sequences with recombinant technology to produce larger polymers such as genes.

DNA ligase is used to join or stitch together two synthetic oligonucleotide fragments. DNA ligase, typically produced from T4-phage-infected *E. coli,* catalyzes the covalent coupling of two oligonucleotide molecules. The synthetic oligonucleotides are first phosphorylated at their 5′-hydroxyl group with T4 polynucleotide kinase, and a phosphodiester bond is formed between the 5′-phosphate and 3′-hydroxyl groups in the presence of ligase catalyst and a complementary template.[115] Many different oligonucleotide fragments can be linked or ligated together to produce the desired sequence having a variety of different restriction (sticky) ends. The newly formed polynucleotide gene can then be cloned by insertion into a plasmid (a small open strand of DNA having the same sticky ends, produced using the same restriction enzyme) and joined by DNA ligase. The gene contained within the plasmid (also containing a suitable antibiotic marker for selection) can be inserted into *E. coli* (or other suitable organism), which can be used to express the desired gene and produce its protein. Genes can also be isolated directly by digestion of native DNA with restriction enzymes. The genes corresponding to the yeast alanine tRNA,[116] insulin A chain,[117] human growth hormone,[118] and interferon[119] are among the many that have been synthesized using this recombinant technology. The use of ligases in the formation and synthesis of biologically active oligonucleotide fragments and genes has been described in detail.[120]

DNA polymerase, a major enzyme in eukaryotic cells, replicates the polynucleotide template by the sequential addition of deoxyribonucleotide monophosphate residue to the 3′-hydroxyl terminus of the primer. Catalytically it interacts with deoxynucleotide triphosphate (dNTP) substrate, the primer, template and divalent cations.

Oligonucleotides of moderate size can be used as template-primers and converted enzymatically into a gene in the presence of DNA polymerase (usually prepared from *E. coli*) and four deoxynucleotide triphosphates. This DNA can then be digested with suitable restriction enzymes and ligated to a similarly restriction-enzyme-digested plasmid. This transformed plasmid containing the desired gene can be identified using complementary synthetic oligonucleotide hybridization probes. AMV reverse transcriptase has been used instead of DNA polymerase in the complete synthesis of enzymatically produced insulin A chain gene.[121]

Application of artificial genes to problems in biochemical and biological systems is increasing. These genes can be used to probe for a defective sequence, to regulate gene expression, to produce proteins, and to recognize the sites of interaction between proteins and nucleic acids. Many proteins are difficult and often impossible to isolate and purify from their natural biological systems. Many of these proteins can be more easily prepared in microorganisms.[116–119,122] An understanding of the gene expression mechanism in *E. coli* has been due in part to the preparation and study of regulatory proteins. Various *E. coli* promoter sequences have been constructed including tyrosine suppressor tRNA promoter,[123] trp promoter,[124] and lambda PR promoter.[125] The interaction of these promoters with DNA polymerase has also been investigated.[123–125]

5.4.3.3. Enzymatic Modification

Site-directed mutagenesis developed rapidly because of the ready availability of restriction nucleases and DNA ligases, which made it possible to prepare specific recombinant DNA molecules.

A wide variety of techniques using site-specific mutagenesis has been reported. The basic approach (Figure 14) uses a synthetic primer that contains a single mismatched codon in the middle of its sequence. Despite this single mismatched site, the primer will usually hybridize sufficiently to its template to permit its extension, thus providing a powerful tool for introducing any type of mutation into a specific target area.[126] An entire DNA sequence can be deleted and replaced (excision and repair) through the action of endonucleases and ligases. Endonucleases have other uses; for example, DNA phosphorylated pyrimidine dimer is recognized and removed by a specific endonuclease enzyme.[127] As the immune system provides protection and resistance against infection, DNA repair processes also perform a protective function by restoring damaged DNA to its original state. Carcinogens such as benzanthracene and benzpyrene, for example, are usually found linked to DNA through the N7 of guanine.[128,129] Chemical modification has also been exploited resulting in

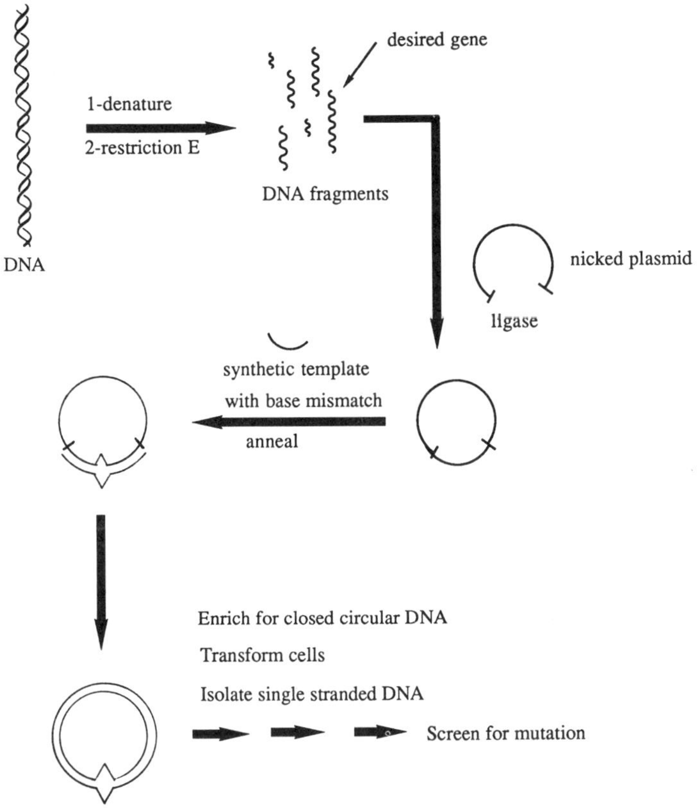

Figure 14. Site-directed mutagenesis. DNA is denatured and broken into small fragments using restriction enzymes. Among these DNA fragments the desired gene is present, and it is inserted into a nicked plasmid. A synthetic template is prepared with a complementary sequence to the desired gene but containing a single base mismatch. This is annealed to the plasmid containing the desired gene and extended to form a closed circular DNA molecule. After a series of steps that enrich for closed circular DNA, transform cells, isolate single-stranded DNA, and screen for mutations, an organism producing a gene product with a single altered amino acid residue can be found.

the development of antiviral and antibacterial drugs that produce modification of a single nucleotide or within nucleic acids.

One important application of chemical and enzymatically synthesized oligonucleotides is as hybridization probes. Nucleic acid hybridization probes offer a powerful technique and a sensitive tool for the detection of pathogens[130–133] and for diagnosing genetic diseases that often cannot be detected by any other method.[134] A nucleotide triphosphate can be labeled with a reporter group such as ^{32}P or biotin and then incorporated into a synthetic nucleic acid sequence through nick translation to produce the required hybridization probe. Alternatively, the intact nucleic acid sequence can also be labeled with the reporter group.[135] The nucleic acids from the sample (target nucleic acids to be examined) are typically immobilized on a solid matrix and probed with a reporter group containing template having a complementary sequence. Histochemical and enzymatic methods are then used to detect sample containing hybridized probe.

5.5. CONCLUSION

This review has attempted to examine the important recent literature on the use of enzymes in the synthesis and modification of biopolymers. It is clear that enzymatic catalysis has a secure place among the methods used by molecular biologists and nucleic acid chemists. Whether the use of enzymes will have the same impact in protein and carbohydrate chemistry remains to be established. The need to improve the regio- and stereoselectivity in peptide and oligosaccharide chemistry represents a potent force dictating the increased use of enzymes. The two major requirements for expanded application of enzymes in these areas are increased enzyme availability and the development of new ways to apply these enzymes to current problems in protein and carbohydrate chemistry.

REFERENCES

1. D. R. Marshak and D. T. Liu, eds., *Therapeutic Peptides and Proteins: Assessing New Technologies*, Banbury Reports #29, Cold Spring Harbor, 1988.
2. L. B. Jaques, *Science 206*, 528 (1979).
3. B. Casu, *Adv. Carbohydr. Chem. Biochem. 43*, 51 (1985).
4. J. F. Kennedy and C. A. White, *Bioactive Carbohydrates*, Ellis Harwood, New York 1983.
5. W. V. Grossman and A. K. Moldave, *Recombinant DNA Technology*, Academic Press, London (1989).
6. J. F. Kennedy, *Enzyme Technology*, VCH Verlagsgesellschaft (1987).
7. C. Laone, J. Tramper, and M. D. Lilly, eds., *Biocatalysis in Organic Media*, Elsevier, Amsterdam (1987).
8. R. J. Linhardt and D. Loganathan, in: *Biomimetic Polymers* (G. Gebelein, ed.), Plenum Press, New York (1990).
9. D. J. Watson, *The DNA Story*, W. H. Freeman and Co., San Francisco (1981).
10. A. J. Zang and J. R. Cech, *Science 231*, 470 (1986).
11. F. M. Ansbel, R. Kingston, D. Moore, J. Seidman, J. Smith, and K. Struhl, eds., *Current Protocol in Molecular Biology*, Wiley/Green, New York (1987).
12. W. E. Cohen and L. Moldave, eds., *Advances in Nucleic Acid Research and Molecular Biology*, New York (1989).

13. M. H. Carthers, *Science 230*, 281 (1985).
14. K. G. Nilsson, *TIBTECH 6*, 256 (1988).
15. H. D. Jakubke, P. Kuhl, and A. Könnecke, *Angew. Chem. Int. Ed. Engl. 24*, 85 (1985).
16. A. Zaks, M. Empie, and A. Gross, *TIBTECH 6*, 272 (1988).
17. G. F. Bickerstaff, *Enzymes in Industry and Medicine*, E. Arnold, ed., (1987).
18. T. J. Ahern, J. I. Casal, G. A. Petsko, and A. M. Klibanov, *Proc. Natl. Acad. Sci. USA 84*, 675 (1987).
19. C.-H. Wong, *Science 244*, 1145 (1989).
20. G. M. Whitesides and C.-H. Wong, *Angew. Chem. Int. Ed. Engl. 24*, 617 (1985).
21. J. S. Dordick, *Enzyme Microb. Technol. 11*, 194 (1989).
22. J. S. Deetz and J. D. Rozzell, *TIBTECH 6*, 15 (1988).
23. P. J. Lea, *Int. Rev. Biochem. 18*, 79 (1978).
24. H. R. V. Arnstein, *Int. Rev. Science, Biochem. Series*, Vol. 7, Butterworths, London (1975).
25. I. K. Inoye and K. Wautab, *J. Chem. Soc., Perkin. Trans*, 1905 (1977).
26. C. Sheehan and G. P. Hess, *J. Am. Chem. Soc. 77*, 1067 (1955).
27. G. W. Anderson, J. E. Zimmerman, and F. M. Callahan, *J. Am. Chem. Soc. 89*, 178 (1967).
28. K. Morihara and T. Oka, *Biochem. J. 163*, 531 (1977).
29. H. D. Law, *The Organic Chemistry of Peptides*, Interscience, Wiley, London (1974).
30. F. Weygand and E. Csendes, *Angew. Chem. 64*, 136 (1952).
31. R. Camble, R. Garner, and G. T. Young, *Nature 217*, 247 (1967).
32. D. F. Elliot and D. Morris, *Chimia 14*, 373 (1960).
33. B. Iselin, *Helv. Chim. Acta 45*, 1510 (1962).
34. M. Mutter and E. Bayer, *Nature 237*, 512 (1972).
35. S. B. H. Kent and R. B. Merrifield, *Peptides*, Brunfeldt, Copenhagen (1981).
36. J. M. Stewart and J. D. Young, *Solid-Phase Peptide Synthesis*, Freeman, San Francisco (1969).
37. R. Arnon, in: *Immunochemistry of Viruses* (Van Regenmortel and A. R. Neurath, eds.), Elsevier (1985).
38. R. Arnon, *TIBS 11*, 521 (1986).
39. R. C. Kennedy, R. D. Hendel, I. Paoletti, J. S. Allen, T. H. Lee, M. Essex, and G. R. Dreesman, *Science 231*, 1551 (1986).
40. J. D. Glass, *Enzyme Microb. Technol. 3*, 2 (1981).
41. J. S. Fruton, *Adv. in Enzymol. 53*, 239 (1982).
42. H. D. Jakubke, P. Kuh., and A. Könnecke, *Angew. Chem. Int. Ed. Eng. 24*, 85 (1985).
43. K. Morihara, *TIBTECH 5*, 164 (1987).
44. M. Homandberg, J. A. Mattis, and M. Laskowski, *Biochemistry 17*, 5220 (1978).
45. I. M. Chaiken and A. Komoriya, *TIBS 3*, 269 (1978).
46. E. Scoffone, R. Racchi, F. Marchiori, L. Moroder, A. Marzotto, and A. M. Tamburro, *J. Am. Chem. Soc. 89*, 5490 (1967).
47. D. E. Harris and R. E. Offord, *Biochem. J. 161*, 21 (1977).
48. F. Borras and R. E. Offord, *Nature 227*, 716 (1970).
49. C. H. Li, J. Blaker, and T. Hayashido, *Biochem. Biophys. Res. Commun. 82*, 217 (1978).
50. M. Pandin, E. A. Paldon, C. Bello, and I. M. Chaiken, *Peptides*, Université de Bruxelles (1976).
51. Y. Isowa, M. Ohmori, and K. Mori, US Patent 4, 436, 925 (1984).
52. K. Oyama, S. Nishimura, Y. Nonaka, K. Kihara, and T. Hashimoto, *J. Org. Chem. 46*, 5241 (1981).
53. R. Saltman, D. Vlach, and P. L. Luisi, *Biopolymers 16*, 631 (1977).
54. M. Bergmann and H. Frankel-Conrat, *J. Biol. Chem. 119*, 707 (1937).

55. R. W. Sealock and M. Laskowski, Jr., *Biochemistry 8*, 3703 (1969).
56. M. Homandberg, J. A. Mattis, and M. Laskowski, Jr., *Biochemistry 17*, 5220 (1978).
57. K. Inouye, K. Wantabe, K. Morihara, Y. Fochino, T. Kanyo, J. Emuro, and S. Sakakibara, *J. Am. Chem. Soc. 101*, 751 (1979).
58. A. Jonezyk and G. H. Gattner, *Hoppe Seyler's Z. Physiol. Chem. 362*, 1591 (1981).
59. C. F. Barbas, J. R. Matos, J. Blair-West, and C.-H. Wang, *J. Am. Chem. Soc. 110*, 5162 (1988).
60. I. K. Bratovanova, I. B. Stoineva, and D. D. Petkov, *Tetrahedron 44*, 3666 (1988).
61. J. Blair-West and C.-H. Wong, *Tetrahedron Lett. 28*, 1629 (1987).
62. A. L. Margolin and A. M. Klibanov, *J. Am. Chem. Soc. 109*, 3802 (1987).
63. T. Nakatsuka, T. Sasaki, and E. T. Kaiser, *J. Am. Chem. Soc. 109*, 3808 (1987).
64. J. Blair-West and C.-H. Wong, *J. Org. Chem. 51*, 2728 (1986).
65. G. Ashwell and E. F. Newfeld, *Cell 12*, 619 (1977).
66. K. B. Lee, D. Loganathan, Z. M. Merchant, and R. J. Linhardt, *Appl. Biochem. and Biotechnol. 53*, 273 (1990).
67. R. W. Stoddart, *The Biosynthesis of Polysaccharides*, Macmillan, New York (1984).
68. R. W. Freisen and S. J. Danishefsky, *J. Am. Chem. Soc. 111*, 6656 (1989).
69. D. R. Mootoo, P. Konradsson, U. Udodong, and B. Fraser-Reid, *J. Am. Chem. Soc. 110*, 5583 (1988).
70. K. G. I. Nilsson, *TIBTECH 6*, 256 (1988).
71. C.-H. Wong, S. L. Haynie and G. W. Whiteside, *J. Org. Chem. 47*, 5416 (1982).
72. J. Thiem and W. Treder, *Angew. Chem. 98*, 1096 (1986).
73. T. A. Bayer, J. E. Sadler, J. I. Rearick, J. C. Paulson, and R. L. Hill, *Adv. Enzymol. 52*, 23 (1981).
74. P. R. Rosevear, H. A. Nunez, and R. Barker, *Biochemistry 21*, 1421 (1982).
75. V. Zehavi, S. Sadeh, and M. Herchman, *Carbohydr. Res. 124*, 23 (1983).
76. K. G. I. Nilsson, *Carbohydr. Res. 188*, 9 (1989).
77. K. Ajisaka, H. Fujimoto, and H. Nishida, *Carbohydr. Res. 180*, 35 (1988).
78. K. Ajisaka, H. Nishida, and H. Fujimoto, *Biotechnol. Lett. 9*, 243 (1987).
79. F. Milliken, P. Smith, C. S. Rose, and P. Cyorgy, *J. Biol. Chem. 217*, 79 (1955).
80. K. G. I. Nilsson, *Carbohydr. Res. 167*, 95 (1987).
81. A. Anon, *Jpn. Bioind. Lett. 2*, 9 (1985).
82. US Patent No. 4, 521, 592 (1985).
83. N. K. Matherson and B. V. McCleary, in: *The Polysaccharide* (G. O. Aspinall, ed.), Vol. 3, Academic Press, New York (1985).
84. J. D. Gregory, T. C. Laurent, and L. Roden, *J. Biol. Chem. 239*, 3312 (1964).
85. R. J. Linhardt, C. L. Cooney, and P. M. Galliher, *Appl. Biochem. and Biotechnol. 12*, 135 (1986).
86. R. J. Linhardt, G. L. Fitzgerald, C. L. Cooney, and R. Langer, *Biochim. Biophys. Acta 702*, 197 (1982).
87. R. J. Linhardt, Z. M. Merchant, K. G. Rice, Y. S. Kim, G. L. Fitzgerald, A. C. Grant, and R. Langer, *Biochemistry 24*, 7805 (1985).
88. R. J. Linhardt, K. G. Rice, and D. M. Cohen, *Biochemistry 28*, 2888 (1989).
89. S. Ishii and T. Yokotsuka, *J. Agr. Food Chem. 20*, 787 (1972).
90. T. N. Wight and R. Mechem, eds., *The Biology of the Extracellular Matrix: Proteoglycans*, Academic Press, New York (1987).
91. K. G. Rice and R. J. Linhardt, *Carbohydr. Res. 190*, 219 (1989).
92. L. M. Mallis, H. M. Wang, D. K. Loganathan, and R. J. Linhardt, *Anal. Chem. 61*, 1453 (1989).
93. D. Loganathan, H. M. Wang, L. M. Mallis, and R. J. Linhardt, *Biochemistry 29*, 4362 (1990).

94. Y. S. Kim, Doctoral Thesis, University of Iowa, August 1988.
95. D. H. Atha, J. C. Lormeau, M. Petitou, and R. D. Rosenberg, *Biochemistry 24*, 6723 (1985).
96. P. Sinay, J. C. Jacquinet, M. Petitou, P. Duchaussoy, I. Lederman, J. Choay, and G. Torri, *Carbohydr. Res. 132*, C5 (1984).
97. L. A. Loeb, P. K. Liu, and M. Fry, *Prog. in Nucleic Acid Res. and Mol. Biol. 33*, 57 (1986).
98. A. Kornberg, *DNA Replication*, Freeman, San Francisco (1980).
99. K. Itakura, T. Hirose, R. Crea, A. D. Riggs, H. Heynker, F. Bolivar, and H. W. Boyer, *Science 198*, 1956 (1977).
100. P. Hodge and D. C. Sherrington, *Polymer-Supported Reaction in Organic Synthesis*, Wiley, New York (1980).
101. G. Alvarado-Urbina, G. M. Sathe, W. C. Liu, M. F. Gillen, and D. P. Duck, *Science 214*, 270 (1981).
102. M. J. Gair, H. W. Matthes, M. Singh, S. Sproat, and R. C. Titmas, *Nucleic Acids Res.*, *10* 1755 (1982).
103. B. C. Reese and R. Ubasawa, *Nucleic Acids Res. 7*, 5 (1980).
104. H. Koster, A. Stumpe, and A. Walter, *Tetrahedron Lett. 24*, 745 (1983).
105. V. Kohli, H. Blocker, and H. Koster, *Tetrahedron Lett. 21*, 2683 (1983).
106. M. D. Matteucci and M. D. Carthers, *J. Am. Chem. Soc. 103*, 3185 (1981).
107. H. G. Khorana, *Science 203*, 614 (1979).
108. J. E. Davis and H. G. Gassen, *Angew. Chem. 22*, 13 (1983).
109. K. Itakura and A. D. Riggs, *Science 209*, 140 (1980).
110. H. O. Smith and D. Nathans, *J. Mol. Biol. 81*, 419 (1973).
111. R. J. Roberts, *Methods Enzymol. 65*, 1 (1980).
112. E. K. Nichols, *Human Gene Therapy*, Harvard University Press, Cambridge (1988).
113. H. G. Gassen and A. Lang, eds., *Chemical and Enzymatic Synthesis of Gene Fragments*, Verlag Chemie, Weinheim (1982).
114. N. Symonds, in: *Biochemistry of Nucleic Acids* (K. Burton, ed.), Vol. 6, p. 165, Butterworths University Park Press, Baltimore (1974).
115. K. Itakura, T. Hirose, R. Crea, R. D. Riggs, and H. L. Hyneker, *Science 198*, 1056 (1977).
116. K. L. Agarwal, H. Buchi, M. H. Caruthers, N. Gupta, H. G. Khorana, K. Kleppe, A. Kumar, E. Ohtsuka, V. Rajbhandary, J. H. Van de Sande, V. Sfaramella, H. Weber, and T. Yamada, *Nature 227*, 27 (1970).
117. D. V. Goeddel, D. G. Kleid, F. Bolivar, H. I. Heynecker, D. C. Yamsura, R. Crea, T. Hirose, A. Kraszewski, K. Itakura, and A. D. Riggs, *Pro. Natl. Acad. Sci. USA 76*, 106 (1979).
118. E. Ohtsuka, Y. Taniyama, S. Iwai, K. Kitano, and S. Miyamoto, *Nucleic Acids Res. 12*, (1984).
119. M. D. Meacok, A. R. Heathcliffe, P. A. Meacock, W. Schuch, D. B. Scalon, T. C. Atkinson, C. R. Newton, and A. F. Markan, *Nature 292*, 756 (1981).
120. P. J. Romaniuk and O. C. Uhlenbeck, *Methods in Enzymol. 100*, 52 (1983).
121. R. C. Scarpulla, S. A. Narang, and R. Wu, *Anal. Biochem.*, *121*, 356 (1982).
122. J. J. Rossi, R. Kierzek, T. Huang, P. Walker, and K. Itakura, *J. Biol. Chem. 257*, 9226 (1982).
123. H. G. Khorana, K. L. Agarwal, P. Besmer, H. Buchi, M. H. Caruthers, P. J. Cashion, M. Fridkin, E. Jay, K. Kleppe, R. Kleppe, A. Kumar, P. C. Loewen, R. C. Miller, K. Minamoto, A. Panet, U. L. Rajbhandary, B. Ramamoorthy, T. Sekiya, T. Takeya, and H. van de Sande, *J. Biol. Chem. 251*, 565 (1976).
124. J. D. Windass, C. R. Newton, J. DeMaeyer-Guignard, V. E. Moore, and A. F. Markham, *Nucleic Acids Res. 10*, 6639 (1982).

125. P. L. Dettaseth, R. A. Goldman, C. L. Cech, and M. H. Caruthers, *Nucleic Acids Res. 11*, 773 (1983).
126. Shortle, D. DiMaio, and D. Nathans, *Annu. Rev. Genet. 15*, 265 (1981).
127. A. Sancar and G. B. Sancar, *Annu. Rev. Biochem. 57*, 29 (1988).
128. S. Neidle, *Nature 283*, 135 (1980).
129. W. A. Haseltine, M. K. Lo, and A. D. D'Andrea, *Science 209*, 929 (1980).
130. J. I. McInnes, S. Dalton, P. D. Vise, and A. J. Robins, *Bio/Technology*, *5* 269 (1987).
131. A. H. Al-Hakim and R. Hull, *Biochem. J. 251*, 935 (1988).
132. R. Hull and A. H. Al-Hakim, *Trends Biotech. 6*, 213 (1988).
133. H. Yokota, K. Yokoo, and Y. Nagata, *Biochim. Biophys. Acta 868*, 45 (1986).
134. R. D. Langer, A. A. Waldrop, and D. C. Wark, *Proc. Natl. Acad. Sci. USA 78*, 6633 (1981).
135. A. H. Al-Hakim and R. Hull, *Nucleic Acids Res. 4*, 9965 (1986).

III

EMERGING BIOCATALYSTS FOR CONVENTIONAL CHEMICAL PROCESSING

6

Lignin-Degrading Enzymes from the Filamentous Fungus *Phanerochaete chrysosporium*

ELIZABETH A. PEASE and MING TIEN

6.1. THE STRUCTURE OF LIGNIN

Lignin is an aromatic polymer which surrounds woody tissue, providing structural rigidity and protection from microbial attack. Next to cellulose, it is the most abundant renewable resource on earth. It is estimated that approximately 25% of the carbon fixed by photosynthesis is incorporated into lignin. The lignin polymer is composed of phenylpropanoid subunits linked together by a variety of bonds resulting in a nonrepeating motif.[1–3] Most biological macromolecules such as cellulose, RNA, DNA, and proteins are largely linear polymers whose subunits are linked together by a repeating bond; thus the mechanism of polymer synthesis and degradation is generally centered around this common bond. Lignin synthesis and degradation differ from this common mechanism since it contains a variety of linkages.[4]

Lignin is a random three-dimensional polymer in which the subunits are linked together by a variety of carbon–carbon or carbon–oxygen

ELIZABETH A. PEASE and MING TIEN • Department of Molecular and Cell Biology, The Pennsylvania State University, University Park, PA 16802.

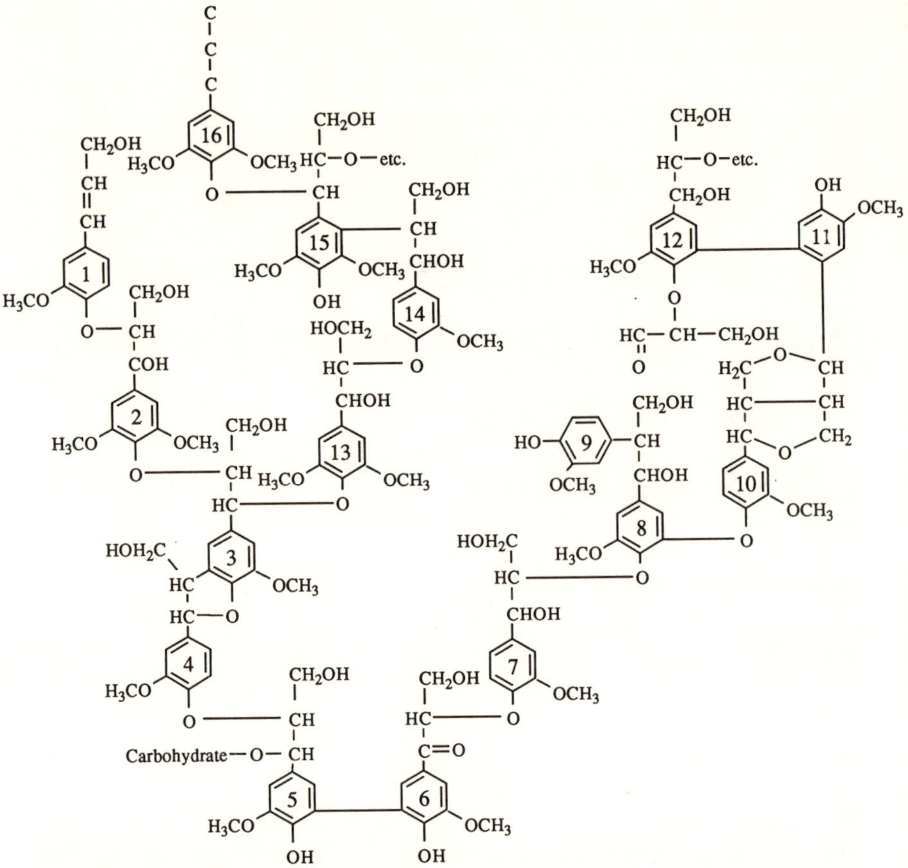

Figure 1. Representative structure of lignin, reproduced from Adler.[4] A variety of linkages are formed between the numbered phenylpropanoid subunits. Subunits 1 and 2 are linked by a β-O-4 type linkage. Subunits 8 and 9 are linked by a β-1 type linkage.

bonds.[2] It is synthesized during cell-wall thickening by the one-electron oxidation of *p*-hydroxy-, methoxy-substituted cinnamyl alcohol units. This one-electron oxidation is peroxidase-catalyzed and produces a radical with four resonance forms:

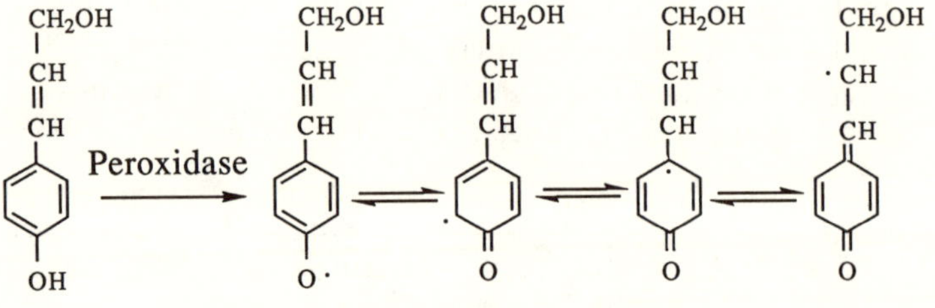

These radicals dimerize and are further oxidized by a one-electron oxidation. Repeated cycles of radical formation and polymerization produce a highly condensed polymer. A representative structure of the lignin polymer is shown in Fig. 1. The predominant linkage found in lignin is of the β-O-4 type (Figure 1). This accounts for approximately 60% of the interunit linkages in spruce lignin.[4] Another linkage is the β-1 type, accounting for approximately 5% of the linkages in spruce lignin (Figure 1).[4] There are more than ten other types of chemical bonds accounting for the remaining linkages in lignin.[4]

6.2. MICROBIAL DEGRADATION OF LIGNIN

Most natural lignin is not degraded to the level of CO_2 but ends up as humic material. This slow degradation rate of lignin plays a major role in controlling the amount of O_2 in the atmosphere. During evolution the partial pressure of molecular oxygen increased in the atmosphere due mainly to the oxidation of H_2O as an electron source by photosynthetic organisms. The increase in molecular oxygen resulted in evolution of aerobes capable of utilizing oxygen as the terminal electron sink in the oxydation of the carbon skeletons fixed from photosynthesis. Thus, if all of the carbons fixed from photosynthesis were completely oxidized, all of the O_2 evolved would be consumed, and O_2 would merely be a trace gas. The high content of atmospheric O_2 has been related to the slow rate of the lignin degradation process, due in part to the recalcitrant nature of lignin. Another important function of lignin is to protect the cellulosic and hemicellulosic material from microbial attack and destruction. The accessibility of these carbohydrates to hydrolytic enzymes produced by pathogenic organisms, particularly fungi, is inversely related to the lignin content of wood tissue. Woody tissue is made up of greater than 25% lignin which surrounds the cellulose and hemicellulose layers forming a protective barrier against microbial degradative enzymes.

Due to the fact that the cellulose and hemicellulose layers are normally inaccessible, only organisms with lignin-degrading capacity can gain access to the carbohydrates. Consequently, this ability gives the invading organism a selective advantage by being able to gain access to and utilize the cellulosic materials as a nutritional source. Although reports have demonstrated bacterial degradation of lignin, this area is not as well characterized as that of fungal decay of lignin.[5] Bacterial degradation of lignin has recently been reviewed by Crawford and Crawford and will not be addressed here.[6] It is widely recognized that the predominant degraders of lignin are pathogenic and saprophytic filamentous fungi. Within this category, the most extensively characterized are the Basidiomycetes which cause the white rot type of wood decay. One of these fungi, *Phanerochaete chrysosporium*, has been extensively studied by many laboratories. Work in the 1970s defined the culture conditions required to activate the ligninolytic response in this fungus. Keyser *et al.* demonstrated that the ligninolytic system of *P. chrysosporium* is expressed in response to nitrogen limitation.[7] This secondary metabolic event can be triggered not only by nitrogen, but also by carbon or sulfar limitation.[8]

The lignin-degrading system of *P. chrysosporium* is highly nonspecific and capable of degrading the lignin polymer to the level of CO_2.[9] Work with the whole polymer by Chen *et al.* detailed some of the products formed from fungal decay of lignin.[10,11] These workers found that the oxygen content of fungal-decayed lignin increases relative to sound lignin. There was also evidence of ring-cleavage products of lignin; however, the predominant reaction was found to be C_α–C_β cleavage producing aromatic aldehydes and acids. This demonstrated that fungal decay of lignin is oxydative, extracellular, in part and nonspecific.[10,11] It was quickly realized

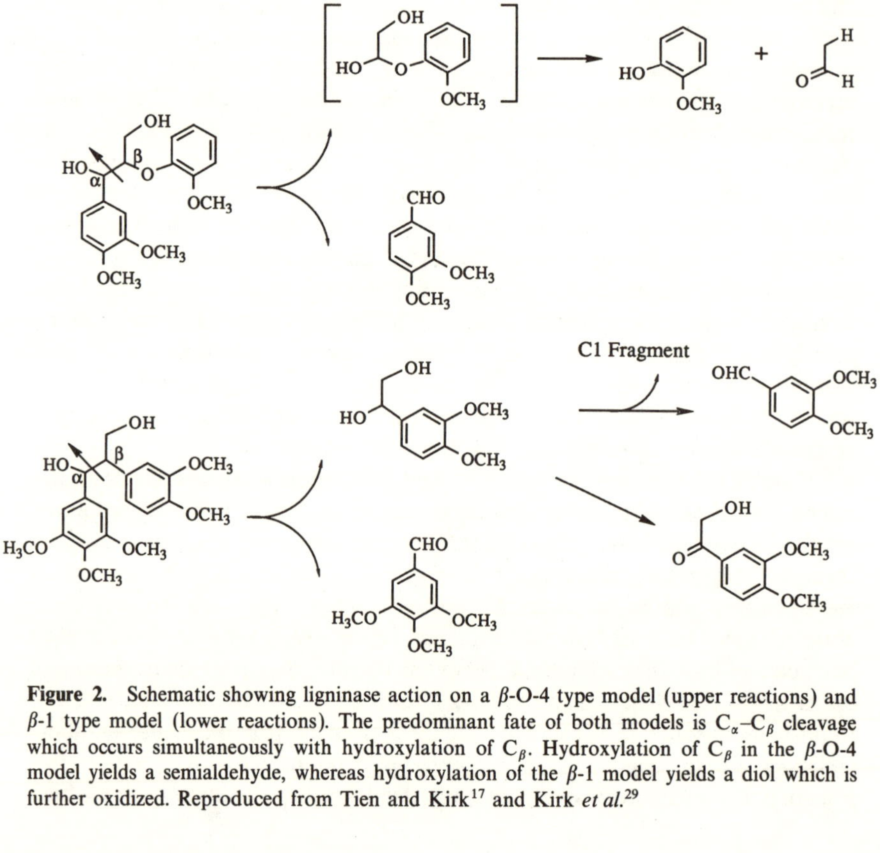

Figure 2. Schematic showing ligninase action on a β-O-4 type model (upper reactions) and β-1 type model (lower reactions). The predominant fate of both models is C_α–C_β cleavage which occurs simultaneously with hydroxylation of C_β. Hydroxylation of C_β in the β-O-4 model yields a semialdehyde, whereas hydroxylation of the β-1 model yields a diol which is further oxidized. Reproduced from Tien and Kirk[17] and Kirk *et al.*[29]

that this process was too complicated to characterize using the whole polymer. Kirk and co-workers proceeded to address the biochemistry of this process by using substructural synthetic models of lignin.[12,13] Models which represent the β-O-4 and β-1 type of substructure were used in whole cultures. Like the lignin polymer, these model compounds were also degraded only when cultures were nutrient starved. The products from the models are shown in Figure 2. As with the natural lignin, the predominant reaction is that of C_α–C_β cleavage. The initial product from the β-1 model is an aromatic aldehyde from the lower ring and a hydroxylated product from the upper portion yielding a diol. The diol is further metabolized to a C_1 product and another aromatic aldehyde. The predominant reaction from the β-O-4 model is also C_α–C_β cleavage yielding an aromatic aldehyde from the lower ring and a semialdehyde from the upper portion which rapidly breaks down to the aldyhyde and alcohol.

6.3. LIGNIN-DEGRADING ENZYMES

The lignin degrading activity of *P. chrysosporium* is catalyzed by the ligninase enzymes (also referred to as lignin peroxidases) found in the extracellular fluid of the cultures.[14,15] The discovery of the enzymes was made possible by the finding of Forney *et al.* that ligninolytic cultures of *P. chrysosporium* produce H_2O_2.[16] Incubation of the lignin model compounds with the extracellular fluid from ligninolytic cultures of *P. chrysosporium* with the addition of H_2O_2 results in C_α–C_β cleavage.[14,15] The extracellular enzyme in the presence of H_2O_2 catalyzes the oxidation of veratryl alcohol to the corresponding aldehyde, which absorbs at 310 nm. Since the alcohol exibits very little absorbance at 310 nm, this oxidation is monitored spectrophotometrically thus providing a convenient assay for the ligninase activity.[17] There are a number of ligninase isozymes which catalyze C_α–C_β cleavage activity or the veratryl alcohol-oxidizing activity.[17] Altogether, over ten extracellular proteins are detected.[18] These isozymes are relatively easy to purify since they are secreted enzymes. A number of purification procedures for these enzymes have been published all of which consist of three basic steps: (i) removal of the mycelium from the extracellular fluid via centrifugation or filtration; (ii) concentration of the extracellular fluid; and (iii) purification by anion exchange chromatography.[19] Although purification on conventional DEAE resin yields pure isozymes,[17] resolution on the commercially-packed HPLC ion exchange resins is much better. Figure 3 shows an elution profile of the extracellular enzymes from ligninolytic cultures of *P. chrysosporium* from the Mono Q resin from Pharmacia using a sodium acetate gradient for elution. The ligninases are

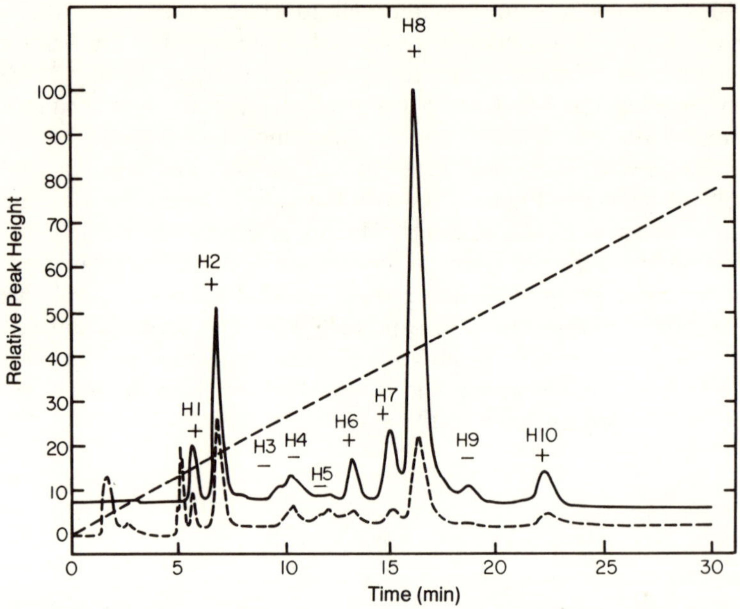

Figure 3. Elution profile of extracellular enzymes of *P. chrysosporium* from a Mono Q column. Full line is absorbance at 409 nm and dashed line is absorbance at 280 nm. Veratryl alcohol-oxidizing activity is marked as positive or negative. Reproduced from Tien and Kirk.[19]

designated here as H1–H10 corresponding to their order of elution. Peaks exhibiting veratryl alcohol-oxidizing activity are marked (+).

In addition to 280 nm absorbance, these proteins also exhibit 409 nm absorbance due to the presence of heme. The hemiproteins can be divided into two categories according to their catalytic activity. One group of isozymes are the lignin peroxidases that all exhibit veratryl alcohol-oxidizing activity. These include H1 (pI = 4.7), H2 (pI = 4.4), H6 (pI = 3.7), H7 (pI = 3.6), H8 (pI = 3.5), and H10 (pI = 3.3).[18] The other group is composed of the manganese-dependent peroxidases that do not have veratryl alcohol-oxidizing activity. These are H3 (pI = 4.9), H4 (pI = 4.5), H5 (pI = 4.2), and H9.[20] These manganese-dependent peroxidases are catalytically distinct from the ligninases.[21,22]

Despite small catalytic differences, the role for the numerous ligninase isozymes is not clearly understood.[18] Overlapping substrate specificity by the isozymes has been observed with all substrates examined.[18] It was believed that the different isozymes may be degradation products of the parent ligninase.[23] This would be consistent with protein recycling during nitrogen starvation. However, this does not appear to be the case. The first indication that the isozymes are not derived from one altered protein came from amino acid sequencing of the amino termini. Despite homologies, the results clearly indicated that the isozymes had different sequences. Immunoprecipitation of *in vitro* translation products from poly(A)$^+$ RNAs of ligninolytic cultures showed multiple bands, also suggesting multiple messages.[24] The cDNA cloning of multiple ligninase genes confirmed the hypothesis that the isozymes are encoded by different genes.[24,25]

Table 1 summarizes some of the physical properties of these hemeproteins. As a family, the ligninase isozymes are structurally related. They are all glycosylated hemeproteins with molecular weights ranging from 38,000 to 46,000.[18,26] Structural homology is also demonstrated by cross-reactivity of polyclonal antibodies and by peptide mapping.[18] Polyclonal antibodies raised to isozyme H8 cross-react with the other lignin peroxidase isozymes.[18,23] However, there is little, if any, reactivity with the manganese-dependent peroxidases.[23,24] Likewise, polyclonal antibodies raised to isozyme H4 cross-react with the other manganese-dependent peroxidases but not with the lignin peroxidases.[20]

The ligninase isozymes catalyze a wide range of H_2O_2-dependent oxidations of aromatic substrates.[27–31] Aside from catalyzing the oxidation of veratryl alcohol, ligninases catalyze (i) C_α–C_β cleavage of lignin model compounds; (ii) the oxidation of benzylic alcohols to the corresponding aldehydes; (iii) C_α–C_β cleavage of dimeric substructural models; (iv) hydroxylation of certain benzylic carbons; (v) phenol dimerization; and (vi) hydroxylation of C_α–C_β double bonds to form the corresponding diol.

Kinetic studies and spectroscopic characterizations of the ligninase isozyme family indicate similarities to other peroxidases and that their catalytic mechanisms are similar.[32–35] The initial catalytic reaction is the

Table 1. Physical Properties of Ligninase Isozymes

Isozyme	Molecular weight	Carbohydrate	Extinction coefficient[a]	pI	Enzyme type[b]
H1	38,000	+	169	4.7	LP
H2	38,000	+	165	4.4	LP
H3	45,000	+	125	4.9	MnP
H4	45,000	+	127	4.5	MnP
H5	45,000	+			MnP
H6	43,000	+	162	3.7	LP
H7	42,000	+	177	3.6	LP
H8	42,000	+	168	3.5	LP
H10	43,000	+	182	3.3	LP

[a] Extinction coefficients given as mM^{-1}.
[b] LP = lignin peroxidase; MnP = Manganese-dependent peroxidase.

two-electron oxidation of resting enzyme by H_2O_2 to form Compound I and H_2O. Compound I oxidizes a substrate molecule by one electron to produce a free radical product and Compound II. Compound II returns to resting enzyme by oxidizing a second molecule of substrate by one electron:

$$\text{Ferric peroxidase} + H_2O_2 \rightarrow \text{Compound I} + H_2O \quad (1)$$

$$\text{Compound I} + A^- \rightarrow \text{Compound II} + A\cdot \quad (2)$$

$$\text{Compound II} + A^- \rightarrow \text{Ferric peroxidase} + A\cdot \quad (3)$$

The chemical mechanism of ligninase catalysis has been elucidated through the work of numerous laboratories.[36–38] The results are consistent with a free-radical mechanism for catalysis. Kersen *et al.* were the first to present experimental evidence that free radicals are formed during ligninase catalysis.[36] These workers also studied ligninase-catalyzed oxidation of methoxybenzene congeners. The one-electron oxidation of some of these congeners produced free radical products stable at low pH and detectable by electron spin resonance (ESR) spectroscopy. ESR signals have been observed for the cation radical of 1,4-dimethoxybenzene and 1,2,3,4- and 1,2,4,5-tetramethoxybenzene. Hammel *et al.* later demonstrated free radical formation from lignin-like model compounds.[37] Cation radicals are also observed in the oxidation of other lignin-like substrates by ligninase, such as in the degradation of a β-1 model dimethoxyhydrobenzoin (DMHB).[37] The predominant reaction of both β-1 models is C_α–C_β cleavage in the alkyl side chain. ESR spectroscopy experiments showed that free radicals are formed during this catalysis. In contrast to the methoxybenzenes, the ESR signal was observed only in the presence of a spin trapping agent such as 5,5-dimethyl-1-pyrroline-*N*-oxide (DMPO). An ESR signal corresponding to carbon-centered radicals was observed in anaerobic incubations of ligninase with H_2O_2, DMHB, and DMPO. Under aerobic conditions, an ESR signal resembling those of peroxy radicals was detected. This indicates that the radicals generated by ligninase can undergo further reactions which result in the propagation of the radical process. Shoemaker *et al.* also proposed a radical mechanism for ligninase.[38] These investigators based their conclusions on the known metabolic fate of lignin model compounds and its similarity to the chemistry of cation radicals.

The free-radical mechanism explains the prominent reactions observed in lignin biodegradation: C_α–C_β cleavage, loss of methoxyls, oxidation of benzylic hydroxyls to ketones, and ring opening. ESR spectroscopy experiments showed that radicals of lignin-like compounds can be relatively stable and long-lived.[36] Because these radicals can diffuse away from the enzyme active site, their fate is influenced by the constituents of their environment (e.g., pH, concentration of oxygen, and other radicals). The degradation of a large number of lignin model compounds by ligninase has been studied by Higuchi *et al.* and Kirk *et al.*[27–31] The product profiles, the stoichiometries, and the kinetics for many of the lignin-like substrates are consistent with cation radical chemistry. If the mechanism with the model compounds is similar to the lignin polymer, the depolymerization occurs through radical-mediated cleavage of carbon–carbon bonds.

One of the unique catalytic abilities of lignin peroxidase in comparison to other peroxidases is the ability to oxidize substrates with higher oxidation–reduction potentials.[39] The oxidation–reduction potential (Em7) values of the ferric/ferrous couple for the lignin and manganese-dependent peroxidases isozymes along with the potentials of several other hemeproteins are summarized in Table 2. When compared to other peroxidases, the Em7 values of the lignin and manganese-dependent peroxidases are more positive (Table 2). The higher Em7 value for the ferric/ferrous couple is consistent with the heme active site of these fungal peroxidases being somewhat more electron-deficient, enabling them to oxidize the substrates with a high oxidation–reduction potential. The Em7 values imply that the oxidation–reduction potentials of Compound I and Compound II which are responsible for oxidizing substrates are also higher for ligninase than other peroxidases. The oxidation–reduction potentials of Compounds I and II formed by horseradish peroxidase (HRP) have both been measured to be approximately 0.95 volts.[46] The oxidation–reduction potentials of both the lignin and manganese-dependent peroxidases are estimated to be higher than this base on their Em7 value and the oxidation–reduction potentials of the substrates which they can oxidize. The Em7 values for these peroxidases are not affected by their respective sub-

Table 2. E_{m7} Values for Lignin Peroxidases, Manganese-Dependent Peroxidases, and Other Hemeproteins

Hemeprotein	E_{m7} (mV)	Reference
Horseradish peroxidase	−278	40, 41
Turnip peroxidase (P_1)	−223	42
Cytochrome c peroxidase	−194	43
Lignin peroxidase H1	−142	44
Chloroperoxidase	−140	40
Lignin peroxidase H8	−137	44
Lignin peroxidase H2	−135	44
Lignin peroxidase H10	−127	44
Turnip peroxidase (P_7)	−100	42
Mn-Dependent peroxidase H4	−93	44
Mn-Dependent peroxidase H3	−88	44
Metmyoglobin	+46	45

strates, veratryl alcohol or Mn(II).[44] These results suggest that substrates probably do not directly interact with the ferric heme-iron; instead only the heme periphery is accessible to the organic substrates.[47,48] This is consistent with the present model for peroxidase catalysis.[47,48]

6.4. MOLECULAR BIOLOGY OF LIGNIN BIODEGRADATION

Several factors initiated the interest in studying the genes encoding the lignin peroxidases. Foremost is overexpression of the enzyme by utilizing one of a number of available molecular biological techniques (see Section 6.5). Another reason for studying the gene is to further characterize the enzymes and their catalytic mechanism. Techniques such as site-directed mutagenesis with subsequent expression of the mutated protein can be used to identify essential structural characteristics.

An important question being addressed using molecular biological approaches is whether the lignin-degrading enzymes are transcribed from different genes or from a single gene that is post-translationally modified. Alternatively, the fungus may utilize both of these mechanisms for protein diversity by having a small number of different genes coding for proteins which are then differently processed. Present data indicate the existence of a multigene family in which the majority of the isozymes are encoded by different genes. These data include diversity among cDNA clones that have been sequenced along with results from Southern blot analysis.[25]

Zhang *et al.* first reported the isolation and identification of four ligninase cDNA clones.[49] Oligonucleotides specific for ligninase coding regions were synthesized and used as probes for cDNA libraries to isolate these clones, which were than analyzed by Southern blot analysis. These four cDNAs were shown to present two distinct ligninase genes by restriction map analysis. Several laboratories have reported on the sequencing of cDNAs encoding the lignin peroxidase genes. The first cDNA sequence reported was for the gene encoding isozyme H8 (λML-1) in 1987 by Tien and Tu,[24] followed by de Boer *et al.*, who reported on cDNAs encoding isozymes H2 (pCLG4) and H10 (pCLG5).[25] Most recently, the presumed sequence for H6 (λML-5) has been published along with an allelic variant of H8 (λML-4) by Andrawis *et al.*[50] The clones demonstrate a high degree of similarity in both the nucleotide and the amino acid sequences. The most similar are H8, H6, and H10, with H2 being least similar to the others. At the nucleotide level H2 is approximately 70% homologous to H6, H8, and H10, whereas H8 is approximately 75% homologous to H10 and 80% homologous to H6. The cDNA sequences encoding two of the manganese-dependent peroxidases have also been reported. Pribnow *et al.* reported the cDNA sequence of isozyme H3 (MnP-1),[51] and Pease *et al.* reported the cDNA sequence of isozyme H4 (λMP-1).[20] The two manganese-dependent peroxidases are also very similar to each other, being approximately 75% similar at the nucleotide level. These isozymes also show some similarity to the lignin peroxidases with approximately 60% nucleotide sequence similarity.

A more meaningful comparison in terms of enzyme structure and function is the amino acid sequence comparison. A summary of similarities based on the amino acid sequence of the various isozymes is shown in Table 3. These similarities are somewhat higher than the expected nucleotide homologies because of the wobble in the genetic code and the fact that the comparison takes into consideration the substitution of similar amino acids as well as identical amino acids.[52] The similarities range from greater than 95% between H6 and H8 to less than 63% between H4 and H6. An even higher degree of amino acid similarity is observed within the active-site residues of the ligninase clones, approaching 100% for certain clones. The proximal and distal histidine and the flanking residues are shown in Figure 4. These residues are conserved among many peroxidases and are thought to be essential for the peroxidase function. The function of peroxidase active-site residues has been discussed elsewhere.[55]

All of the full-length cDNAs isolated thus far are approximately 1300 bp in size, ranging from 1285 to 1312 bp.[20,24,25,50,51] All deduced amino acid sequences contain a leader sequence that is proteolytically cleaved as a signal for secretion. This leader sequence ranges from 21 to 28 amino acids. The lignin peroxidases contain leader sequences of 27 or 28 amino acids and have a conserved Lys-Arg dibasic cleavage site.[24] The manganese-dependent peroxidases, on the other hand, have shorter signal sequences of either 21 or 24 amino acids, and the cleavage site is not conserved.[20,51] All of the isozymes are glycoproteins, and the cDNA sequences exhibit several potential glycosylation sites with the conserved eukaryotic consensus sequence of Asn-X-Thr/Ser.[56] Most eukaryotic genes also contain the consensus sequence for polyadenylation AATAAA approximately

Table 3. Percent Similarity of Predicted Amino Acid Sequences of Lignin Peroxidases

Isozyme	H2	H6	H8	H10	H4	H3
H2		82.2%	84.1%	78.6%	62.9%	65.7%
H6	82.2%		95.2%	86.5%	63.4%	66.4%
H8	84.1%	95.2%		89.5%	65.2%	67.1%
H10	78.6%	86.5%	89.5%		63.5%	66.7%
H4	62.9%	63.4%	65.2%	63.5%		89.4%
H3	65.7%	66.4%	67.1%	66.7%	89.4%	

			Distal Histidine					Proximal Histidine						ref
H2	41	E A L	R	M V	F H D	S I A	......E I E T	V	W L L S A	H	S I A	181		25
H6	40	E S I	R	L V	F H D	S I A	......E L E L	V	W M L S A	H	S V A	179		50
H8	40	E S I	R	L V	F H D	S I A	......E L E L	V	W M L S A	H	S V A	179		24
H10	40	E S I	R	L V	F H D	S I A	......E L E L	V	W M L S A	H	S V A	178		25
H3	39	E V I	R	L T	F H D	A I A	......P F E V	V	S L L A S	H	S V A	176		51
H4	39	E V I	R	L T	F H D	A I A	......P F E V	V	S L L A S	H	T V A	176		20
HRP	35	S I L	R	L H	F H D	C F V	......S S D L	V	A L S G G	H	T F G	173		53
CCP	45	V L V	R	L A	W H T	S G T	......D R E V	V	A L M G A	H	A L G	177		54

Figure 4. Comparison of amino residues of lignin peroxidases with other peroxidases at regions near proximal and distal histidine. The sequences for the lignin peroxidases H2, H6, H8, and H10 and for the manganese-dependent peroxidases, H3 and H4, are shown along with two commonly known peroxidases—horseradish peroxidase (HRP) and cytochrome c peroxidase (CCP). Conserved residues are enclosed in boxes. The amino acid sequences are from the references listed.

11–13 bp from the poly(A) sequence.[57] There is not a well defined nucleotide consensus sequence for polyadenylation found in *P. chrysosporium.* The sequence AAATAT is seen in H8 approximately 12 base pairs upstream of the polyadenylation site and similar, however not identical, sequences appear in some of the other cDNA clones.[20,24,25,50,51]

There have been a number of genomic ligninase sequences published. The first to appear in the literature was a brief report containing the sequence of isozyme H8 by Smith *et al.*[58] This was quickly followed by Brown *et al.* with four homologous sequences (LIG 1, 2, 3, and 4).[59] LIG 1 was a full-length sequence that was highly homologous to the genomic sequence for H8 reported by Smith and co-workers and the H8 cDNA reported by Tien and Tu.[24] LIG 2 was the most homologous to H8; however, the sequence was not reported in its entirety. It should be mentioned that these clones were from a different strain than the H8 clones previously reported. Asada *et al.* and Walther *et al.* reported other genomic sequences that also were homologous to H8 but not identical.[60,61] Again, these were not identified as corresponding to any of the other cDNA sequences previously published. Some of these clones may be allelic variants. Schalch *et al.* characterized two clones 96% similar to H8 genomic clone.[62] By differential hybridizations and sequence analysis they concluded that these are not allelic variants. Andrawis *et al.* reported the isolation and sequence of both the cDNA and genomic DNA sequence of an allelic variant of the previously reported H8 isozyme.[50] These sequences are all highly homologous and contain conserved regions within them. This raises the obvious question of whether these clones are really different and if so why are there so many variations of one gene? There are a few possible explanations for the existence of highly homologous sequences. One is that *P. chrysosposium* strains ME446 and BKM-F-1767 are heterokaryotic and many of the clones may be allelic variants. The other explanation is that since some of the clones came from BKM-F-1767 and some from ME446, there could be variations in the enzymes between the strains. Finally, these could represent significant functional differences in the enzymes.

One characteristic that all of the genes have in common is that they contain introns. Comparison of the genomic sequence for isozyme H8 with the cDNA sequence indicates that there are eight intervening sequences and nine coding regions. The gene structure is shown in Figure 5. The introns are of a consistent size, being approximately 50 bp. The exon sizes appear to be the variable component of the *P. chrysosporium* genomic clones, varying in length from 9 to 424 bp in size. The genes contain the commonly accepted eukaryotic transcriptional upstream regulatory sequences, including a TATA box and CAAT box at positions −75 and −107, respectively.[61,63] The initiation codon is contained within the sequence GACCATGG, which has been implicated in efficient recognition of the correct ATG for initiation.[64] These sequences as yet have not been shown to be active in the transcription process in *P. chrysosporium*, and there is no information available concerning the regulatory elements involved in ligninase transcription.

Another question that remains unanswered is how the levels of protein and activity (defined by veratryl alcohol oxidation in the presence of H_2O_2) are regulated. The lignin peroxidases are expressed solely during secondary metabolism under nutrient limitation. The majority of the work has been done on nitrogen-limited cultures, but some information is available about response to carbon-limited conditions. In nitrogen-limiting conditions the normal pattern of isozyme expression is that the enzyme activity is detected on day 3 and increases up to day 6.[23,24] If the cultures are carbon-limited, a different pattern of expression is exhibited, with the activity appearing and peaking on day 3 and disappearing by day 6.[23,63]

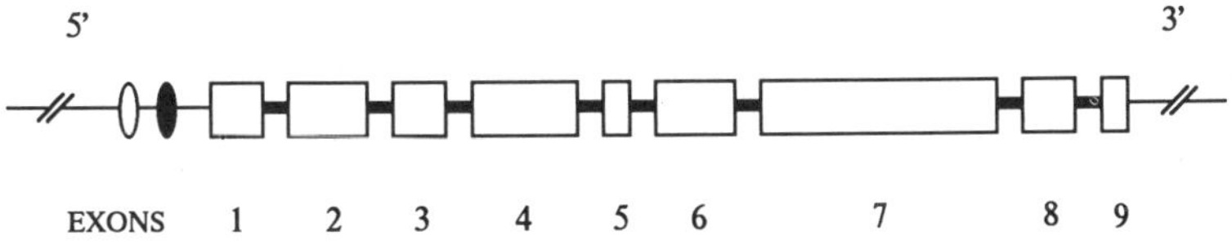

Figure 5. Ligninase isozyme H8 gene structure. The genomic clone of H8 is represented here. The thin line represents noncoding regions, the squares represent exons, the thicker lines represent the introns, the filled oval represents the TATA box sequence at position -75 with respect to the ATG start codon, and the open oval represent the CAAT sequence at position -107 with respect to the ATG.

These patterns of enzyme activity are the result of both the level of expression of the isozymes and the specific isozymes that are being expressed. Only H2 is seen under carbon limitation, but all ten isozymes are seen under nitrogen limitation, with H8 and H2 predominating (see Figure 3).[63] The activity was correlated to the RNA abundance in several experiments.[34,63] The RNA content was analyzed by Northern blot analysis using probes specific for particular isozymes. These results indicated that the message for each isozyme parallels the appearance and activity of that isozyme. Therefore, it has been concluded that regulation takes place at the transcriptional level. It was observed that the manganese-dependent peroxidases are also regulated at the transcriptional level by comparing the enzyme activity (defined by phenol red oxidation[65]), protein levels, and RNA abundance.[20] The time course of expression, however, is different from that of the lignin peroxidases. The manganese-dependent peroxidases are expressed early in secondary metabolism. Under nitrogen limitation, these isozymes appear and peak on day 3 and then gradually decrease through day 6.[20] The managense-dependent peroxidases are not detectable when the fungus is subjected to carbon limitation.[23,63] This set of data raises many interesting and important questions about the regulation, a key point currently under study by numerous groups.

6.5. HETEROLOGOUS AND HOMOLOGOUS EXPRESSION OF LIGNINASES

To assess the potential applications of ligninase (see Sections 6.6 and 6.7) and to further characterize the enzyme, significantly higher quantities of enzyme will have to be available. One of the most useful applications of molecular biology is heterologous expression of the ligninase. This would allow large quantities of enzyme to be produced for analysis and possible direct applications. A large number of laboratories have been working on this problem, but the results have been disappointing so far. Farrell and co-workers reported the expression of ligninase in *E. coli* and reconstitution of active enzyme.[66] Although this process is feasible for basic studies, it is not a viable alternative for industrial application. Downstream processing adds a high cost to production. Our laboratory has also worked on heterologous expression in *E. coli* for some time with very little success. The enzyme is expressed as inactive inclusion bodies and is approximately 5% of the total cellular protein. Our reconstitution attempts have been neither reliable nor effective.[67] These attempts are similar to the procedures used by Farrell and co-workers and are similar to previously published techniques.[66,68] The method entails solubilization of the inclusion bodies with urea and β-mercaptoethanol followed by dialysis in the presence of heme. Although expression in *E. coli* can be a valuable experimental tool for studying enzyme function, it shows little promise as a practical source of enzyme. Expression in other heterologous systems has also not resulted in active ligninase preparations. Our laboratory has attempted expression in yeast. Again, the ligninase peptide is synthesized, but the enzyme is not active. Attempts to express the enzyme in related fungi such as *Trichoderma* have also been unsuccessful.[69] Heterologous expression is presently more of a technology than a science; thus no reasons for these failures are readily apparent.

Overexpression of the ligninases in *P. chrysosposium* shows much promise in light of the failures with the other systems. This would require a transformation system for *P. chrysosporium*. Introduction of ligninase genes behind strong constitutive promoters into *P. chrysosporium* could lead to a deregulated, ligninase-overproducing strain that would be useful for both basic and applied research. By taking advantage of these techniques, the genome could be altered to create a strain that exhibits superior ability to biopulp and degrade environmental waste. A transformation system would also permit in-depth studies on the mechanism of ligninase catalysis through site-directed mutagenesis and detailed studies on the regulation of the ligninase genes. At present, two transformation systems have been reported. The selectable marker used by Alice *et al.* is based on converting an adenine auxotrophic strain to a prototroph by integration of the gene for an adenine biosynthetic enzyme from *Schizophyllum commune* into the *P. chrysosporium* genome.[70] Randall *et al.* used the antibiotic G418 and the kanamycin resistance gene for a selectable marker by integration of a plasmid containing the resistance gene to confer antibiotic resistance.[71] Both of these systems are modeled after fungi and should prove interesting upon optimization.

6.6. POTENTIAL APPLICATIONS OF THE LIGNINASES IN BIOPULPING

An important application for the ligninases is in the pulp and paper industry. The use of lignin-degrading fungi or isolated ligninases for the treatment of pulp could potentially provide greater selectivity than chemical processes in the removal of lignin. Studies with model compounds and pulps indicate that the ligninase is highly selective for aromatic substrates. Carbohydrates are not degraded by the ligninase; hence less destruction of the cellulose and hemicellulose fibers would occur than with current chemical processes. This would result in a high-quality paper product with increased strength. Presently much of the research in biopulping is occurring at the Forest Product Laboratories in Madison, Wisconsin, under the guidance of Gary Leatham and T. Kent Kirk.

The use of a lignin-degrading organism or isolated ligninase could also potentially lower the energy demands for the pulping process. Thermal-mechanical pulping, widely used in Europe, is an energy-consuming process. Kirk and co-workers have shown that a small decrease in the lignin content caused by fungal pretreatment can significantly decrease the energy requirement in this type of pulping.[72] In comparison to the kraft pulping process, the use of ligninase in degrading lignin would proceed at much lower temperatures and require lower levels of chemicals. The pulping and bleaching process used today produces a large volume of pollutants. These lignin-derived pollutants are chlorinated and highly toxic due to the chemicals used in the pulping process. The use of ligninase can potentially eliminate the need for many of these chemicals and lower the pollution volume.

Tien and Kirk initially demonstrated the depolymerization of lignin by ligninases.[14] However, the use of ligninases for biopulping has not shown much promise. These studies indicate that the ligninase does not sufficiently degrade milled-wood or the lignin kraft pulp. This has caused some skepticism concerning the ability of ligninase to degrade lignin. However, much of these studies have been performed in industrial laboratories, and it is difficult to assess what routes have been investigated. Furthermore, these studies are preliminary, and negative data should always be viewed with caution. Much more research needs to be done to understand how the enzyme attacks its native substrate. Leisola and co-workers have recently shown that ligninase can actually catalyze the polymerization of lignin.[73] It is difficult to surmise the significance of these results. One speculates whether this is a trick nature is playing on the *in vitro* system. This observation begs the question of whether polymerization of lignin by ligninases has any *in vivo* significance. Does the fungus have mechanisms to counteract the polymerization process? Nevertheless, it is a problem which has to be circumvented before industrial application of ligninases is possible in the pulping process. The role of the multiple ligninases in lignin degradation also needs to be clarified. Which of these enzymes are necessary and whether other enzymes or cofactors are involved in lignin degradation are questions which need to be answered before ligninase pulping is a reality.

The discovery of ligninases and their mechanism of action has stimulated research into biomimetic systems. The rationale for such research is to synthesize an efficient low-molecular-weight lignin-degrading catalyst which can selectively remove lignin. These synthetic enzymes have taken the form of porphyrin complexes modified with an electron-withdrawing substituent.[74] This electron-withdrawing substituent increases the oxidation–reduction potential of the porphyrin iron. This in turn imparts to the complex the ability to oxidize lignin-like substrates with high oxidation–reduction potentials. Although these complexes have potential, selectivity and high rates of degradation have yet to be demonstrated.

6.7. POTENTIAL APPLICATIONS OF LIGNINASES IN METABOLISM OF ENVIRONMENTAL POLLUTANTS

Work by Kirk and Chang established that the ligninolytic system of *P. chrysosporium* decolorizes the chlorinated aromatic-containing E1 effluent of pulping mills.[75] More recent results reveal dehalogenation of components in the effluent.[76] The ability of *P. chrysosporium* to degrade lignin and chlorinated lignin-derived by-products of the kraft pulping process led Bumpus *et al.* and Eaton to investigate whether it would degrade recalcitrant environmental pollutants.[77,78] Eaton studied the metabolism of polychlorinated biphenyls (PCBs).[78] Bumpus *et al.* studied a wider range of

Table 4. Compounds Degraded by *P. chrysosporium*

Polycyclic aromatics	*Chlorinated aromatics*
Benzo[*a*]pyrene	4-Chlorobenzoic acid
Biphenyl	Dichlorobenzoic acid
2-Methylnaphthalene	2,4,6-Trichlorobenzoic acid
Phenanthrene	4,5-Dichloroguaiacol
Benzo[*a*]anthracene	6-Chlorovanillin
Pyrene	4,5,6-Trichloroguaiacol
Anthracene	Tetrachloroguaiacol
Perylene	Pentachloroguaiacol
Dibenzo[*p*]dioxin	3-Chloroaniline
	3,4-Dichloroaniline
Triphenylmethane dyes	2,4,5-Trichlorophenoxyacetic acid
Crystal violet	*Polycyclic chlorinated aromatics*
Pararosaniline	
Cresol red	DDT (1,1,1-trichloro-2,2-bis(4-chlorophenyl)ethane
Bromophenyl blue	2,3,7,8-Tetrachlorodibenzo-*p*-dioxin
Ethyl violet	3,4,3′,4′-Tetrachlorobiphenyl
Malachite green	2,4,5,2′,4,5′-Hexachlorobiphenyl
Brilliant green	Aroclor 1254
	Aroclor 1242
Biopolymers	2-Chlorodibenzo[*p*]dioxin
Lignin	Dicofol (2,2,2-trichloro-1,1-bis(4-chlorophenyl)ethanol
Cellulose	
Kraft lignin	*Chlorinated alkyhalides*
3-Chloroaniline-lignin conjugates	Lindane
3,4-Dichloroaniline-lignin conjugates	Chlordane

pollutants including PCB congeners 3,4,3′,4′-tetrachlorobiphenyl and 2,4,5,2′,4′,5′-hexachlorobiphenyl in addition to 2,3,7,8-tetrachlorodibenzo-*p*-dioxin (2,3,7,8-TCDD), 1,1-bis(4-chlorophenyl)-2,2,2-trichloroethane (DDT), benzo(*a*)pyrene, and lindane.[77] ^{14}C-labeled substrates were shown to be metabolized to $^{14}CO_2$ in all cases, indicating a degradative effect by the fungus. The metabolism of these compounds, like lignin, occurred after the cells entered secondary metabolism brought about by nitrogen starvation. Despite early skepticism that one fungus could degrade such a large number of chemicals, much of the work by Bumpus *et al.* has been repeated in other laboratories. Table 4 lists some of the fifty chemicals that are degraded by *P. chrysosporium.* It is this ability to degrade such a large number of compounds that makes *P. chrysosporium* potentially useful in waste-treatment systems.

Since the metabolism of these compounds by the fungus appeared to correlate with the lignin biodegradation, the effect of the enzyme alone was investigated. The involvement of ligninase in degrading many of these compounds has been demonstrated. Haemmerli *et al.* have shown that the ligninase metabolizes benzo(*a*)pyrene to yield cation free radicals which undergo further decomposition to produce oxygenated products.[79] Aust and co-workers, who recently demonstrated that pentachlorophenol is also degraded by the ligninase, are continuing to screen for other compounds with are metabolized by the ligninase.[80] These observations indicate that the enzyme is capable of catalyzing these degradation reactions and may prove to be promising in certain applications.

6.8. CONCLUDING REMARKS

There are many possible applications for the lignin peroxidases produced by *P. chrysosporium*, ranging from degradation of environmental pollutants to aiding in the process of making paper. Availability of the enzyme is the key to assessing the feasibility of the potential applications that have been mentioned. At present it is difficult to produce large amounts of the enzyme. These problems are being addressed by a number of laboratories encompassing a wide range of techniques and disciplines. Many of the questions involving the ligninase catalytic properties and expression problems should be answered soon, and the enzyme will then be used in the above-mentioned applications.

REFERENCES

1. J. M. Harkin, in: *Oxidative Coupling of Phenols* (W. I. Taylor and A. R. Battersby, eds.), Marcel Dekker, New York (1967).

2. K. V. Sarkanen and C. H. Ludwig, *Lignins: Occurrences, Formation, Structure, and Reactions*, Wiley-Interscience, New York (1971).
3. K. Freudenberg, in: *Constitution and Biosynthesis of Lignin* (A. C. Neish and K. Freudenberg, eds.), Springer, New York (1971).
4. E. Adler, *Wood Sci. Technol. 11*, 169 (1977).
5. T. K. Kirk and R. L. Farrell, *Annu. Rev. Microbiol. 41*, 465 (1987).
6. R. L. Crawford and D. L. Crawford, *Enzyme Microb. Technol. 6*, 434 (1984).
7. P. Keyser, T. K. Kirk, and J. G. Zeikus, *J. Bacteriol. 135*, 790 (1978).
8. T. W. Jeffries, S. Choi, and T. K. Kirk, *Appl. Environ. Microbiol. 42*, 290 (1981).
9. T. K. Kirk, W. J. Connors, R. D. Bleam, W. F. Hackett, and J. G. Zeikus, *Proc. Natl. Acad. Sci. USA 72*, 2515 (1975).
10. C.-L. Chen and H.-M. Chang, *Holzforschung 36*, 3 (1982).
11. C.-L. Chen, H.-M. Chang, and T. K. Kirk, *J. Wood Chem. Technol. 3*, 35 (1983).
12. T. K. Kirk and F. Nakatsubo, *Biochim. Biophys. Acta 756*, 376 (1983).
13. F. Nakatsubo, I. D. Reid, and T. K. Kirk, *Biochim. Biophys. Acta 719*, 284 (1982).
14. M. Tien and T. K. Kirk, *Science 221*, 661 (1983).
15. J. K. Glenn, M. A. Morgan, M. B. Mayfield, M. Kumahara, and M. H. Gold, *Biochem. Biophys. Res. Commun. 114*, 1077 (1983).
16. L. J. Forney, C. A. Reddy, M. Tien, and S. D. Aust, *J. Biol. Chem. 257*, 11455 (1982).
17. M. Tien and T. K. Kirk, *Proc. Natl. Acad. Sci. USA 81*, 2280 (1984).
18. R. L. Farrell, K. E. Murtagh, M. Tien, M. D. Mozuch, and T. K. Kirk, *Enzyme Microb. Technol. 11*, 322 (1989).
19. M. Tien and T. K. Kirk, *Methods in Enzymol. 161*, 238 (1988).
20. E. A. Pease, A. Andrawis, and M. Tien, *J. Biol. Chem. 264*, 13531 (1989).
21. A. Paszczynski, V-B. Huynh, and R. Crawford, *FEMS Microbiol. Lett. 29*, 37 (1985).
22. J. K. Glenn and M. H. Gold, *Arch. Biochem. Biophys. 242*, 329 (1985).
23. M. S. A. Leisola, B. Kozulic, F. Meussdoerffer, and A. Fiechter, *J. Biol. Chem. 262*, 419 (1987).
24. M. Tien and C.-P. D. Tu, *Nature 326*, 520 (1987).
25. H. A. de Boer, Y. Z. Zhang, C. Collins, and C. A. Reddy, *Gene 60*, 93 (1987).
26. M. Tien, *CRC Critical Reviews in Microbiology 15*, 141 (1987).
27. T. Higuchi, in: *Biosynthesis and Biodegradation of Wood Components*, Academic Press, Orlando, Florida 557 (1985).
28. T. Kamaya and T. Higuchi, *FEMS Microbiol. Lett. 22*, 89 (1984).
29. T. K. Kirk, M. Tien, P. J. Kersten, M. D. Mozuch, and B. Kalyanaraman, *Biochem. J. 236*, 279 (1986).
30. S. Kawai, T. Umezawa, and T. Higuchi, *Appl. Environ. Microbiol. 50*, 1505 (1985).
31. T. Umezawa and T. Higuchi, *FEBS Lett. 182*, 257 (1985).
32. M. Tien, T. K. Kirk, C. Bull, and J. A. Fee, *J. Biol. Chem. 261*, 1687 (1986).
33. V. Renganathan and M. Gold, *Biochemistry 25*, 1626 (1986).
34. A. Andrawis, K. A. Johnson, and M. Tien, *J. Biol. Chem. 263*, 1195 (1988).
35. D. Kuila, M. Tien, J. A. Fee, and M. R. Ondrias, *Biochemistry 24*, 3394 (1986).
36. P. J. Kersten, M. Tien, B. Kalyanaraman, and T. K. Kirk, *J. Biol. Chem. 260*, 2609 (1985).
37. K. E. Hammel, M. Tien, B. Kalyanaraman, and T. K. Kirk, *J. Biol. Chem. 260*, 8348 (1985).
38. H. E. Shoemaker, P. J. Harvey, R. M. Bowen, and J. M. Palmer, *FEBS Lett. 183*, 13 (1985).
39. K. E. Hammel, B. Kalyanaraman, and T. K. Kirk, *J. Biol. Chem. 261*, 16948 (1986).
40. R. Makino, R. Chiang, and L. P. Hager, *Biochemistry 15*, 4748 (1976).
41. H. Yamada, R. Makino, and I. Yamazaki, *Arch. Biochem. Biophys. 169*, 344 (1975).
42. J. Ricard, G. Mazza, and R. J. P. Williams, *Eur. J. Biochem. 28*, 566 (1972).

43. C. W. Conroy, P. Tyma, P. H. Daum, and J. E. Erman, *Biochim. Biophys. Acta 537*, 62 (1978).
44. C. D. Millis, D. Cai, M. T. Stankovich, and M. Tien, *Biochemistry*, in press (1989).
45. J. F. Taylor and V. E. Morgan, *J. Biol. Chem. 144*, 15 (1942).
46. Y. Hayashi and I. Yamazaki, *J. Biol. Chem. 254*, 9101 (1979).
47. M. A. Ator and P. Ortiz de Montellano, *J. Biol. Chem. 262*, 1542 (1987).
48. M. A. Ator, S. K. David, and P. Ortiz de Montellano, *J. Biol. Chem. 262*, 14954 (1987).
49. Y. Z. Zhang, G. J. Zylstra, R. H. Olsen, and C. A. Reddy, *Biochem. Biophys. Res. Commun. 137*, 649 (1986).
50. A. Andrawis, E. Pease, I. Kuan, E. Holzbaur, and M. Tien, *Biochem. Biophys. Res. Commun. 162*, 673 (1989).
51. D. Pribnow, M. B. Mayfield, V. J. Nipper, J. A. Brown, and M. H. Gold, *J. Biol. Chem. 264*, 5036 (1989).
52. D. J. Lipman and W. R. Pearson, *Science 227*, 1435 (1985).
53. K. G. Welinder, *FEBS Lett. 72*, 19 (1976).
54. J. Kaput, S. Goltz, and G. J. Blobel, *J. Biol. Chem. 257*, 15054 (1982).
55. T. L. Poulos and J. Kraut, *J. Biol. Chem. 255*, 8199 (1980).
56. A. Neuberger, A. Gottshalk, R. D. Marshal, and R. D. Spiro, in: *The Glycoproteins: Their Composition, Structure and Function, Part A* (A. Gottschalk, ed.), p. 450, Elsevier, Amsterdam (1972).
57. N. J. Proudfoot and G. G. Brownbee, *Nature 263*, 211 (1976).
58. T. L. Smith, H. Schalch, J. Gaskell, S. Covert, and D. Cullen, *Nucleic Acids Res. 16*, 1219 (1988).
59. A. Brown, P. F. G. Sims, U. Raeder, and P. Broda, *Gene 73*, 77 (1988).
60. Y. Asada, Y. Kimura, M. Kuwahara, A. Tsukamoto, K. Koide, A. Oke, and M. Takanami, *Appl. Microbiol. Biotechnol. 29*, 469 (1988).
61. I. Walther, M. Kalin, J. Reiser, F. Suter, B. Fritsche, M. Saloheimo, M. Leisola, T. Teeri, J. K. C. Knowles, and A. Feichter, *Gene 70*, 127 (1988).
62. H. Schalch, J. Gaskell, T. Smith, and D. Cullen, *Mol. Cell Biol. 9*, 2743 (1989).
63. E. L. F. Holzbaur, A. Andrawis, and M. Tien, *Biochem. Biophys. Res. Commun. 155*, 626 (1988).
64. M. Kozak, *Nucleic Acids Res. 9*, 5233 (1981).
65. M. Kuwahara, J. K. Glenn, M. A. Morgan, and M. H. Gold, *FEBS Lett. 169*, 247 (1984).
66. R. L. Farrell, P. Gelep, A. Anilionis, K. Javaherian, T. E. Maione, and J. R. Rusche, European Patent Application number 87,810,516.2.
67. A. Andrawis, E. A. Pease, and M. Tien, in: *Biotechnology in Pulp and Paper Manufacturing: Applications and Fundamental Investigation* (T. K. Kirk and H.-M. Chang, eds.), Butterworth Publishers (1989).
68. M. Gribskov and R. R. Burgess, *Gene 26*, 109 (1983).
69. T. T. Teeri, P. Lehtovaara, M. Penttila, M. Saloheimo, and J. K. C. Knowles, *The Third Chemical Congress of North America*, Toronto, Ontario, Canada, June 5–11 (1988).
70. M. Alic, J. R. Kornegay, D. Pribnow, and M. H. Gold, *Appl. Environ. Microbiol. 55*, 406 (1989).
71. T. Randall, T. R. Rao, and C. A. Reddy, *Biochem. Biophys. Res. Commun. 161*, 720 (1989).
72. K.-E. Eriksson and T. K. Kirk, in: *Comprehensive Biotechnology* (C. L. Cooney and A. E. Humphrey, eds.), Pergamon, Toronto (1985).
73. S. D. Haemmerli, M. S. A. Leisola, and A. Feichter, *FEMS Microbiol. Lett. 35*, 33 (1985).
74. F. Çui, D. Dolphin, T. Wijeseker, R. Farrell, and P. Skerker, in: *Biotechnology in Pulp and Paper Manufacturing: Applications and Fundamental Investigation* (T. K. Kirk and H.-M. Chang, eds.), Butterworth Publishers (1989).
75. T. K. Kirk and H.-M. Chang, *Enzyme Microb. Technol. 3*, 189 (1981).

76. V.-B. Huynh, H.-M. Chang, and T. W. Joyce, *Tappi J. 68*, 98 (1985).
77. J. A. Bumpus, M. Tien, D. Wright, and S. D. Aust, *Science 228*, 1434 (1985).
78. D. C. Eaton, *Enzyme Microb. Technol. 7*, 194 (1985).
79. S. D. Haemmerli, M. S. A. Leisola, D. Sanglard, and A. Fiechter, *J. Biol. Chem. 261*, 6900 (1986).
80. G. J. Mileski, J. A. Bumpus, M. Jurek, and S. D. Aust, *Appl. Environ. Microbiol. 54*, 2885 (1988).

7

Microbial Biocatalysis for the Alteration of Fossil Fuels

MICHAEL E. McILWAIN and PATRICK R. DUGAN

7.1. INTRODUCTION

Fossil fuels, including coal, oil, oil shales, tar sands, and natural gas, are the world's predominant consumable energy sources. The United States annually consumes approximately one billion tons of coal, 6.2 billion barrels of oil, and more than 18 billion cubic feet of natural gas.[1] The strategic importance of these fuels has resulted in renewed interest in using biological catalysis—both whole cells and cell-free extracts—to recover, convert, process, and supplement these fossil fuels. Although the first reports of the ability of microorganisms to interact with fossil fuels date back to the late 1940s and early 1950s, large-scale technologies have not yet appeared. This fact is not due to the lack of action by microorganisms, but rather to (a) the slow rate of biological metabolic processes as compared to conventional chemical process rates, (b) a general lack of knowledge of specific metabolic and biochemical reactions, and (c) the slowness with which the potential for bioprocesses is accepted by the energy-producing community. Recent advances in genetics and biochemistry now permit modification of microorganisms and their metabolic rates, and this new development has stimulated a resurgence of interest in this area.

MICHAEL E. McILWAIN and PATRICK R. DUGAN • Center for Bioprocessing and Biotechnology Group, Idaho National Engineering Laboratory, EG & G Idaho, Inc., P.O. Box 1625, Idaho Falls, ID 83415.

This chapter considers biocatalytic interactions with constituents of fossil fuels. Microbial systems will be shown to have potential for removal of sulfur from coal, solubilizing low-rank (relatively low energy, high oxygen content) coals, and converting coal-derived materials such as syngas and solubilized coal polymers to higher-value-added liquids and gases. The utility of microbial materials to enhance the recovery of oil from oil-bearing reservoirs is also considered. An overview discussion of this type would not be complete without a discussion of the impact of biocatalysis on the processing of fossil fuels.

7.2. MICROBIAL COAL DESULFURIZATION

Coal desulfurization is desirable because combustion of high sulfur coal is partially responsible for the production and subsequent evolution of oxides of sulfur that are converted in the atmosphere to acids that contribute to atmospheric acid precipitation. These acids are widely thought to have adverse environmental effects.[2] Since coal is the result of fossilization of a wide variety of once living organisms and other organic material deposited over long periods, it can be characterized as a heterogeneous material. Sulfur in coal is present in both organic and inorganic compounds. The inorganic forms are primarily mineral sulfides, such as iron pyrite and marcasite. Minor amounts of sulfate and elemental sulfur are sometimes present, but these materials are generally thought to be the result of pyrite conversion during mining, washing, and storage of mined coal.[3,4] Organically bound sulfur in certain bituminous coals has been characterized as sulfur containing aromatics (such as thiophene, benzothiophene, and dibenzothiophene derivatives), disulfide groups linking alkyl and cyclic subunits, thiols, and thiosulfonates.[5]

A wide variety of microorganisms have been identified that have the ability to oxidize sulfur-containing compounds comparable to those commonly identified in coals. Ideally, the optimum microorganism-based process for desulfurization is one in which the organism derives its energy solely from the oxidation of the reduced sulfur accompanied by the release of sulfur as soluble sulfate or via hydrolysis of the sulfide followed by its release as hydrogen sulfide. Many organisms contain biocatalysts (enzymes) that are active with sulfur-containing species; however, few have the ability to release these sulfur species without loss of substrate organic carbon.

7.2.1. Pyritic Sulfur

Certain microorganisms are able to use both the sulfur and iron contained in pyrite to provide metabolic energy. The net effect of the oxidative bacterial metabolism is the conversion of water-insoluble iron pyrite into sulfuric acid and iron(III) hydroxysulfate complexes. The generalized chemical reaction for this process is shown in Eq. (1):

$$FeS_2 + 3O_2 + 2H_2O \rightarrow 4H^+ + 2SO_4^= + Fe^{3+} \tag{1}$$

The ferric ions formed in Eq. (1), in the presence of *Thiobacillus* bacteria react with water to produce ferric hydroxysulfate complexes (such as the formation of jarosite shown in Eq. (2)) and additional hydrogen ions; and in the absence of thiobacilli, ferric hydroxides and oxides such as magnetite plus hydrogen and sulfate ions are formed as shown in Eq. (3).

$$FeS_2 + O_2 + H_2O \xrightarrow{\textit{Thiobacillus}\ \text{sp.}} H^+ + SO_4^= + \underset{\text{Jarosite}}{[Fe_6(OH)_{12}(SO_4)_4]^=} \tag{2}$$

$$FeS_2 + O_2 + H_2O \longrightarrow H^+ + SO_4^= + \underset{\text{Magnetite}}{Fe_2O_3} \tag{3}$$

Environmental problems associated with acidic water drainage from coal piles and coal mines were the original source for the isolation of *Thiobacillus ferrooxidans*.[6] The early research relating to acidophillic bacteria found a variety of other microbial types associated with these environments.[7,8] Of these organisms, *T. ferrooxidans* and *Leptospirillum ferrooxidans* have been the only microbes identified as directly affecting the rate of pyrite oxidation.[9,10] The remaining organisms have been suggested to serve the role of removing autotoxic metabolic by-products, providing trace organic nutrients, assisting in the removal of oxide materials situated on pyrite surfaces that impede microbial access to the pyrite, and accelerating the conversion of partially reduced sulfur to sulfate.[11–14]

Various geothermal springs have also provided organisms with the ability to oxidize pyrite at elevated temperatures (i.e., greater than 50 °C).[15–20] Of these, *Sulfolobus acidocalderius* and *Sulfolobus brierlyi* have been reported to promote coal pyrite oxidation.[21–22] These organisms were found to oxidize both iron (Fe^{2+}) and sulfur and to grow on simple organic compounds and yeast extract. That is, they are facultative autotrophic bacteria. Other recently identified thermophilic bacteria have also been shown to metabolize the elemental sulfur found in coal arising from exposure of coal pyrite to air.[4]

T. ferrooxidans, a bacterium for which several different strains have been described, is the most widely studied of these pyrite-oxidizing organisms. It grows in the pH range from about 1.5 to 4.5 with optimum growth of various isolates in the range of 1.5 to 3.5. This organism oxidizes ferrous iron using molecular oxygen, but it will conserve energy by oxidation of elemental sulfur and reduced sulfur compounds (i.e., there are approximately 46 kJ released per mole of Fe^{2+} oxidized to Fe^{3+}, whereas the oxidation of a mole of S^{2-} to SO_4^{2-} releases about 670 kJ). The low pH of the medium in which these organisms grow provides a gradient in hydrogen ion concentration that serves as one mechanism to drive membrane phosphorylation of ADP to ATP and in turn provides energy for fixation of carbon dioxide.[23] The reduction of molecular oxygen and subsequent formation of water serves as the final electron acceptor for the oxidation of both sulfide and ferrous iron.[24] The oxidation of sulfide to sulfate, which releases a total of eight electrons, is a highly integrated process involving both soluble and membrane-bound enzymatic components. The oxidation of ferrous to ferric iron also appears to be associated with the microbial membrane. Electron micrographs of pyrite that has been immersed in solutions containing *T. ferrooxidans* show organisms dispersed on the surface of the pyrite crystals. As the cell density of the immersion solution increases, the number of cells attached to the pyrite surface also increases. At low cell densities, the cells are attached to the crystal along the length of the cell. When cell densities are sufficient to saturate the surface with organisms, the cell alignment appears to change to an end on attachment.[25] The biochemistry of *T. ferrooxidans* seems to have become highly adapted to the acidic environment it produces and the chemical nature of sulfur-containing minerals.

7.2.1.1. Factors that Influence Desulfurization

The factors that influence the rate and extent of coal desulfurization (pyrite) can be separated into those that involve physical, chemical, and biological characteristics of the bioprocess. The physical parameters include temperature, coal particle size, and pulp density. Chemical properties can vary depending on the organism(s) selected for the process, but most important of these are pH, gases, nutrients, and toxicants. Finally, the biological factors, such as cell density, resistance to toxicants, and strain characteristics must be considered. It is apparent that these factors are interrelated. Attempts to evaluate a microbial system or process for coal desulfurization must evaluate each of these properties and optimize all of these combined to obtain the greatest degree of desulfurization while minimizing the residence time, water usage, and amount of additives.[26]

a. Physical Factors. In terms of physical properties, the reaction mixture temperature needs to be controlled to match the specific organism type. For example, *T. ferrooxidans*, a mesophile, grows best in the temperature range of 30 to 35 °C. *Sulfolobus* species and other thermophiles grow at higher temperatures and can oxidize pyrite at temperatures as high as 85 °C. Particle size of the feed coal is also a factor in determining the rate and ultimate amount of inorganic sulfur removed. Since pyrite oxidation depends on exposing pyrite either to contact the microbe or the solution containing Fe^{3+} ions, which serve as oxidants, the coal must be ground to a size sufficiently small to expose the surface of the pyrite. In a recent study, the conclusion is presented that oxidation of coal pyrite by Fe^{3+} is negligible and that microbes play the main role in pyrite oxidation.[27] If one selects a coal with a nominal pyrite inclusion size of 10 μm, geometric calculations suggest that the nominal coal particle size must be approximately 50 μm to obtain optimum pyrite removal.[28] Coal grinding is one of the more significant costs associated with any coal desulfurization process. Grinding costs increase with the need to decrease coal particle size, and care must be taken to grind the coal only enough to obtain the desired amount of sulfur removal. The pulp density, or weight of coal per weight of process water, is another important consideration. Ideally, the process should have the highest pulp density mechanically maintainable to minimize the amount of process water. Many investigators have reported that the optimum pulp density relative to the rate of pyritic sulfur converted to sulfate per unit weight of coal by *T. ferrooxidans* in shake flask experiments is in the range of 10 to 20 wt%. However, the total amount of pyritic sulfur converted to sulfate per unit time has been demonstrated to increase up to about 40% pulp density (i.e., there is more pyrite to be removed as the pulp density increases.). At higher pulp densities (up to 50%) the ultimate pyrite removal rate per unit weight of coal declines while the total amount of sulfur removed per unit time per volume of water increases.[2, 29] The decline in ultimate pyrite removal appears to be associated with the oxygen transport in shake flasks. When similar studies are performed in highly aerated bioreactors, ultimate pyrite removal can be maintained up to pulp densities as high as 50%.[30]

b. Biological Factors. Biological factors are important in the removal of pyrite from coal. As discussed previously, the organism strain or type or mixture of organisms employed can influence the amount of sulfur removed. However, better results are obtained if the culture is allowed to adapt to the type of coal to be processed.[29, 31, 32] This adaptation process seems to select a population of organisms resistant to toxic materials contained in the coal or able to attack a specific variety of coal pyrite which is known to contain various metal impurities. It should be noted that pyrite samples vary tremendously with respect to amount and mix of elemental impurities. Further, adaptation of the culture appears to induce metal resistance. In the case of arsenic and mercury contained in certain coals, the metal may be reduced, combined with organic compounds, and/or volatilized from the medium.[33] The cell density is also important. The optimum condition is one that has sufficient cell density in the reactor to just saturate the surface of the exposed coal pyrite. The organisms that carry out pyrite desulfurization are aerobic chemoautotrophs requiring oxygen and carbon dioxide. The oxidation of pyrite depends on the presence of dissolved gases. It was shown previously that the partial pressure of air in the reactor can influence desulfurization by *T. ferrooxidans*. However, hyperbaric oxygen creates excessive dissolved oxygen levels (and presumably toxic oxygen by-products such as superoxide and peroxide) that are toxic to *T. ferrooxidans*.[34]

c. Chemical Factors. The above discussion of gases leads naturally to the topic of chemical properties. Nutrients, such as mineral salts containing nitrogen or phosphorus, are often found to enhance the rate of pyrite oxidation, especially if the coal is deficient in these materials.[34] Certain organisms employed for coal desulfurization require media with low pH values for optimum desulfurization. Further, during pyrite oxidation hydrogen ions are produced. If the pH is allowed to decline below approximately 1.7, the oxidation of pyrite will be retarded for many strains of *T. ferrooxidans* and at values below 1.5 the oxidation is inhibited. It is important that pH be monitored and controlled at an optimum value for the culture mixture used in the process. Finally, various types of by-products can accumulate in the medium which influences the desulfurization of coal. These by-products can act as metabolic inhibitors, or they can build up on the pyrite surface, which restricts the access of either the organisms or the oxygen to the pyrite.

7.2.1.2. Processes to Remove Pyrite

Control of the chemical properties is always vital to ultimate process performance. The number of parameters to be controlled is usually reflected in the complexity of the process. Process complexity translates directly into the cost of the product. In the case of coal these costs in terms of value added must always be considered, since the value added to coal is currently considered to be low. Therefore, current research concerned with microbial pyrite removal from coal is seeking to find ways to limit the number of parameters that need to be controlled.

A number of process schemes have been proposed for microbial pyrite removal from coal. They range from designs using stirred reactors to leaching of coal piles.[35–39] Depending on the specific process, the number of unit operations needed to perform desulfurization will vary. The features that most designs have in common include coal handling, organism preparation, a bioreactor, dewatering and waste water clean-up units (see Figure 1). The simplest process concept for large-scale desulfurization is heap leaching in which a coal pile composed of granular coal is con-

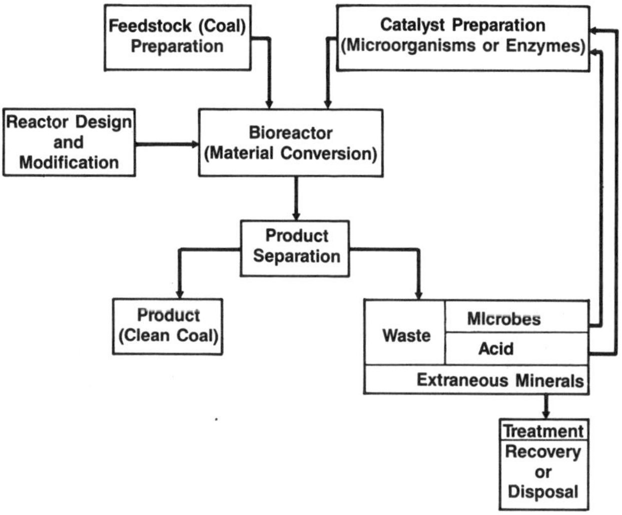

Figure 1. Typical biological process for coal conversion.

tinuously aerated from the bottom or within the coal pile. Organisms, nutrients, and acidified water are trickled or sprayed onto the top of the pile. A continuous operation is achieved by addition of coal either to the top of the pile with removal from the bottom or addition to one side of a pile with removal from the opposite side. Water, spent nutrients, iron, and sulfate are removed from the bottom, treated with lime to control metals, sulfate, and pH, and microbes recycled back into the system. A series of cascade Pachuca tanks has also been proposed as a replacement for the coal pile. Each tank is aerated. As the coal passes from tank to tank it becomes progressively desulfurized. Slurry pipelines, holding ponds, barges, and tanker ships have also been proposed as bioreactors. Work at the INEL is currently investigating the potential of aerated troughs, fluidized beds, and conveyed coal beds as possible methods of continuously processing coal.

A slightly different approach to microbial coal desulfurization is the concept of using microorganisms to promote the floatation cleaning of coal.[40–42] Organisms in suspension are mixed either during coal wet grinding or following grinding to precondition the coal prior to froth floatation. Pyrite oxidizing microbes attach to the pyrite surfaces, and since the microbes enhance the water wetting of pyrite, the particles with the greatest amount of pyrite and least amount of coal are the most wettable, and this results in improved separation efficiency. Results to date for microbial-assisted froth floatation suggest that microbes or their products assist the separation of pyrite from coal. However, similar if not improved results can be achieved by addition of chemical surfactants.[41, 42]

7.2.2. Organosulfur Removal

Several investigations have shown that microorganisms have the ability to convert organically bound sulfur to sulfate when organosulfur compounds similar to those found in coal and petroleum are used.[43–46] Early efforts in this area found that many soil microbes have the ability to metabolize organosulfur compounds. The primary mode of microbial attack is by oxidation of the aromatic ring structure resulting in ring cleavage followed by oxidative metabolism of the terminal carbon units (see Figure 2a).[44]

An organism isolated from soil was found to also exhibit an alternate pathway for degradation of thiophenic molecules (Figure 2b). When the primary pathway involving ring cleavage was blocked, the organism, designated CB1, was able to directly attack the carbon–sulfur bond resulting in the suggested production of dihydroxybiphenyl and sulfate.[46]

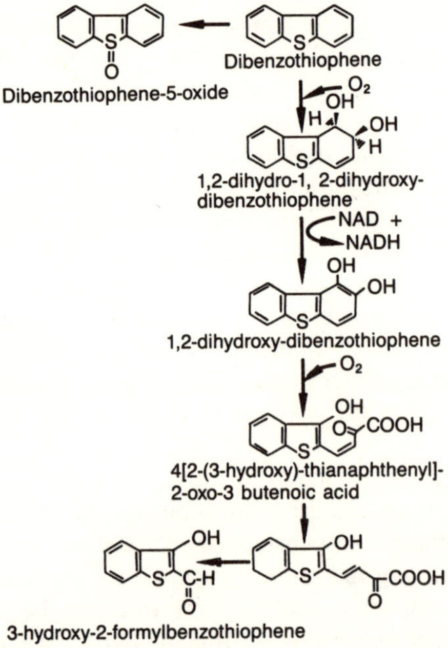

Figure 2a. Aerobic degradation of dibenzothiophene by pseudomonads *via* carbon skeleton attack (see Refs. 44 and 45).

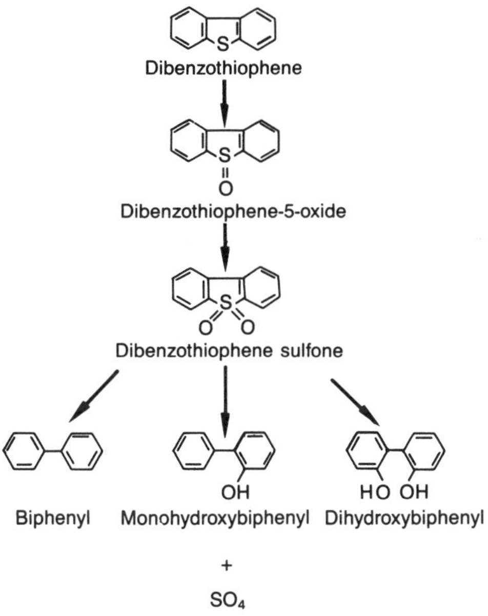

Figure 2b. Proposed mechanisms of direct oxidation of the ring sulfur of dibenzothiophene resulting in release of sulfate (see Ref. 46).

This finding represented a major advance toward a mild process for organic sulfur removal. More recent studies have shown that a great number of different soil organisms are capable of catalyzing the removal of sulfur from organosulfur containing chemicals via this pathway. One of the organisms (a *Rhodococcus* Species) appears to have a pathway that provides the sulfur for cellular constituents from dibenzothiophene and other organosulfur substrates when the organism is in a sulfate-limited or deficient environment. A mixed culture isolate (designated IGTS7) has been shown to remove more than 90% of the 2.43% organic sulfur contained in a +9 mesh ground Illinois #6 coal sample.[47] The coal contact time required for this removal was greater than 200 days. These results are preliminary, and with additional research, it is anticipated that the contact time can be dramatically reduced, or that the enzyme activity can be increased by use of genetic manipulation. Currently, a micrococcal organism has also been isolated with a similar ability to degrade aromatic carbon–sulfur bonds. While researchers are continuing to isolate organisms from various environments, genetic experiments are currently attempting to induce carbon–sulfur bond cleavage in strains of *Escherichia coli.* This area of research is rapidly advancing.

7.3. COAL CONVERSION

Prior to the discovery of domestic petroleum deposits, coal served as a source of hydrocarbon liquids, chemical feedstocks, and solvents. With the advent of declining domestic petroleum reserves, coal has again been considered as a source for production of these hydrocarbon materials. However, coal-derived chemicals are not cost competitive with world petroleum as sources of fuel and bulk chemicals. In order to reduce the cost of coal-derived chemicals and to open new markets for coal, new technologies are needed. This need for new technologies is one of the driving forces for microbial conversion processes.

The conversion of coal using microorganisms can be divided into two topics: solubilization of coal and synthesis of higher-value-added materials from coal-derived material. The term "coal solubilization" is used to describe the conversion of coal solids to water-soluble materials. Application of microbes for synthesis involves the use of either solubilized coal product or effluents from other coal conversion processes to produce various types of alcohols, acids, or a methane gas.

7.3.1. Coal Solubilization

A variety of studies have shown that microorganisms have the ability to promote solubilization, depolymerization, and oxidation of coal and coal-derived materials.[48, 49] Most studies have either used lower-rank coals (coals having Btu content of less than 8000/ton), or coals that have undergone chemical pretreatment. These pretreatments are oxidative in nature and produce humic acid like components as starting materials. Some "young" lignite coals can contain as much as 35 to 70% lignin-like constituents. It is, therefore, appropriate to examine some of the chemistry associated with microbial interaction with lignin to aid in understanding how microorganisms are interacting with coal.

White rot fungi, such as *Phanerochaete chrysosporium* and *Polyporus versicolor*, are known to be responsible for decomposition of woody plant tissues. The mechanism by which these fungi decompose the lignin component of woody tissue involves enzymes known as ligninases and laccases.[50, 51] The mode of action of these enzymes is associated with a nonspecific hydrolysis and ring cleavage of the phenylpropyl units in lignin. Figure 3a shows one of the proposed mechanisms for cleavage of lignin. Figure 3b illustrates an etherase-based mechanism for ligninolytic activity shown by *Poria subacida.*[52] In general, the chemistry of microorganism attack on lignin appears to be associated with either the C_α or C_β carbons of the phenylpropyl subunits. However, each organism type seems to have developed a unique set of enzymes to degrade lignin. For example, *P.*

chrysosporium has been shown to have both an arylglycerol-β-aryl etherase (Figure 4a) and a 1-α and 4-β ligninase (Figure 4b).

Although no accurate model of lignite coal or oxidized lignite currently exists, it is known that phenylpropyl units are not common in higher-rank coals.[53] At present, it is uncertain as to the extent, if any, of ligninase- or laccase-catalyzed reactions associated with microbial degradation of coal. It is important, however, to illustrate how versatile microorganisms can be when interacting with relatively recalcitrant materials such as lignin and lignite.

Suggestions that microbial action could occur on coal were reported as early as the 1960s.[54, 55] This line of investigation was not further pursued until quite recently when two groups of workers almost simultaneously

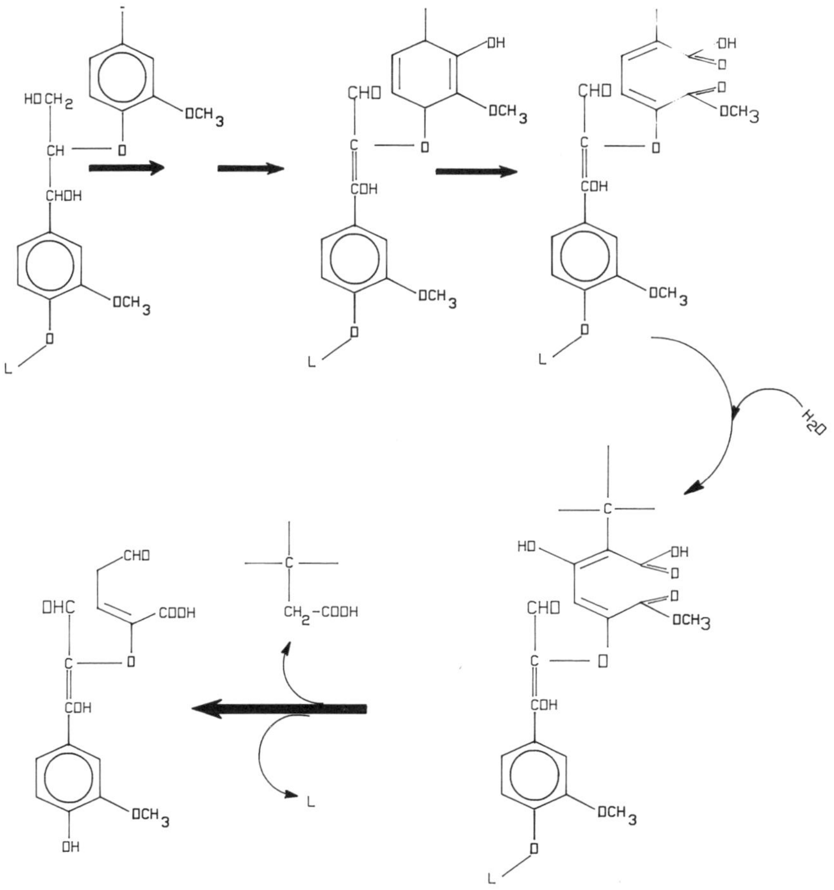

Figure 3a. Mechanism of biological degradation of lignin.

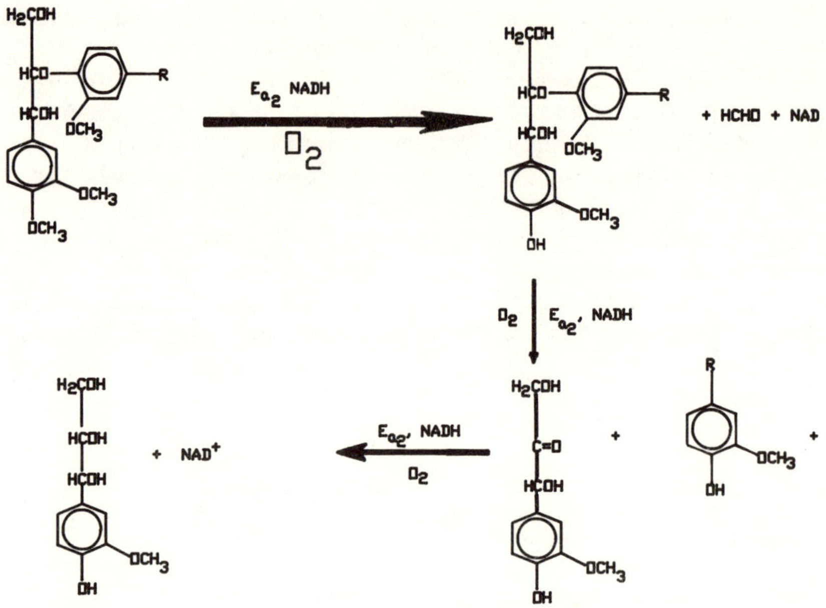

Figure 3b. Hypothetical scheme for cleavage of ether bonds by *Poria subacida.*

reported that filamentous fungi produced coal-derived liquid materials.[48, 56] When the hyphae (fungi filaments) were examined under a microscope, the newly germinated hyphae had no affinity for coal. As the culture matured, greater and greater numbers of coal particles were observed attached to the hyphae. Further, at the points of contact with the coal, the hyphae appeared dark or black, indicating microbial uptake of coal-derived material. Additional studies confirmed that organic-containing liquids were produced by numerous fungi when the fungi were grown in the presence of low-rank coals.[57–59] Table 1 lists some of these organisms.

Based on the confirmation that organisms could live and grow on coal, research began to focus on how to enhance solubilization and to determine what by-product materials were being produced. When a range of coal types were examined, the greatest degree of solubility was found to be associated with a lignite coal called leonardite.[60] Comparison of chemical compositions of coals showed that the concentration of oxygen in leonardite was considerably higher than that found for other coals. This finding led several groups to investigate the impact of oxidative pretreatments on the coals in relation to microbial action.[61, 62] The oxidative procedures that were evaluated ranged from heating the coal in air to treatments with oxidizing agents, such as ozone, hydrogen peroxide, potassium permanganate, and nitric acid. Results proved that oxidative pretreatments

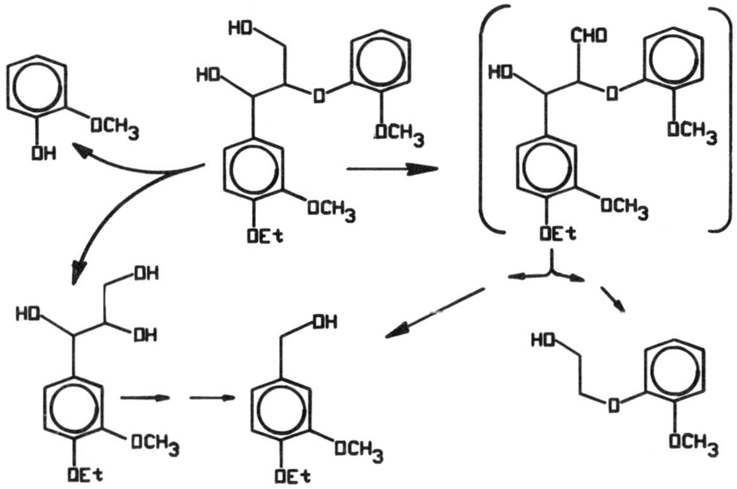

Figure 4a. Arylglycerol-β-aryl etherase.

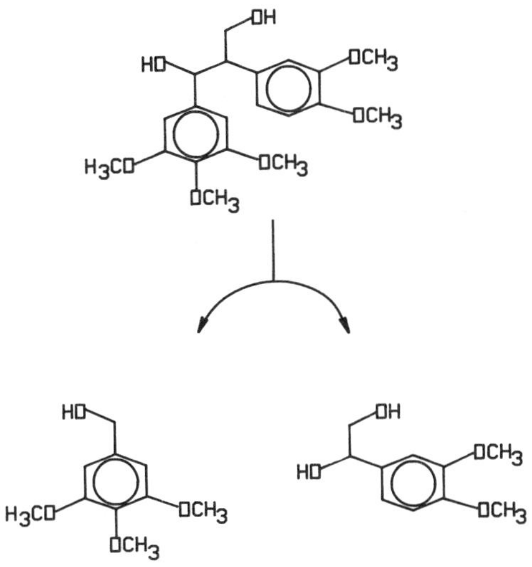

Figure 4b. 1-α and 4-β ligninase reaction.

improved subsequent biological solubilization and showed that the extent of solubilization was improved for low-rank coals. These findings also suggested that as the oxidative strength of the pretreatment increased, microbial degradation was enhanced. When Texas lignite is exposed to 8*N* nitric acid, an exothermic reaction occurs. This reaction results in approximately 35 to 40% of the coal being lost either during the reaction

Table 1. Fungae and Filamentous Bacteria Examined for Solubilization Activity

Ascomycetes sp.	*Aspergillus* sp.
Candida sp. (ML-13, ACL-13)	*Coprinus comatus*
Coriolus hirsutus	CP-1 and CP-2
Cunninghamella sp. (YML-21)	*Ganoderma lucidum*
Geosmithia argellacea	*Lentinus edodes*
Lenzites trabea	*Neurospora sitophila*
Paecilomyces TLi	*Penicillium* sp. (RML-5)
Penicillium waksmanii (ML-20)	*Phanerochaete chrysosporium*
Pleurotus ostreatus	*Pleurotus sapidus*
Polyporus monticola	*Polyporus versicolor*
Rhizopus arrhizus	*Sporothix* sp.
Streptomyces flavovirens	*Streptomyces setonii*
Streptomyces viridosporus	

or during the coal washing needed to remove excess acid. Comparison of Fourier Transform Infrared spectra for samples of treated and untreated Texas lignite indicated that nitric acid results in a significant amount of coal oxidation, and nitration of the coal material may also occur. Pretreatment with 8*N* nitric acid for 48 h resulted in the lignite being almost completely solubilized.[63, 64] Pretreatment was also found to accelerate the onset of microbial action. When untreated coal was placed on an established fungal matte, the time required for the formation of black droplets could be greater than a week. A similar test performed with 8*N* nitric-acid-pretreated coal produced black droplets within two hours after introduction of the coal to the fungal matte. It was also demonstrated that 2*N* nitric acid removed most of the calcium, strontium, and several other metallic components from Illinois #6 coal during pretreatment. These metals are believed to contribute to holding the coal organic matrix in its solid configuration.[65]

As the amount of biologically produced material increased from isolated drops to milliliters of material, researchers began to examine the material being produced. The material was found to be highly polar and soluble in water up to greater than 20 wt%.[63] Solubility was shown to be highly pH-dependent, and a brown-to-black solid could be precipitated on re-acidification of the microbially produced liquid.[60] Chemical analysis of this precipitate showed relatively little difference between the original lignite and microbial product except that the amounts of nitrogen and phosphorus in the microbial product were elevated. Spectrochemical analysis of the microbial product also indicated an increase of oxygen functionalities.

Data indicates that coal is not being directly converted to an organic liquid, but is being dissolved in water. The solubilization process is greatly enhanced by either using a weathered coal or performing an oxidative pretreatment on the coal. Coal types that are most receptive to microbial solubilization tend to be of low rank, as lignites, for example; although some subbituminous and bituminous coals have been reported to be partially solubilized. The water-soluble coal material appears to be large, highly charged macromolecules. Combined analytical data on the characterization of the microbial produced material indicates that it is a high-molecular-weight polycarboxylated aromatic polymer. Current research activities are focusing on issues related to a) how pretreatment enhances or promotes microbial solubilization, b) whether microorganisms secrete material that can react with the coal polymer to decrease its molecular weight, and c) whether enzymes isolated from lignin-degrading organisms (e.g., ligninases, laccases, peroxidases, etherases, or others) can degrade the coal polymer.

Several studies found that the resulting nitric-acid-treated coal is readily solubilized in alkaline solutions.[59, 62] If the pretreated coal is washed with a series of buffer solutions of varying pH value, a comparison of the final extraction buffer pH and the accumulated weight of material dissolved suggests that acidic groups in the coal are being neutralized. Maximal coal solubilization always occurs when the final extraction buffer pH value ranges from 5.2 or higher. The resulting neutralization of coal acidic groups seems to promote increased water solubility. This base- or buffer-solubilized coal can be reprecipitated using either mineral acids or metal solutions (e.g., calcium, aluminum, iron, and so on).

Since the culture pH value for many fungal and bacterial cultures which were effective in solubilizing acid treated coal fell in the range of 5.0 to 6.0, researchers began to question whether base-promoted solubilization could also be responsible for microorganism-promoted solubilization of coal. Several *Streptomyces* species known to promote lignin degradation and coal solubilization were used to test this hypothesis. The pH of liquid cultures was monitored at intervals following introduction of acid-pretreated coal. The culture pH was observed to initially decrease due to the residual acid in the coal and then gradually increase with time. Comparison of the extent of coal solubilization with the rise of culture pH indicated that culture pH was correlated with coal solubilization and suggested that one mechanism responsible for solubilization of acid-pretreated coal was related to microbial production of an alkaline material. More recent results[59, 66] indicated that the production of alkaline material(s) by fungi is also related to the degree of coal solubilization. These organic bases are neutralizing the acid groups on the coal, which increases its hydrophilic nature and thereby enhances its water solubility.

Studies have also been initiated to determine whether fungi and bacteria are responsible for additional degradation of the solubilized coal.

Studies using *Streptomyces viridosporus*, *Streptomyces setonii*, and *Streptomyces flavovirens* suggest that the solubilized coal is being altered due to microbial attack. This conclusion is based on changes observed in the ultraviolet spectra of culture solutions and on changes in the infrared spectra of acid precipitated materials. One fungal organism (designated RWL-5) has been found to promote sufficient alteration in the solubilized coal that the polymer will not precipitate upon re-acidification. Results also show that products obtained from different fungi grown on the same coal are not composed of exactly the same materials.[67] High-pressure liquid chromatographs of the microbial products show peaks which do not elute at the same retention times. These results suggest that microorganisms do have an ability to attack solubilized coal polymers.

Cell-free enzyme extracts have also been observed to promote coal solubilization and degradation of the resulting solubilized polymer.[63,67,68] These studies are still preliminary and do not as yet support a definitive conclusion. Several researchers report that certain laccases and peroxidases have the ability to enhance the rate of coal solubilization.[68] A purified lignin peroxidase from *P. chrysosporium*, acting on previously base-solubilized extracts of North Dakota lignite and German subbituminous coal, produced major decreases in the high-molecular-weight coal polymer peak as determined by gel filtration.[68] This decrease in molecular weight was shown to be dependent on the polymer to enzyme ratio. The decrease in the high-molecular-weight polymer peak was accompanied by the appearance of both higher- and lower-molecular-weight peaks. No monomeric peaks were observed. Experiments were performed to determine whether the polymer was being degraded or the polymer subunits were just being dispersed and re-aggregated. The latter explanation could account for the appearance of higher-molecular-weight fragments. However, a similar re-aggregation of chemically degraded fragments could also be responsible for the observed results. Additional research is required in this area before any definite conclusion concerning cell-free enzymatic degradation of the solubilized coal polymer can be drawn.

7.3.2. Conversion of Coal-Derived Materials

In addition to coal solids and solubilized coal, gases produced during thermal gasification of coal are ideally suited as microbial metabolites. Many of the unwanted compounds present in liquids produced during coal liquefaction can also serve as carbon sources for microorganisms. Although the amount of research associated with conversion of coal-derived materials is limited, microbial synthesis, conversion, and selective removal of components from these materials may provide future fuel supplements and chemical feedstocks that are presently supplied by petroleum.

Solubilized coal polymers are currently being explored as substrates for microbial synthesis of value-added chemicals. Microorganisms are being isolated that have the ability to use these solubilized coal polymers as metabolic carbon sources for production of various organic acids, alcohols, ketones, and possibly polysaccharides. Attempts are being made to use solubilized coal as a carbon source for anaerobic production of methane. To satisfy the sizable volume of gas needed to fuel a power generator, extremely large bioreactors will be required, and natural caverns and salt domes are currently being considered as potential bioreactors.[69] The experimental studies associated with this "methane from coal" concept are attempting to identify and isolate microorganisms or mixtures of microbes that can digest coal polymers, produce methane, and be maintained under conditions typical of the proposed bioreactors.

7.3.2.1. Conversion of Coal Synthesis Gas

The primary thrust of early research on microbial conversion of coal synthesis gas was to determine the feasibility of converting gas mixtures containing hydrogen, carbon monoxide, and carbon dioxide to methane. Initial studies employed organisms collected from sewage treatment plants and various types of animal wastes.[70] A mixed culture was obtained which was capable of converting approximately 95% of the carbon monoxide and hydrogen to products in approximately two hours. The products obtained were primarily methane with small amounts of acetate. Attempts to improve the conversion of carbon monoxide to methane led to screening large numbers of organisms. In the process, two organisms were identified, *Peptostreptococcus productus* and *Acetobacterium woodii*, which convert carbon monoxide and water to acetate and carbon dioxide. When a test gas composed of 65% carbon monoxide, 22% hydrogen, 11% carbon dioxide, and 2% methane was introduced to a previously acclimated culture of *P. productus*, approximately 90% of the carbon monoxide was converted to acetate in 0.7 h. Analysis of the products also showed the presence of small amounts of other organic acids and alcohols. The sensitivity of *P. productus* to hydrogen sulfide and carbon disulfide, common hydrogenation catalyst poisons, was also evaluated and found not to inhibit acetate production when the concentration of these gases was maintained at levels typical of coal gasification. In an attempt to produce more acids and alcohols using the mixed culture, various types of methane formation blocking agents were investigated.[71] Preliminary screening studies reported that addition of blocking agents resulted in production of small quantities of ethanol, butanol, and methanol, as well as acetate. A pure culture isolated from this mixed consortium was capable of producing only ethanol and acetate.

Current efforts in this area are continuing to investigate the production of low-molecular-weight alcohols from both mixed and pure cultures grown in the presence of gasifier products and are encouraging as to the potential for microbial processes to convert carbon monoxide to low-molecular-weight organic acids and alcohols.[72] Studies using pure isolates have reported that product distributions can be controlled while maintaining process efficiency. An organism tentatively identified as a *Clostridium* sp. required a small amount of yeast extract in the growth medium and agitation to enhance conversion of carbon monoxide to ethanol. An ethanol concentration of approximately 4.3 g per liter has been obtained using this *Clostridium* sp., and efforts are being made to increase the ethanol concentration and to examine bioreactor designs that enhance the transport of carbon monoxide to the organism. All of the above-mentioned microbial conversions are dependent on hydrogen (or other suitable sources of metabolic energy), and ultimate success of the process is likely to be limited by the availability of hydrogen or other suitable reductants.

7.3.2.2. Selective Treatment of Coal-Derived Materials

Current research in this area can be described as being exploratory and fundamental in nature. The primary focus is to identify microbial pathways to remove unwanted materials from coal-derived liquids (compared to water-soluble products). When coal is liquefied, the resulting coal liquid contains significant amounts of nitrogen and sulfur-containing components. To remove these materials conventional chemical processes rely on hydrogenation to transform nitrogen and sulfur-containing components to ammonia and hydrogen sulfide, respectively. Hydrogen consumption is significant and extremely costly. In the case of nitrogen-containing compounds, efforts are currently underway to identify organisms that have the ability to metabolize these materials. Reports to date indicate that microorganisms are very selective as to the nitrogen compound metabolized.[73] In the case of organic sulfur removal, microorganisms have been demonstrated to have the ability to promote aqueous separation of organosulfur-containing species from hydrocarbon liquids. Organisms are currently available which convert aromatically bound sulfur to water-soluble compounds. A major obstacle to be addressed is how to contact the organism with the coal liquid. Organisms typically function in aqueous phases either on particle surfaces or in bulk liquids. Since coal liquids do not readily dissolve in water, new bioreactor types designed to handle immiscible liquids are required. Efforts to design these new bioreactors are currently underway.

7.4. MICROBIAL-ENHANCED OIL RECOVERY

Petroleum is the predominant source of hydrocarbon transportation fuels and will remain so for the foreseeable future. Since the United States' petroleum production peaked in 1968 and demand has exceeded production since the late 1940s, the country has become more dependent on imported petroleum.[73] The United States currently imports more than 50% of its petroleum which has a significant impact on the trade deficit and has raised concern regarding national security in times of shortage. One of the greatest challenges to be addressed during the next decade is solving the problem associated with supply of transportation fuels.

Although the United States has massive discovered oil deposits, it has been estimated that approximately two-thirds of all oil discovered still remains in the reservoir. The objective of enhanced oil recovery (EOR) is to produce this residual oil. Microbial-enhanced oil recovery (MEOR) is similar to EOR processes that employ chemical agents to assist in the liberation of oil trapped in reservoir rock after primary production has ceased.[75] Microorganisms are known to produce surface active materials, such as emulsifiers and surfactants.[75] Several biologically synthesized surfactants are currently being employed for oil field applications.[77, 78] Certain microorganisms are able to produce extracellular polymers, and these materials can serve as both water viscosity modifiers and porosity plugs to improve secondary recovery (waterflood sweep efficiency). During metabolism, polysaccharides, sugars, and hydrocarbons can be converted to acids and alcohols. Acids and alcohols can act to reduce the viscosity of petroleum, enhance the performance of surfactants, and react with carbonaceous rocks. In addition, certain organic acids are strong chelating agents for Group IIa metal ions such as calcium. In the reservoir, these organic acids can displace oil constituents bonded to clays and carbonates. Finally, many reservoirs have been overproduced during primary production and no reservoir pressure remains. Microbial metabolism can result in release of carbon dioxide, nitrogen, hydrogen, and methane, which can partially repressurize the reservoir.

MEOR has advantages in relation to other types of chemical-assisted EOR methods. The most significant of these are *in situ* production of materials, environmental compatibility of these materials, and low cost of reagents. Chemicals directly added to the well act the moment they enter the reservoir, and the treated region of the reservoir is that which is directly adjacent to the well bore. Since microbes generate metabolic products and extracellular materials over a longer duration (when the appropriate conditions or nutrients are present), wide dissemination of microbes could result in greater dissemination of polymers and surface active materials than can be achieved through well bore injection. This could result in improved oil recovery. The primary cost difference between MEOR and chemical EOR is associated with the cost of reagents. MEOR reagents can be comprised of wastes or by-products which are subsequently converted by microorganisms to relatively inexpensive chemicals.

Numerous field evaluations of MEOR have been reported since the mid-1940s.[79] Results of these investigations vary, and few have conclusively demonstrated enhanced oil production.[79] The primary reason for the poor results seems to be associated with a lack of understanding of the oil-bearing reservoir and its hydrology and a lack of suitable experimental controls, rather than with failure of the organism to function. Most field evaluations were based on trial-and-error methods and were not supported by extensive laboratory studies. Recent investigations have made an effort to first confirm in the laboratory the hydrodynamics of the reservoir, the compatibility of organisms with oil and brine, and selection of an appropriate type of microorganism. Results reported for recent field evaluations suggest short-term improvements in production; long-term production gains are currently being evaluated.[81]

Research associated with MEOR continues to investigate the use of microbes and their metabolic products. Biosurfactants capable of reducing the surface tension of various types of crude oil are being evaluated. Biopolymers are being studied as to their ability to improve volumetric sweep efficiency. Thermophilic bacteria are being isolated to develop cultures that will function in the higher temperature environments associated with many deep petroleum deposits. Studies associated with gas generation, selective plugging, the degradation of hydrocarbons, and the role of microbes as agents that induce corrosion of equipment are continuing. Finally, fundamental studies to determine the function of microorganisms under reservoir conditions and their interactions with crude oil and rock are currently underway.

Although MEOR is the oldest of the microbial fossil fuel-processing techniques, a considerable amount of fundamental information concerning the role and function of microorganisms in reservoirs is needed. MEOR currently suffers from attempts to employ this technology based on empirical understanding and without a firm technical data base. It is to be hoped that once a better fundamental data base is obtained, the full potential of MEOR can be realized.

7.5. CONCLUSION

Biological processing of fossil fuels can be considered to be an emerging technology. Although various microorganism-based concepts for processing or recovery of fossil fuels have been investigated for more than

40 years, the concepts have neither been fully demonstrated to be commercially viable, nor are they at a sufficient stage of development to be evaluated on a large scale. Other concepts require more fundamental and applied research to be performed. Microbial removal of organic sulfur constituents from coal currently needs more information relating to the genetics of organisms and the biochemistry associated with aromatic sulfur metabolism. Coal solubilization using bacteria and fungi has just begun to investigate the viability of microbial degradation of coal polymers. More data is needed on relevant biochemistry, as well as an expanded collection of microorganisms with the ability to interact with the coal polymers. Microbial conversion of coal-derived materials has been shown to be viable; however, there is a need for new bioreactor types to facilitate processing large volumes of solids and gaseous substrates. Microbial-enhanced oil recovery requires an improved understanding of how microorganisms interact with both petroleum and reservoir rock under various reservoir conditions. Further, a greater variety of microorganism types need to be evaluated in relation to their ability to interact with petroleum materials. The microbiology of petroleum reservoirs is starting to be explored, and the data obtained from these studies will certainly benefit MEOR development. Although a considerable amount of information needs to be obtained relating to biological processing of all fossil fuels, the results of past and current studies suggest that the area has exciting potential. Given time and continued support for these activities, new technologies will begin to be utilized by the fossil fuel industry. Based on the current state of development, it can be predicted that coal desulfurization, production of methane and alcohols, and enhanced oil recovery will be aided by biological processes.

ACKNOWLEDGMENTS

This work was supported by the Advanced Research and Technology Program managed by the Office of Fossil Energy under Contract No. DE-AC07-76ID01570 from the US Department of Energy to EG & G Idaho, Inc. at the Idaho National Engineering Laboratory.

REFERENCES

1. Monthly Energy Review, IEA/D0E-0035, January (1989).
2. P. R. Dugan, in: *Processing and Utilization of High Sulfur Coals, Coal Science and Technology*, Vol. 9, p. 101, Elsevier, Amsterdam (1985).
3. J. E. Duran, S. R. Mahassay, and L. M. Strock, *Fuel 65*, 1167 (1986).
4. R. N. Schicho, S. Brown, G. Olsen, E. Parks, and R. Kelly, *Fuel 68*, 1368 (1989).

5. A. Attar, *Hydrocarbon Processing*, January, 175 (1979).
6. A. R. Colmer and M. E. Hinkle, *Science 106*, 253 (1947).
7. Z. M. Zarubina, N. N. Lyalikova, and Ye. I. Shmuk, *Inz. Akad. Nauk SSSR, Otd. Tekh. Nauk, Metall. Topl. 1*, 117 (1959).
8. D. Ashmeed, *Colliery Guardian 190*, 694 (1955).
9. M. P. Silverman, M. H. Rogoff, and I. Wender, *Fuel 42*, 113 (1963).
10. V. V. Balashova, I. Y. Vedinina, G. E. Markosan, and G. A. Zavarin, *Microbiol. 43*, (1974).
11. P. R. Dugan, C. B. MacMillan, and R. M. Pfister, *J. Bacteriol. 101*, 973 (1970).
12. S. N. Groudev and F. N. Genchev, *Compt. Rendus Bulgarian Acad. Sci. 32*, 353 (1979).
13. A. P. Harrison, *Ann. Rev. Microbiol. 38*, 265 (1984).
14. J. H. Tuttle, P. R. Dugan, and W. A. Appel, *Appl. and Environ. Microbiol. 33*, 459 (1977).
15. C. L. Brierley, Ph. D. Thesis, Montana State University, *Dissertation Abstracts 67*, 1342 (1966).
16. C. L. Brierley and J. A. Brierley, *Can. J. Microbiol. 19*, 183 (1973).
17. J. A. Brierley, P. R. Norris, D. P. Kelly, and N. W. LeRoux, *European J. Appl. Microbiol. Biotech. 5*, 291 (1981).
18. T. D. Brock, K. M. Brock, R. T. Belly, and R. L. Weiss, *Arch. Microbiol. 84*, 54 (1972).
19. R. Guay and M. Silver, *Can. J. Microbiol. 21*, 281 (1975).
20. B. B. Bohlool and T. D. Brock, *Appl. Microbiol. 28*, 11 (1974).
21. F. Kargi and J. M. Robinson, *Appl. Environ. Microbiol. 24*, 2115 (1982).
22. F. Kargi and J. M. Robinson, *Biotech. Bioeng. 26*, 687 (1984).
23. W. A. Apel, P. R. Dugan, and J. H. Tuttle, *J. Bacteriol. 142*, 295 (1980).
24. M. P. Silverman and H. L. Ehrlich, *Adv. Appl. Microbiol. 4*, 153 (1964).
25. M. Rodriquez-Leiva and H. Tributsch, *Arch. Microbiol. 149*, 401 (1988).
26. G. F. Andrews, in: *Biotechnology in Mineral and Metal Processing*, Ch. 12, Society of Mining Engineers, Inc., Colorado (1989).
27. W. Uhl, H.-J. Hone, M. Beyer, and J. Klein, *Biotech. Bioeng. 34*, 1341 (1989).
28. G. F. Andrews, *Proceeding of the 4th Biological Processing of Fossil Fuels Workshop*, Tyson's Corners, VA, August (1989).
29. P. R. Dugan, in: *Microbial Chemoautotrophy* (W. R. Strohl and O. H. Tuovinen, eds.), The Ohio State University Press, Columbus (1984).
30. G. F. Andrews, M. Dauch, and T. Harrison, *Biotechnol. Bioeng. 32*, 813 (1988).
31. P. R. Dugan and W. A. Apel, *Metallurgical Applications of Bacteria*, Academic Press, New York (1978).
32. M. Elzeky and Y. A. Attia, in: *Biotechnology in Mineral and Metal Processing* (B. J. Scheiner, F. M. Doyle, and S. K. Kawatra, eds.), Ch. 18, Society of Mining Engineers, Inc., Colorado (1989).
33. G. J. Olsen, F. E. Brinckman, K. L. Jewett, and J. S. Thayer, in: *Biotechnology in Mineral and Metal Processing* (B. J. Scheiner, F. M. Doyle, and S. K. Kawatra, eds.), Ch. 6, Society of Mining Engineers, Inc., Colorado (1989).
34. G. J. Olsen and F. E. Brinckman, *Proceedings, Biological Treatment of Coals Workshop*, Herndon, VA, June (1986).
35. E. Beier, *Processing and Utilization of High Sulfur Coals, Coal Science and Technology*, Vol. 9, Elsevier, Amsterdam, p. 653 (1985).
36. F. Kargi, *Enzyme Microb. Technology 4*, 13 (1982).
37. P. Bos, T. F. Huber, K. Luyben, and J. G. Kuenen, *Resources, Conservation and Recycling 1*, 279 (1988).
38. R. D. Sproull, H. J. Francis, C. R. Krishna, and D. J. Dodge, *Proceedings, Biological Treatment of Coals Workshop*, Herndon, VA, July, p. 83 (1986).
39. C. Rai, *Biotech. Progress 1*, 200 (1985).

40. J. D. Isbister, G. Anspach, and A. DeSouza, unpublished results (1987).
41. A. J. Davis and A. S. Atkins, *Resources, Conservation and Recycling 1*, 223 (1988).
42. Y. A. Attia, *Proceeding, Bioprocessing of Coals Workshop-III*, Tyson's Corners, VA, August, p. 199 (1988).
43. D. A. Malik, *Process Biochemistry 10*, September (1978).
44. K. Yamada, Y. Minoda, K. Kodama, S. Nakatani, and T. Akasaki, *Agric. Biol. Chem. 32*, 840 (1968).
45. D. J. Monticello and W. R. Finnerty, *Annu. Rev. Microbiol. 39*, 371 (1985).
46. J. D. Isbister and E. A. Kobylinski, in: *Processing and Utilization of High Sulfur Coals, Coal Science and Technology*, Vol. 9, Elsevier, Amsterdam (1985).
47. J. J. Kilbane, *Proceeding, Bioprocessing of Coals Workshop-III*, Tyson's Corners, VA, p. 156 (1988).
48. M. S. Cohen and P. D. Gabriele, *Appl. Environ. Microbiol. 44*, 23 (1982).
49. C. D. Scott, G. W. Strandberg, and S. N. Lewis, *Biotech. Progress 2*, 131 (1986).
50. T. K. Kirk and H. H. Chang, in: *Microbial Degradation of Organic Compounds* (D. T. Gibson, ed.), Marcel Dekker, New York (1986).
51. T. K. Kirk and H. H. Chang, *Holzforschung 29*, 56 (1975).
52. T. Fukuzumi and T. Shibamoto, *Nippon Mokuzai Gakkaishi 11*, 45 (1965).
53. R. Narayan, in: *Proceedings: First Annual Workshop on Biologic Processing of Coals*, Electric Power Research Institute, Monterey, CA, February (1986).
54. J. A. Korburger, *Proc. W. Va. Acad. Sci. 36*, 26 (1964).
55. M. H. Ragoff, *Microbiology of Coal*, US Bureau of Mines, Circular No 8075.
56. R. M. Fakoussa, Ph.D. Thesis, University of Bonn (1981).
57. B. W. Wilson, R. M. Bean, J. A. Franz, B. L. Thomas, M. S. Cohen, H. Aronson, and E. T. Gray, Jr., *Energy and Fuels 1*, 80 (1987).
58. H. B. Ward, *Syst. Appl. Microbiol. 6*, 236 (1985).
59. G. W. Strandberg and S. N. Lewis, *Biotechnol. Bioeng. Symp. 17*, 153 (1986).
60. B. W. Wilson, R. M. Bean, J. Pyne, L. I. Stewart, and J. Fredrickson, in: *Proceedings, Biological Treatment of Coals Workshop*, Herndon, VA, June, p. 114 (1986).
61. M. Dahlberg, in: *Proceedings, Biological Treatment of Coals Workshop*, Herndon, VA, June, p. 172 (1986).
62. D. R. Quigley, J. R. Wey, C. R. Breckenridge, and H. J. Hatcher, in: *Proceedings, Biological Treatment of Coals Workshop*, Vienna, VA, July, p. 151 (1987).
63. C. D. Scott and S. N. Lewis, in: *Biotechnology Applied to Fossil Fuels* (D. L. Wise, ed.), CRC Press, Cleveland, OH (1988).
64. R. E. Wyza, A. E. Desouza, and J. D. Isbister, in: *Proceedings, Biological Treatment of Coals Workshop*, Vienna, VA, July, p. 119 (1987).
65. P. R. Dugan, J. Wey, D. Stoner, and D. Quigley, in: *Resources, Conservation and Recycling*, Elsevier (1990).
66. G. W. Strandberg and S. N. Lewis, *J. Indust. Microbiol. 1*, 371 (1987).
67. B. W. Wilson and M. S. Cohen, *Proceedings: First Annual Workshop on Biologic Processing of Coals*, Electric Power Research Institute, Monterey, CA, February (1987).
68. L. Wondrack, M. Szanto, and W. A. Wood, *Appl. Biochem. Biotech. 20/21*, 765 (1989).
69. E. Kern, in: *Proceedings: First Annual Workshop on Biologic Processing of Coals*, Electric Power Research Institute, Monterey, CA, February, Ch. 12, p. 13 (1987).
70. J. L. Gaddy, in: *Proceedings: Biological Treatment of Coals Workshop*, Herndon, VA, July, p. 194 (1986).
71. E. C. Clausen and J. L. Gaddy, in: *Proceedings: Biological Processing of Coals Workshop*, Vienna, VA, August, p. 89 (1987).
72. J. L. Vega, S. Prieto, B. B. Elmore, E. C. Clausen, and J. L. Gaddy, *Appl. Biochem. Biotech. 20/21*, 781 (1989).

73. J. Aislabie and R. M. Atlas, *Appl. Biochem. Biotech. 20/21*, 799 (1989).
74. US Department of Energy, in: *Federal Oil Research: A Strategy for Maximizing the Productivity of Known US Oil*, NTIS-PR-360, August (1989).
75. R. J. Smith and A. G. Collins, in: *Microbes and Oil Recovery*, Vol. 1, International Bioresources Journal (J. E. Zajic and E. C. Donaldson, eds.), Bioresources Publishers, El Paso, TX, p. 1 (1985).
76. J. E. Zajic, A. Gurrola, and J. Mueler, in: *Microbes and Oil Recovery*, Vol. 1, International Bioresources Journal (J. E. Zajic and E. C. Donaldson, eds.), Bioresources Publishers, El Paso, TX, p. 151 (1985).
77. W. R. Finnerty, M. E. Singer, F. Ohene, and H. Attaway, in: *Microbes and Oil Recovery*, Vol. 1, International Bioresources Journal (J. E. Zajic and E. C. Donaldson, eds.), Bioresources Publishers, El Paso, TX, p. 276 (1985).
78. W. R. Finnerty, *Applications of Microbial Processes to Viscosity Reduction of Heavy Crude Oil—Final report*, US Department of Energy, Bartlesville, OK, DOE/BC/10507-41, May (1987).
79. V. Moses and D. G. Springham, in: *Bacteria and the Enhancement of Oil Recovery*, Elsevier, Barking Essex, England (1982).
80. D. O. Hitzman, in: *Proceedings of the 1982 International Conference on Microbial-Enhanced Oil Recovery*, Afton, OK, DOE Conf-8205140, p. 162, May (1982).
81. R. S. Bryant, T. E. Burchfield, D. M. Dennis, and D. O. Hitzman, paper at the SPE/DOE 6th International Symposium on Enhanced Oil Recovery, Tulsa, OK, SPE-17341, April (1988).

8

Enzymes in Organic Solvents

ALEKSEY ZAKS

8.1. INTRODUCTION

The high potential of biocatalysis for a variety of industrial applications is widely recognized. For centuries various enzyme-based transformations have been used in wine and bread making, milk clotting, and beer brewing. Added to this list are recent advances in food technology such as processes for the production of high-fructose corn syrup, chillproofing of beer, hydrolysis of lactose, and saccharification of starch-containing raw materials. The food industry is now the largest consumer of enzymes, accounting for over 50% of the $445 million enzyme market.[1] The use of enzymes as practical catalysts in the chemical industry, however, has been limited. This is mainly attributed to their high cost, relative instability in harsh industrial environments, and strong competition with current and well established chemical processes. However, applied enzymology has always been an area of high expectation. Unique properties of enzymes such as their superior regio- and stereoselectivity, ability to catalyze a variety of reactions under mild conditions, nontoxicity, and biodegradability make them potentially useful in a variety of applications.

Recent advances in molecular biology, protein crystallography, and genetic engineering have resulted in the production of new enzymes with increased stability and modified substrate specificity. Moreover, chymosin, bacterial α-amylase, a number of lipases, and subtilisin have already received approval for industrial applications.[1]

ALEKSEY ZAKS • Enzytech, Inc., 64 Sidney St., Cambridge, MA 02139.

Until now, in industry enzymes have been used in predominantly aqueous environments. However, because of the low solubility of many organic compounds in water (fats, oils, fatty acids and alcohols, aromatics, steroids, etc.), undesirable side reactions (hydrolysis, acyl migration, nucleophilic addition of hydroxide ion), and unfavorable thermodynamic equilibrium, water is a poor reaction medium for the majority of chemical transformations.

In order to overcome these caveats, a number of approaches have been developed. These include the use of enzymes in aqueous solutions containing a water-miscible organic cosolvent,[2,3] aqueous-organic biphasic mixtures,[4,5] reversed micelles,[6,7] and, finally, anhydrous organic solvents.[8–10]

A variety of synthetic reactions including oxidoreductions,[5,12] epoxidations,[13] isomerizations,[14] and synthesis of peptide and ester bonds[15] have been carried out efficiently in aqueous reaction mixtures containing organic solvents. The addition of cosolvents to the reaction medium increases the solubility of substrates usually having little or no effect on the substrate specificity of enzymes. There are some exceptions, however. It has been reported that the ratio of esterase to amidase activity of trypsin, chymotrypsin, papain, and subtilisin can be varied by adding of organic solvents to the reaction medium.[16] For example, in 40–60% of DMF, DMSO or dioxane, the enzymes express no amidase activity, whereas the esterase activity remains significant. Based on this finding, a preparative synthesis of various peptides and penicillin precursors has been demonstrated.[16]

In the case of micellar enzymology the reaction medium consists of a nonpolar organic solvent. Enzyme molecules are dissolved in "water pools" separated from the organic solvent by a shell formed by the polar heads of surfactant molecules.[6] Both the molar ratio of water to surfactant and hydrophobicity of the enzyme determines the strength of interactions between the protein surface and the inner wall of the reversed micelle. These interactions are thought to be responsible for the observed changes in the kinetic properties of some enzymes in reversed micelles.[17] The two approaches described above have a number of advantages over traditional enzymatic reactions in aqueous solutions including increased substrate solubility and often more favorable thermodynamic equilibrium. However, enzyme instability in the presence of organic cosolvents and the necessity to separate the products from surfactant molecules (in the case of micellar enzymology) limit the practical use of these methods.

The most recent advances in enzymatic conversions in water-restricted environments is the use of enzymes in predominantly nonaqueous media including anhydrous organic solvents. The investigation of the kinetic and structural behavior of enzymes in these extreme environments has already resulted in a number of exciting discoveries that have opened new directions in the industrial applications of enzymes. In this review I will focus on this approach.

8.2. EFFECT OF WATER AND ORGANIC SOLVENTS ON ENZYMATIC CATALYSIS IN NONAQUEOUS MEDIA

The nature of the organic solvent and amount of water in the reaction medium are two fundamental factors that determine the behavior of enzymes in water-restricted environments. Water is absolutely essential for enzymatic activity. Indeed, it participates in all noncovalent interactions maintaining the protein in its native conformation and plays a crucial role in enzyme dynamics. Though the crucial role of water has been recognized for many years, the *minimum* amount of water necessary for enzyme to function has been determined only for a few enzymes. Thus far, lysozyme[18–21] and chymotrypsin[22] have been used in most studies. Dehydrated protein powders were gradually hydrated by exposing them to air of different humidity and the molecular events occurring within the protein molecules have been observed by a number of spectroscopic techniques. It was noted that the rehydration process significantly increased the mobility of the protein matrix which, at a certain level of hydration, (in the case of lysozyme it corresponded to 0.2 g water per g protein) allows enzymatic activity. As the amount of water on the protein increased, enzymatic activity and the protein's motional properties also increased. At the level of 0.38 g water/g protein, the lysozyme molecules became fully hydrated and the enzymatic activity reached 10% of that in water.[19]

These elegant studies on dry powders of enzymes reveal the minimum amount of water needed for catalysis. Unfortunately, these results cannot be directly applied to enzyme-catalyzed reactions in nonaqueous media because they were obtained in the absence of an organic solvent. The presence of an organic solvent significantly complicates experimental approaches. Organic solvents affect an enzymatic reaction in various ways. First, the solvent affects the distribution of water between an enzyme and a reaction medium. Second, an organic solvent may directly interact with the enzyme, adversely affecting its native catalytically active conformation and thereby inactivating it. Finally, partitioning of substrates and/or products of the reaction between the active site of an enzyme and medium may influence a number of kinetic and thermodynamic parameters of the process. The resulting effect of an organic medium on an enzymatic process is most often a combination of the above factors.

The impact of water on the kinetic behavior of enzymes in water-restricted environments is best observed when enzymes are directly suspended in organic solvents instead of being immobilized on a support. This approach has been described in a number of reports.[23, 24, 26, 46] The activities of horse liver alcohol dehydrogenase, mushroom polyphenol oxidase, and yeast alcohol oxidase have been measured in a variety of organic solvents and plotted as a function of the concentration of water in the reaction medium (Figure 1).[23] All three enzymes are catalytically active in a variety of organic solvents. The enzymatic activity in every solvent tested strongly increases as the water content in the solvent is raised. For example, alcohol oxidase activity in ethyl acetate increases 1000-fold as the concentration of water in the solvent is changed from 0.5 to 1.1%. The absolute amount of water required for catalysis for each enzyme varies significantly from one solvent to another. The activity of alcohol dehydrogenase, for example, in the presence of 0.5% water in hydrophobic isopropyl ether is more than three orders of magnitude higher than in hydrophilic methyl acetate or acetonitrile at the same water concentration. Despite these large variations, under optimal conditions all three enzymes reveal activities comparable to that in water: the activity of alcohol oxidase in *tert*-amyl alcohol containing 10% water is 21% of that in aqueous buffer; in 97% octanol the activity of polyphenol oxidase reaches 40% of its activity in water; and, the activity of alcohol dehydrogenase in 99.5% isopropyl ether is one quarter of that in aqueous solution.[23]

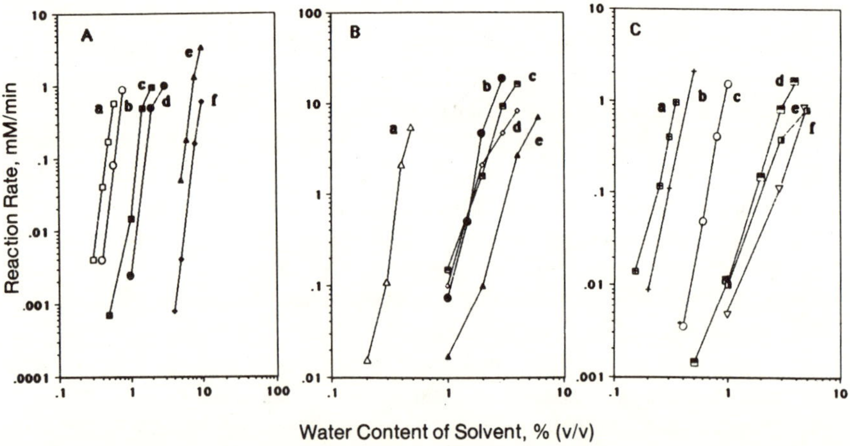

Figure 1. Dependence of enzymatic activity on the concentration of water in various organic solvents. (A) Yeast alcohol oxidase in (a) ethyl ether; (b) butyl acetate; (c) ethyl acetate; (d) *n*-octanol; (e) *tert*-amyl alcohol; (f) *sec*-butanol. (B) Mushroom polyphenol oxidase in (a) hexyl acetate; (b) *n*-octanol; (c) methyl acetate; (d) *n*-butanol; (e) *tert*-amyl alcohol. (C) Horse liver alcohol dehydrogenase in (a) methylene chloride; (b) isopropyl ether; (c) butyl acetate; (d) methyl acetate; (e) tetrahydrofuran; (f) acetonitrile.[23]

Apparently (Figure 1) enzymes suspended in hydrophobic solvents require much less water for activity than those suspended in hydrophilic ones. In the former media, water tends to partition onto the enzyme so that as little as 1% water in the solvent yields up to 30% water on the protein. Addition of water to hydrophilic solvents results in a moderate increase in the concentration of water on the protein (Figure 2A). It was hypothesized[23] that it is the difference in the distribution of water between the enzyme and the medium in different solvents that is responsible for such different enzymatic activity. The dependence of enzymatic activity on the amount of water bound to the enzyme in various organic solvents (Figure 2B) supported this assumption. Regardless of the solvent's hydrophobicity, the dependence fits roughly the same pattern: the higher the water content on the enzyme, the greater the enzymatic activity.

It seems reasonable to assume that it is the concentration of water *on the enzyme* that has the most significant effect on its activity in a particular solvent. Most likely, enzymatic activity in organic media is determined not so much by interactions of the solvent with the enzyme *per se* but by those with water on the enzyme.[23] The loss of this essential water has a strong detrimental effect decreasing activity by more than three orders of

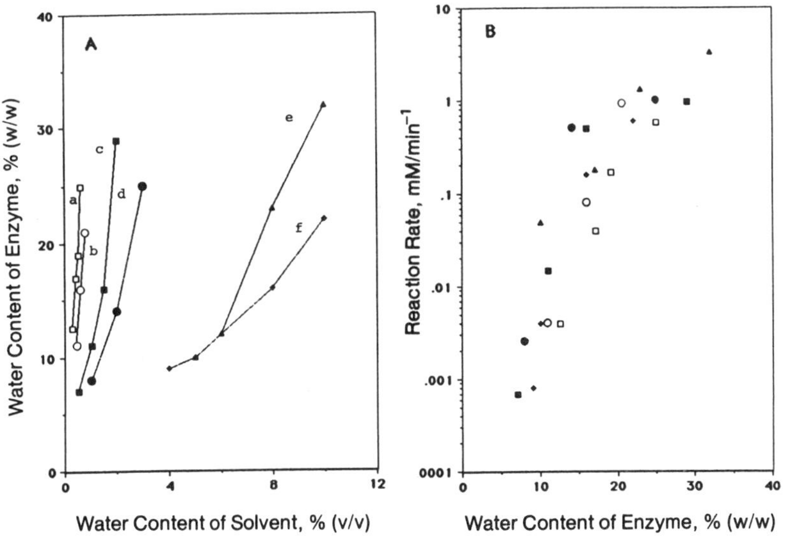

Figure 2. (A) Dependence of the amount of water bound to yeast alcohol oxidase on the water content of the solvent in which the enzyme is suspended: (a) ethyl ether; (b) butyl acetate; (c) ethyl acetate; (d) *n*-octanol; (e) *tert*-amyl alcohol; (f) *sec*-butanol. (B) Dependence of alcohol oxidase activity in various organic solvents on the amount of water bound to the enzyme: (□) ethyl ether; (○) butyl acetate; (■) ethyl acetate; (●) *n*-octanol; (▲) *tert*-amyl alcohol; (◆) *sec*-butanol.[23]

magnitude. When water is replenished, however, the catalytic activity is restored.

It has been demonstrated that this essential water can be substituted to some degree by other compounds.[23] The presence of 1% formamide in the reaction medium increased the activity of alcohol dehydrogenase in butyl acetate 15-fold over the activity in the absence of additive. Three percent formamide in octanol affords a 35-fold increase in the rate of oxidation of 4-methylcatechol catalyzed by tyrosinase. Presumably, due to their high potential to form hydrogen bonds the above additives can mimic the interactions of water with the protein thereby restoring enzymatic activity. They may also act as a barrier preventing unfavorable interactions between the protein and the solvent.

The search for a quantitative correlation between a suitable solvent property and the activity of a biocatalyst is a highly challenging problem in the area of nonaqueous enzymology. Since hydrophobic solvents do not distort the layer of essential water around the enzyme as much as hydrophilic counterparts, enzymatic activity in hydrophobic solvents should be higher than that in hydrophilic ones. It has been shown that a majority of enzymes behave according to the above prediction.[23–26] Some enzymes, however, were found to be more sensitive to the nature of organic solvents than others. For example, while the activity of porcine pancreatic lipase is comparable in most solvents tested, that of yeast lipase varies by three orders of magnitude.[24] Moreover, the activity of subtilisin in the more hydrophilic solvents *tert*-amyl alcohol and acetone is higher than that in hydrophobic toluene, butyl ether, or carbon tetrachloride.[26]

To estimate the polarity of solvents Brink and Tramper suggested using of the Hildebrand solubility parameters.[27] However, correlation of this parameter with the rates of a number of microbial epoxidations in various water-immiscible solvents was poor. Laane *et al.*[25] proposed the use of the log P value of the solvent (partition coefficient of a given solvent in octanol-water system). Though the latter parameter is considered more useful in determining solvent suitability, the correlation between log P and enzymatic activity has been demonstrated for only a few enzymes.[25] It is hypothesized[28] that the solubility of a given solvent in water is a better parameter for describing the ability of a solvent to distort the layer of essential water. Unfortunately, because of the lack of data on the solubility of organics in water, this apparently reasonable correlation remains unproven.

8.3. ROLE OF SUPPORT

One of the advantages of nonaqueous enzymology is that there is no general need to immobilize an enzyme. Enzymes are insoluble in most organic solvents and therefore can be successfully used in a powdered state. After the reaction is complete the catalyst is easily removed by filtration or centrifugation. However, if the use of an immobilized enzyme is preferred, the choice of the support becomes of primary importance. The support influences the microenvironment of the enzyme and the distribution of water in the system, thereby affecting the enzyme's activity and stability. It has been shown[9, 29] that the activity of a number of enzymes depends strongly on the nature of the support. In the case of chymotrypsin and alcohol dehydrogenase, for example, there appears to be competition for the water in the system between the enzyme, support, and solvent. When the concentration of water in the reaction mixture was constant, the activity of the enzymes immobilized on different supports decreased with increasing affinity of the support for water.[29] Indeed, hydrophilic supports adsorb water. As a result the enzyme becomes insufficiently hydrated and looses its activity. Interestingly, when reaction rates were measured at several water contents, the activity of the preparation immobilized on a hydrophilic support was lower in all cases, indicating the presence of direct interactions between an enzyme and a support.[29]

The nature of the support may also affect the distribution of a substrate between an enzyme and the reaction medium. Experiments with polymer-entrapped enzymes[30] have shown that the increase in content of hydrophobic residues resulted in increased activity of the catalyst toward hydrophobic substrates.

The use of support may be particularly beneficial with highly active enzymes. The insolubility of these enzymes in organic solvents may result in intraparticle diffusional limitations. This is eliminated by dispersing enzymes on materials with a large surface area. It has been shown that the rate of horseradish peroxidase-catalyzed oxidation of *o*-phenylenediamine increased 160-fold when the enzyme was deposited onto 75–150 μm glass beads.[11]

Internal diffusional limitations are also eliminated by solubilization of enzymes in organic solvents. This is achieved by covalent attachment of polyethylene glycol derivatives to accessible lysine residues on the surface of enzymes. Using this technique several lipases,[30] chymotrypsin,[31] catalase,[32] and horseradish peroxidase[33] have been made soluble in a number of organic solvents with the retention of catalytic activity.

8.4. PROPERTIES OF ENZYMES IN ORGANIC SOLVENTS

8.4.1. Substrate Specificity

Binding energy is the driving force for all molecular recognitions and enzyme catalysis. Intermolecular interactions determine the strength of binding between ligands and receptors and are responsible for the binding of substrates to enzymes, The binding energies in biological systems are always determined by the differences in energy between molecules in the solution interacting with the solvent and molecules in a complex interacting with each other. Thus, when the aqueous reaction medium is substituted by an organic solvent the free energy of enzyme–substrate binding changes affecting both substrate specificity and activity of the catalysts. This change in substrate preference upon the transition from aqueous to nonaqueous phases has been demonstrated with chymotrypsin and subtilisin.[34] For these enzymes the major driving force of enzyme–-substrate binding is hydrophobic interaction between the side chains of amino acids and the binding pocket of the enzyme. Since the presence of water is imperative for hydrophobic interactions, the substrate preference of these enzymes in organic solvents differs significantly from that in water. As opposed to the situation in water, hydrophilic *N*-Ac-L-Ser-OMe and *N*-Ac-L-His-OMe are better substrates in octane than more hydrophobic *N*-Ac-L-Phe-OMe.

One of the most striking features of enzymatic catalysis in organic solvents is the possibility to affect specificity and activity of enzymes in a predictable manner. This phenomenon—inhibitor-induced activation—was first observed for chymotrypsin[26] and investigated further in detail for subtilisin.[35] It has been found that in nonaqueous media, activity of the enzyme lyophilized from aqueous slutions containing ligand is up to 100-fold higher than that of the enzyme lyophilized in their absence (Table 1). It is hypothesized[26, 35] that the presence of the ligand (usually a competitive inhibitor) locks the enzyme in a more favorable conformation resembling the enzyme–substrate complex. Due to the high rigidity of enzymes in anhydrous organic solvents this conformation is retained even after the ligand is removed. Subtilisin and chymotrypsin function in this conformation more efficiently in the restrictive environment of an organic solvent. Ligand-induced memory is observed only in anhydrous organic solvents: addition of even small amounts of water to the reaction medium apparently loosens up protein molecules and irreversibly destroys the enzyme's "memory." Ligand-induced alteration of subtilisin also affects enzyme substrate specificity and stability.[35]

The use of nonaqueous solvents results in the possibility of using proteolytic enzymes in a number of synthetic reactions where water, as the reactant, is substituted by other nucleophiles. The chemoselectivity of enzymatic acylation of heterofunctional compounds, such as aminoalcohols, can be controlled by the nature of the acyl moiety.[36] The rate of enzymatic acylation of hydroxyl group of 6-amino-1-hexanol with 2-chloroethyl butyrate catalyzed by *Aspergillus niger* lipase, was 37 times greater than that of the amino group. In contrast, when 2-chloroethyl ester

Table 1. Enhancement of Subtilisin Activity in Octane Induced by Various Ligands[35]

Ligand	Substrate	Rate enhancement[a]
N-Ac-Tyr-NH_2	*N*-Ac-Ala-OMe	55
	N-Ac-Tyr-OEt	11
	N-Ac-Phe-OEt	3
	N-Ac-Trp-OEt	25
N-Ac-Trp-NH_2	*N*-Ac-Ala-OMe	96
	N-Ac-Tyr-OEt	12
	N-Ac-Phe-OEt	3
	N-Ac-Trp-OEt	16
2-Naphthol	*N*-Ac-Ala-OMe	24
	N-Ac-Tyr-OEt	21
Indole	*N*-Ac-Ala-OMe	8
	N-Ac-Tyr-OEt	15
Phenylboronic acid	*N*-Ac-Ala-OMe	6
	N-Ac-Tyr-OEt	10

[a] Defined as the ratio of enzymatic transesterification of a given ester catalyzed by ligand-activated subtilisin to that by subtilisin lyophilized without ligand.

of *N*-acetyl-L-phenylalanine was used as the acylating agent, the reactivity of the hydroxyl group was less than one-fifth of that of the amino group. This reversal of substrate specificity was also observed for porcine pancreatic lipase and *Pseudomonas* sp. lipoprotein lipase. Though the mechanism of this phenomenon is not yet understood, the ability to control the chemoselectivity of enzymatic reactions is likely to be of great practical importance.

8.4.2. Stereospecificity

In general, enzymes placed in organic solvents retain their stereoselectivity and can be used successfully for preparative synthesis. However, upon the transition from water to an organic solvent as the reaction medium, the enantioselectivity of some enzymes changes.[37, 37a] Lipase-catalyzed esterifications in organic solvents, for example, are often more enantionselective than the corresponding hydrolytic reactions in water. On the other hand, the enantioselectivity factor defined as $(k_{cat}/K_m)_L/(k_{cat}/K_m)_D$ for subtilisin Carlsberg, elastase, α-lytic protease, subtilisin BPN′, α-chymotrypsin, and trypsin in nonaqueous solvents is 1–2 orders of magnitude lower than in water[37] (Table 2). Moreover, in the latter case enantioselectivity depended strongly on the hydrophobicity of the reaction medium being highest in hydrophilic solvents such as DMF and acetonitrile and lowest in hydrophobic solvents (methylene chloride,

Table 2. Enantioselectivity of Various Proteolytic Enzymes in Water and in Butyl Ether[37]

Enzyme	Amino acid substrate	$({}^{k}\text{cat}/{}^{K}\text{m})_L/({}^{k}\text{cat}/{}^{K}\text{m})_D$ in water	in butyl ether
Subtilisin Carlsberg	*N*-Ac-Ala-OEtCl	1,800	4.4
Subtilisin Carlsberg	*N*-Ac-Phe-OEtCl	15,000	5.4
Elastase	*N*-Ac-Ala-OEtCl	>1,000	4.5
α-lytic protease	*N*-Ac-Ala-OEtCl	>10,000	8.3
Subtilisin BPN′	*N*-Ac-Phe-OEtCl	16,000	7.3
α-Chymotrypsin	*N*-Ac-Ala-OEtCl	710	3.2
Trypsin	*N*-Ac-Phe-OEtCl	>4,000	3.2

butyl ether, or cyclohexane). This phenomenon was explained by differences in the free energy of transfer of water molecules from the active center of the enzymes into different solvents upon binding of the L- and D-isomers.[37] It was hypothesized that productive binding of the L-isomer released more water molecules than that of the D-isomer. Taking into account that the release of water to hydrophobic solvents is more energetically unfavorable than to hydrophilic ones, and assuming that this release is the rate-limiting step, the authors concluded that the rate of the reaction with the L-isomer should be diminished to a greater degree than with the D-isomer. As a result enantioselectivity in hydrophobic solvents decreased.

8.4.3. Structure

The knowledge of protein conformations in environments of organic solvents is crucial for the understanding and greater utilization of enzymatic catalysis in organic solvents. There have been a number of early studies on the effect of organic solvents on protein structure.[38–41] Unfortunately, the insolubility of proteins in a majority of organic solvents significantly limited the number of solvents tested. Solvents that dissolve proteins such as DMSO, DMF, acidic methanol, and 2-chloroethanol were used in most studies. It has been shown on a number of examples that these solvents directly interact with proteins causing significant changes in their structure. In contrast, other organic solvents such as polyols (including glycerol, 2-methyl-2,4-pentanediol, and sugar-containing solutions) were shown to stabilize proteins.[42,43] Until recently the effect of other hydrophilic solvents and hydrophobic water-immiscible solvents has not been investigated to any significant extent.

A number of kinetic experiments reveal that enzymes in some organic solvents exhibit activities comparable to those in water.[9, 23, 45, 75] This indicates that their structure in these solvents is not radically different from that in water. Important information on the transition state structure has been recently obtained by applying Hammet analysis for subtilisin-catalyzed transesterifications in nonaqueous media.[44] In all solvents tested (including water-immiscible butyl ether and *tert*-amyl alcohol, and water-miscible acetone, tetrahydrofuran, and acetonitrile) ρ-values were between 0.72 and 0.93 (which falls within less than a 15% range and is close to that in water). The similarities of ρ-values in nonaqueous media indicate that the microenvironment of the transition state of subtilisin-catalyzed reactions is the same in organic solvents and in water. It also suggests that the enzyme's active center is shielded from the bulk solvent.

The first direct evidence that the structure of the active site of enzymes in organic solvents is similar to that in aqueous solutions was recently obtained by solid-state ^{15}NMR analysis.[45] The unique tautomeric structure and hydrogen bonding of the active center His-36 imidazole of α-lytic protease from *Lysobacter enzymogenes* suspended in anhydrous acetone and octane was preserved, indicating that the enzyme remains intact in these solvents. Furthermore, there was no evidence of appreciable structural heterogeneity in the population of enzyme active centers in these solvents. In contrast, dimethyl sulfoxide, the only solvent tested in which α-lytic protease was inactive, was found to disrupt the catalytic triad.

8.4.4. Thermostability

A number of reports describe the use of dry organic solvents as a reaction media that significantly increases enzyme stability and, in particular, their thermostability.[26, 35, 46–48] Indeed, processes which lead to the irreversible inactivation of enzymes (formation of incorrect structures, disulfide bonds interchange, deamidation of asparagine and glutamine residues, and hydrolysis of peptide bonds) require water and therefore do not occur in water-free environments such as organic solvents. Moreover, decreased water activity reduces the mobility of the protein molecule and consequently prevents partial unfolding which is the first step of the thermoinactivation process. A dramatic increase in the stability of porcine pancreatic lipase in a number of organic solvents relative to that in water is one example.[46] Whereas in an aqueous buffer at 100 °C the lipase is inactivated almost instantly, the half-life of the enzyme is dry tributyrin at this temperature is nearly 26 h. The amount of water and the nature of the solvent have a strong effect on the stability of enzymes: enzymes placed in anhydrous hydrophobic solvents exhibit the highest stability.

8.5. APPLICATIONS OF ENZYMES IN ORGANIC SOLVENTS

8.5.1. Hydrolases

Lipases represent a group of enzymes that is most widely used for transformations in organic solvents. Although in nature lipases catalyze the hydrolysis of triglycerides using water as a substrate, they also accept other nucleophiles such as alcohols, amines, thioesters, and oximes. Thus hydrolysis is substituted by more desirable synthetic reactions.

8.5.1.1. Modifications of Triglycerides

Lipases are used in a variety of esterification and interesterification reactions for the production of triglycerides with the desired physical properties.[49–51] The major advantage of using lipases stems from the high regiospecificity of the enzymes resulting in a product with more defined chemical composition than that obtained by traditional chemical processes. A number of lipases have been successfully used in regiospecific trans-esterifications of various oils. A process for the production of a cocoa butter substitute via lipase-based regiospecific esterification of palm oil with stearic acid is near commercialization.

8.5.1.2. Synthesis of Optically Active Compounds

Use of lipases for preparative synthesis of optically active compounds is an area of considerable interest. Porcine pancreatic and *Candida cylindracea* lipases efficiently catalyze esterification of racemic bromo-propionic acid with butanol.[52] Only the L-isomer is esterified. (S)-(−)-2-bromopropionic acid (an intermediate in the production of the potent herbicide (R)-2-phenoxypropionic acid) is obtained in 75% of theoretical yield and 99.6% enantiomeric excess (ee). *C. cylindracea* lipase is also useful for stereoselective esterification of menthol. The enzyme was immobilized by entrapment with urethane prepolymers and used in cyclohexane and isooctane for esterification of racemic menthol with 5-phenylvaleric acid.[53] As a result only the L-isomer was esterified. *Pseudomonas fluorescence* lipase adsorbed on Celite was used for resolutions of optically active alcohols from racemates.[56] Acetic, propionic, or butyric anhydrides were employed as acylating agents. The lipase is catalytically active in a variety of organic solvents including benzene, hexane, methylene chloride, and chloroform. The enzyme preferentially utilizes the R-enantiomer, so that the remaining alcohol is enriched in the S-form and the ester produced is enriched in the R-form.[54] *P. fluorescence* lipase is reported to catalyze irreversible and highly enantioselective acylation of 2-halo-1-arylethanols with vinyl esters.[55] As a result, a number of optically active esters can be obtained in high yield and purity.

C. cylindracea lipase was also used as a catalyst for preparative resolution of α-substituted cyclohexanols.[56] Powdered lipase catalyzes the enantioselective synthesis of the ester of lauric acid and different racemic 2-substituted cyclohexanols in organic solvents. The enzyme is specific for the alcohols of R-configuration. *C. cylindracea* lipase-catalyzed acylation of primary allenic alcohols in hexane results in only a partial resolution of enantiomers.[57] The best results (70% optical purity at 12% yield) are obtained with more hindered alcohols.

Asymmetric esterifications and transesterifications catalyzed by hydrolytic enzymes in nonaqueous solvents have been used frequently for resolutions of racemic alcohols, acids, and their derivatives. Unfortunately, this methodology has never been successful for the resolution of racemic amines because the enzymes acted equally well on both isomers. It has been found recently that enantioselectivity of a number of proteolytic enzymes can be markedly affected by the organic solvent.[37, 58] For example, the enantioselectivity factor of subtilisin in the acylation of α-methylbenzylamine varies from near unity in hexane to 7.7 in 3-methyl-3-pentanol.[58] This degree of enantioselectivity was sufficient for kinetic resolutions of racemates. As a result, a number of optically active amides and amines have been prepared on a millimole scale with high enantiomeric excess.

8.5.1.3. Lactonizations and Polycondensations

In addition to the reactions described above, lipases catalyze intramolecular esterifications in organic solvents. Using this technology macrocyclic lactones have been produced from free hydroxy acids[59] and their methyl esters.[60] Macrocyclic lactones were also produced via direct condensation of diacids with diols, catalyzed by lipases in nonaqueous media.[61] In most cases examined, the major products of this reaction were mono- and dilactones accompanied by linear oligomeric esters.

Lipases have also been used to catalyze stereoselective oligomerizations in organic solvents.[62] For instance, the reaction between racemic bis-(2-chloroethyl)-2,5-dibromoadipate and 1,6-hexadiol catalyzed by a variety of lipases in toluene yields optically active trimers and pentamers.

8.5.1.4. Modifications of Steroids, Sugars, and Sugar-Related Compounds

Lipase-catalyzed acylations are not limited to simple alcohols. Lipases efficiently catalyze the regioselective acylation of a number of steroids, sugars, and sugar derivatives.[63–67] Gram quantities of monoacylated sugars have been produced by the reaction between trichloroethyl carboxylates and various sugars in anhydrous pyridine. Interestingly, whereas some lipases acylate exclusively the C-3 hydroxyl group in C-6 protected glucose, others display an overwhelming preference toward the C-2 hydroxyl group.[64] Similar results were observed in lipase-catalyzed acylations of sugars with vinyl esters.[66] Vinyl and isopropenyl esters have been particularly useful for lipase-catalyzed acylation of a number of hydroxy compounds including glycerol and serinol derivatives, ferrocenylethanol, and other alcohols. Due to the high reactivity of these esters, reactions are faster and easier to optimize. In addition, the alcohol product rapidly tautomerizes to volatile acetaldehyde or acetone, making the process irreversible.[66]

Lipase-catalyzed transesterifications have also been performed on a wide range of sugar alcohols and vegetable oils. The products of these reactions were primary monoesters. These enzymatically produced sugar alcohol esters were found to be excellent surfactants far superior to their chemically produced counterparts.[67]

The scope of enzyme-catalyzed regioselective acylations in anhydrous organic solvents was extended even further to hydroxy steroids. It was found[68] that certain hydrolytic enzymes discriminate between secondary hydroxyl groups in the A-ring and the D-ring of steroids (Figure 3). For example, *Chromobacterium viscosum* lipase and *Bacillus subtilis* protease (subtilisin) esterify the model hydroxy steroid (I) in dry acetone with high and opposite regioselectivity. While lipase reacts exclusively with the OH group in the C-3 position, subtilisin displays a marked preference for the C-17 hydroxyl (Figure 3). On the basis of these findings, gram quantities

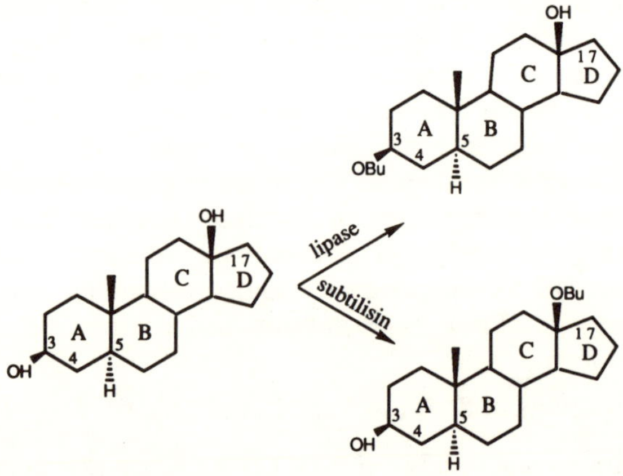

Figure 3. Regioselective acylation of dihydroxy steroids with trifluoroethyl butyrate catalyzed by lipase and subtilisin in anhydrous acetone.[68]

of a number of 3β and 17β monobutyryl steroids were obtained in good yield.

Sugars are reasonably well soluble in only few very hydrophilic organic solvents such as pyridine, formamide, and dimethylformamide, i.e., solvents which are the least appropriate for lipases.[24] For instance, out of a dozen commercially available lipases, only those from porcine pancreas and *C. viscosum*, were active in pyridine and none exhibited any appreciable activity in dimethylformamide. The methodology of enzyme-catalyzed modifications of sugars in organic solvents was greatly extended by the finding that the proteolytic enzyme subtilisin is catalytically active in anhydrous dimethylformamide.[69] By taking advantage of the unique dissolving potency of DMF and the broad substrate specificity of subtilisin, a number of carbohydrates and other sugar-containing compounds including maltose, cellobiose, lactose, sucrose riboflavin, salicin, adenosine, and uridine have been regioselectively acylated.

8.5.1.5. Synthesis of Peptides

Due to the high stereo- and regioselectivity of enzymes, mild reaction conditions which do not result in racemization, and only limited need for protective groups, enzymes are widely used for preparative peptide synthesis.[70] Under physiological conditions, however, the equilibrium in enzyme-catalyzed reactions lies far in the direction of peptide bond hydrolysis. There are several strategies that enable one to bypass unfavorable thermodynamic equilibrium. The one considered most successful is the reduction of water activity in the reaction system by substituting water with organic solvents. Also, the presence of organic solvents lowers the dielectric constant of the reaction mixture, therefore reducing the degree of ionization of carboxyl and amino groups. Depending on the particular requirements of the system the concentration of an organic solvent in the reaction mixture may range from 10 to nearly 100%. For example, the synthesis of an aspartame precursor was successfully carried out by thermolysin-catalyzed coupling of *N*-(benzyloxycarbonyl)-L-aspartic acid and L-phenylalanine methyl ester in ethyl acetate containing less than 3% water.[71] Similarly, chymotrypsin immobilized on chitin is reported to catalyze synthesis of *N*-acetyl-tyrosine-glycineamide in 95% acetonitrile.[72]

The major limitation of protease-catalyzed peptide synthesis is undesirable hydrolysis of the growing polypeptide chain. Lipase-catalyzed peptide synthesis in organic solvents is a successful alternative.[73,74] The use of organic solvents instead of water as the reaction medium favors the equilibrium of the reaction toward synthesis and increases the solubility of substrates. In addition, lipases, in contrast to proteolytic enzymes, have a broad specificity allowing for the incorporation of D-amino acids.

8.5.2. Oxidoreductases

8.5.2.1. Alcohol Dehydrogenase

Alcohol dehydrogenases are widely used in the preparative production of optically active alcohols and ketones. However, the relative instability of dehydrogenases and their cofactors in aqueous environments and strong enzyme inhibition by both substrates and products limit their use for large-scale synthesis. The use of organic solvents as media for alcohol dehydrogenase-catalyzed reactions helps eliminate some of these hurdles. Horse liver alcohol dehydrogenase (HLADH) and yeast alcohol dehydrogenase (YADH) are catalytically active in a number of organic solvents.[9, 23, 75] These enzymes can be used either in the free form[23] or immobilized on an appropriate support.[9, 75] The nature of the support strongly influences enzymatic activity. For example, yeast alcohol dehydrogenase immobilized on unmodified glass has very low activity whereas glycophase glass gives superior results.[9] Since neither the cofactors nor the enzymes are soluble in most organic solvents, they have to be brought together prior to placement in the organic medium. This is accomplished by dissolving them in water where they bind to each other due to their high affinity. The solution is then lyophilized and the resultant powder placed into an organic medium containing substrates. It has been reported that when ethanol is used as a recycling system, a cofactor turnover number of 10^6 is obtained.[75] HLADH is not only active in organic solvents but it also retains its stereoselectivity. Using HLADH as a catalyst, a variety of optically active alcohols and ketones (ee 95–100%) have been prepared on 1–10 mmole scale.[75]

8.5.2.2. Polyphenol Oxidase

Polyphenol oxidase-catalyzed hydroxylation of phenols is of great interest for practical organic chemistry primarily because of the high regioselectivity of the reaction. Attempts to use polyphenol oxidase as a practical catalyst however have not been successful due to the instability of polyphenol oxidase and *o*-quinones in water. These complications have been overcome by using polyphenol oxidase in organic solvents.[76] Polyphenol oxidase was immobilized on glass beads affording a larger surface area, placed in chloroform, and used as a catalyst for regiospecific oxidation of a number of phenols. The substrate specificity of the enzyme in chloroform was found to be similar to that in water. The quinones formed during the reaction were reduced to catecholes by conventional chemical means thus affording a net regioselective hydroxylation of phenols.[76]

8.5.2.3. Horseradish Peroxidase

Horseradish peroxidase efficiently catalyzes a variety of oxidations in a number of different organic solvents.[77] The rate of oxidation of *p*-anisidine with hydrogen peroxide in toluene, ether, and ethyl acetate is as high as 20% of that observed in aqueous solutions. Due to its high activity and stability in nonaqueous media, horseradish peroxidase was found useful in a variety of applications. For example, it was used in a coupled enzymatic system with cholesterol oxidase for the quantitative analysis of cholesterol in toluene.[78] Also, a variety of polymers have been produced by horseradish peroxidase-catalyzed coupling of phenols.[77] The enzyme polymerizes a number of phenols in dioxane, acetone, dimethylformamide, and methylformate with the solvent content up to 95%. The molecular weight of a polymer can be controlled by varying the concentration of water in the reaction medium. Depending upon experimental conditions, polymers with average molecular weights from 400 to 2.6×10^4 D have been obtained.[77]

8.6. CONCLUSIONS

Enzymatic catalysis in organic solvents is a new and rapidly growing area of enzymology. Although the first reports on catalytic activity of chymotrypsin[79] and xanthine oxidase[80, 81] suspended in nonaqueous media appeared as early as twenty years ago, interest in this field has experienced significant growth only in the past five years. It has been a common belief that enzymes are not catalytically active in organic solvents. Critical examination of the literature indicates that the conclusion that organic media are not suitable for enzyme-catalyzed processes was based on experiments which employed a very limited number of enzymes. The strong denaturing effect of such hydrophilic water-miscible solvents as acidic methanol, DMSO, formamide, and 2-chloroethanol, which are known to inactivate proteins even at low concentrations,[38–41] has been generalized and accepted as a common feature of nonaqueous media. Ironically, because of the very fact that hydrophobic solvents can hardly dissolve any water, they have not been used as reaction media for enzymatic processes. Now it has become generally accepted that hydrophobic, water-immiscible solvents are greatly superior to hydrophilic water-miscible solvents as enzymatic reaction media.

It has been demonstrated in recent years that a great variety of enzymes are catalytically active in organic solvents. The enzymes studied thus far are diverse in their action (oxidoreductases, hydrolases, isomerases), origin (bacteria, molds, yeasts, plants, and mammals), molecular weights, and subunit composition. Despite this diversity a number of similarities in their behavior is observed. Enzymes are active in a variety of solvents and require only a small amount of water. Due to the low water activity, enzymes are highly thermostable in dry organic solvents and can be used for months without loss of activity. Similar to their action in aqueous solutions, enzymes in organic solvents are substrate-specific and stereo- and regioselective. All these data strongly indicate that the ability of enzymes to catalyze reactions in organic solvents is a general phenomenon.

Enzymatic transformations have a great potential in food, pharmaceutical, and specialty chemical industries. The discovery and dynamic advancement of nonaqueous enzymology should turn this potential into a reality.

ACKNOWLEDGMENTS

I thank Drs. Radha Iyengar and Arthur McEvily for their critical reading of the manuscript and helpful advice.

REFERENCES

1. C. G. Greenwald and J. M. Nystrom, *Spectrum, Food Industry*, Arthur D. Little Decision Resources (1988).
2. L. G. Butler, *Enzyme Microb. Technol. 1*, 253 (1979).
3. S. Fukui and A. Tanaka, *Adv. Biochem. Eng. 29*, 1 (1984).
4. M. D. Lilly, *J. Chem. Tech. Biotechnol. 32*, 162 (1982).
5. G. Carrea, *Trends Biotechnol. 2*, 102 (1984).
6. P. L. Luisi, *Angew. Chem. 24*, 439 (1985).
7. P. L. Luisi and C. Laane, *Trends Biotechnol. 4*, 153 (1986).
8. A. M. Klibanov, *Chemtech. 16*, 354 (1986).
9. J. S. Deetz and J. D. Rozzell, *Trends Biotechnol. 6*, 15 (1988).
10. A. Zaks and A. J. Russell, *J. Biotechnol. 8*, 259 (1988).
11. J. Dordick, *Enzyme Microb. Technol. 11*, 194 (1989).
12. G. Carrea, S. Riva, R. Bovara, and P. Pasta, *Enzyme Microb. Technol. 10*, 333 (1988).
13. R. D. Schwartz and C. J. McCoy, *Appl. Environ. Microbiol. 34*, 47 (1977).
14. B. C. Buckland, P. Dunnill, and M. D. Lilly, *Biotechnol. Bioeng. 17*, 815 (1975).
15. G. A. Homandberg, J. A. Mattis, and M. Laskowski, Jr., *Biochemistry 17*, 5220 (1978).
16. C. F. Barbas, J. R. Matos, J. B. West, and C.-H. Wong, *J. Amer. Chem. Soc. 110*, 5162 (1988).
17. K. Martinek, A. V. Levashov, Y. L. Khmelnitski, N. L. Klyachko, and I. V. Berezin, *Science 218*, 889 (1982).
18. J. A. McCammon, B. R. Gelin, and M. Karplus, *Nature 267*, 585 (1977).
19. J. A. Rupley, E. Gratton, and G. Careri, *Trends Biochem. Sci. 8*, 18 (1983).
20. G. Careri, E. Gratton, P. H. Yang, and J. A. Rupley, *Nature 284*, 572 (1980).
21. J. E. Schinkel, N. W. Downer, and J. A. Rupley, *Biochemistry 24*, 352 (1985).
22. S. Bone, *Biochim. Biophys. Acta 916*, 128 (1987).

23. A. Zaks and A. M. Klibanov, *J. Biol. Chem. 263*, 8017 (1988).
24. A. Zaks and A. M. Klibanov, *Proc. Natl. Acad. Sci. USA 82*, 3192 (1985).
25. C. Laane, S. Boeren, K. Vos, and C. Veeger, *Biotechnol. Bioeng. 30*, 81 (1987).
26. A. Zaks and A. M. Klibanov, *J. Biol. Chem. 263*, 3194 (1988).
27. L. E. S. Brink and J. Tramper, *Biotechnol. Bioeng. 27*, 1258 (1985).
28. P. J. Halling, *Trends Biotechnol. 7*, 50 (1989).
29. M. Reslow, P. Adlercreutz, and B. Mattiasson, *Eur. J. Biochem. 172*, 573 (1988).
30. K. Takahashi, Y. Kodera, T. Yoshimoto, A. Ajima, A. Matsushima, and Y. Inada, *Biochem. Biophys. Res. Commun. 131*, 532 (1985).
31. A. Matsushima, M. Okada, and Y. Inada, *FEBS. Lett. 178*, 275 (1984).
32. K. Takahashi, A. Ajima, T. Yoshimoto, and Y. Inada, *Biochem. Biophys. Res. Commun. 125*, 761 (1984).
33. K. Takahashi, H. Nishimura, T. Yoshimoto, Y. Saito, and Y. Inada, *Biochem. Biophys. Res. Commun. 121*, 261 (1984).
34. A. Zaks and A. M. Klibanov, *J. Am. Chem. Soc. 108*, 2767 (1986).
35. A. J. Russell and A. M. Klibanov, *J. Biol. Chem. 263*, 11624 (1988).
36. N. Chinsky, A. L. Margolin, and A. M. Klibanov, *J. Amer. Chem. Soc. 111*, 386 (1989).
37. T. Sakurai, A. L. Margolin, A. J. Russell, and A. M. Klibanov, *J. Amer. Chem. Soc. 110*, 7236 (1988).
37a. C.-S. Chen and C. J. Sih, *Angew. Chem. Int. Ed. Engl. 28*, 695 (1989).
38. S. J. Singer, *Adv. Protein Chem. 17*, 1 (1962).
39. N. J. Solli and T. T. Herskovits, *Anal. Biochem. 54*, 370 (1973).
40. M. Winkler and P. Doty, *Biochim. Biophys. Acta 54*, 488 (1961).
41. T. T. Herskovits, B. Gadegbeku, and H. Jaellet, *J. Biol. Chem. 245*, 2588 (1970).
42. E. P. Pittz and S. Timasheff, *Biochemistry 17*, 615 (1978).
43. T. Arakawa and S. Timasheff, *Biochemistry 21*, 6536 (1982).
44. L. T. Kanerva and A. M. Klibanov, *J. Amer. Chem. Soc. 111*, 6864 (1989).
45. P. A. Burke, S. O. Smith, W. W. Bachovchin, and A. M. Klibanov, *J. Amer. Chem. Soc. 111*, 8290 (1989).
46. A. Zaks and A. M. Klibanov, *Science 224*, 1249 (1984).
47. G. Ayala, M. Tuena de Gomez-Puyou, A. Gomez-Puyou, and A. Darzon, *FEBS Lett. 203*, 41 (1986).
47a. G. Garza-Ramos, A. Darszon, M. Tuena de Gomez-Puyou, and A. Gomez-Puyou, *Biochemistry 28*, 3177 (1989).
48. C. J. Wheeler and R. Croteau, *Arch. Biochem. Biophys. 248*, 429 (1986).
49. M. Reslow, P. Adlercreutz, and B. Mattiasson, *Appl. Microbiol. Biotechnol. 26*, 1 (1987).
50. A. R. Macrae, *J. Am. Oil Chem. Soc. 60*, 243 (1983).
51. A. P. Ison, P. Dunnill, and M. D. Lilly, *Enzyme Microb. Technol. 10*, 47 (1988).
52. G. Kirchner, M. P. Scollar, and A. M. Klibanov, *J. Am. Chem. Soc. 107*, 7072 (1985).
53. S. Koshiro, K. Sonomoto, A. Tanaka, and S. Fukui, *J. Biotechnol. 2*, 47 (1985).
54. D. Bianchi, P. Cesti, and E. Battistel, *J. Org. Chem. 53*, 5531 (1988).
55. J. Hiratake, M. Inagaki, T. Nishioka, and J. Oda, *J. Org. Chem. 53*, 6130 (1988).
56. G. Langrand, J. Baratti, G. Buono, and C. Triantaphylides, *Tetrahedron Lett. 27*, 29 (1986).
57. G. Gil, E. Ferre, A. Meou, J. Le Petit, and C. Triantaphylides, *Tetrahedron Lett. 28*, 1647 (1987).
58. H. Kitaguchi, P. A. Fitzpatrick, J. E. Huber, and A. M. Klibanov, *J. Amer. Chem. Soc. 111*, 3094 (1989).
59. A. L. Gutman, K. Zuobi, and A. Boltansky, *Tetrahedron Lett. 28*, 3861 (1987).
60. A. Makita, T. Nihira, and Y. Yamada, *Tetrahedron Lett. 28*, 805 (1987).
61. Z.-W. Guo and C. J. Sih, *J. Amer. Chem. Soc. 110*, 1999 (1988).

62. A. L. Margolin, J.-Y. Crenne, and A. M. Klibanov, *Tetrahedron Lett. 28*, 1607 (1987).
63. M. Therisod and A. M. Klibanov, *J. Am. Chem. Soc. 108*, 5638 (1986).
64. M. Therisod and A. M. Klibanov, *J. Am. Chem. Soc. 109*, 3977 (1987).
65. W. J. Hennen, H. M. Sweers, Y.-F. Wang, and C.-H. Wong, *J. Org. Chem. 53*, 4939 (1988).
66. Y.-F. Wang, J. J. Lalonde, M. Momongan, D. E. Bergbreiter, and C.-H. Wong, *J. Amer. Chem. Soc. 110*, 7200 (1988).
67. J. Chopineau, F. D. McCafferty, M. Therisod, and A. M. Klibanov, *Biotechnol. Bioeng. 31*, 208 (1988).
68. S. Riva and A. Klibanov, *J. Amer. Chem. Soc. 110*, 3291 (1988).
69. S. Riva, J. Chopineau, A. P. G. Kieboom, and A. Klibanov, *J. Amer. Chem. Soc. 110*, 584 (1988).
70. W. Kullmann, *Enzymatic Peptide Synthesis*, CRC Press (1987).
71. K. Oyama, S. Nishimura, Y. Nonaka, K. Kihara, and T. Hashimoto, *J. Org. Chem. 46*, 5241 (1981).
72. H. Kise, A. Hayakawa, and H. Noritomi, *Biotechnol. Lett. 9*, 543 (1987).
73. A. L. Margolin and A. M. Klibanov, *J. Am. Chem. Soc. 109*, 3802 (1987).
74. J. B. West and C.-H. Wong, *Tetrahedron Lett. 28*, 1629 (1987).
75. J. Grunwald, B. Wirz, M. P. Scollar, and A. M. Klibanov, *J. Amer. Chem. Soc. 108*, 6732 (1986).
76. R. Z. Kazandjian and A. M. Klibanov, *J. Amer. Chem. Soc. 107*, 5448 (1986).
77. J. S. Dordick, M. A. Marletta, and A. M. Klibanov, *Biotechnol. Bioeng. 30*, 31 (1987).
78. R. Z. Kazandjian, J. S. Dordick, and A. M. Klibanov, *Biotechnol. Bioeng. 28*, 417 (1986).
79. F. R. Dastoli and S. Price, *Arch. Biochem. Biophys. 118*, 163 (1967).
80. F. R. Dastoli and S. Price, *Arch. Biochem. Biophys. 122*, 289 (1967).
81. F. R. Dastoli, N. A. Musto, and S. Price, *Arch. Biochem. Biophys. 115*, 44 (1966).

9

Catalysis by Alcohol Dehydrogenases in Organic Solvents

JEFFREY S. DEETZ and J. DAVID ROZZELL

9.1. INTRODUCTION

Modern organic synthesis has been advanced in large part through an increased understanding of solvent effects and by the development of sophisticated reaction catalysts. However, some of the best-designed catalysts, i.e., naturally-occurring enzymes, have been generally excluded from use in organic synthesis because of their perceived requirement for a totally aqueous reaction solvent. Over recent years, reports documenting the ability of enzymes to function under nearly anhydrous conditions may reconcile this dilemma[1–8] and lead to a significant expansion of the industrial applications of enzymes.

Removing the constraints imposed by aqueous solutions has a number of potential advantages[9]:

1. Enzymatic reactions can be carried out on substrates which have limited solubility in water or which are water-labile.
2. Costs of product recovery may be reduced because water, which has a high heat of vaporization, does not have to be removed.

JEFFREY S. DEETZ and J. DAVID ROZZELL • Genetics Institute, Inc., 87 Cambridge Park Drive, Cambridge, MA 02140.

3. The threat of microbial contamination of the reaction mixture can be greatly reduced.
4. Thermodynamic equilibria, particularly those involving water, can be shifted to favor certain reactions (e.g., transesterifications, esterifications, and amide formation), which proceed only to a very low extent in aqueous solutions.
5. Immobilization of enzymes can be accomplished by relatively simple and inexpensive procedures such as adsorption to inert supports.

A few reactions catalyzed by enzymes in organic milieu are now nearing commercialization. One of the most important is the upgrading of triglyceride oils by the lipase-catalyzed incorporation of certain fatty acyl moieties into relatively inexpensive fats. As an example, increasing the stearic acid content of palm oil gives rise to a cocoa butter-like fat, and this process is being developed by several companies, including Unilever.[14] A second process under development is the resolution of 2-halopropionic acids by stereoselective esterification in organic solvents such as hexane or butyl acetate.[15]

These first examples of enzymatic catalysis in organic media focused on the use of hydrolytic enzymes. However, the use of more complex enzymes was a natural extrapolation of this early work. Alcohol dehydrogenase (ADH) was chosen as an example of a more complex enzymatic reaction system.

9.2. MODEL SYSTEMS FOR CATALYSIS BY ALCOHOL DEHYDROGENASES

Alcohol dehydrogenase catalyzes the oxidation of alcohols and the reduction of carbonyl compounds such as aldehydes and ketones; many of these are poorly soluble in water, but very soluble in organic solvents. The enzyme requires a nicotinamide cofactor for catalytic activity; the cofactor mediates the hydride transfer from the alcohol to the carbonyl compound. The reaction can be viewed as two independent half-reactions: the oxidation of an alcohol with concomitant reduction of the nicotinamide cofactor followed by reduction of a carbonyl compound and simultaneous reoxidation of the cofactor. The successive oxidation and reduction steps recycle the cofactor for participation in multiple catalytic events. The general reaction scheme combining both oxidation and half-reactions is shown in Figure 1 using cinnamyl alcohol and octanal as coreactants.

Because of the requirement for recycling the cofactor with ADH enzymes, the complete reaction requires the use of a starting substrate pair

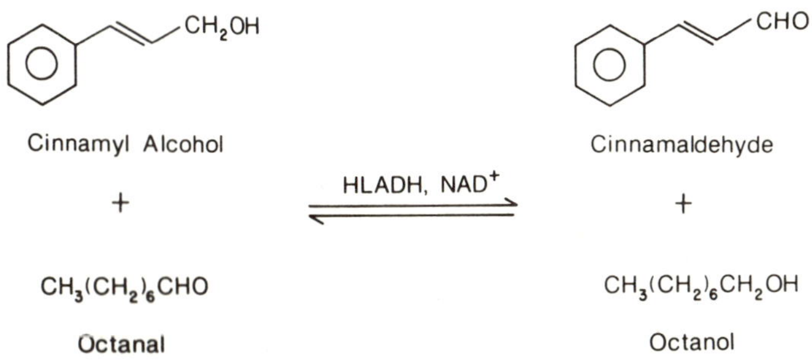

Figure 1. General reaction used for the assay of the activity of alcohol dehydrogenase in organic solvents.

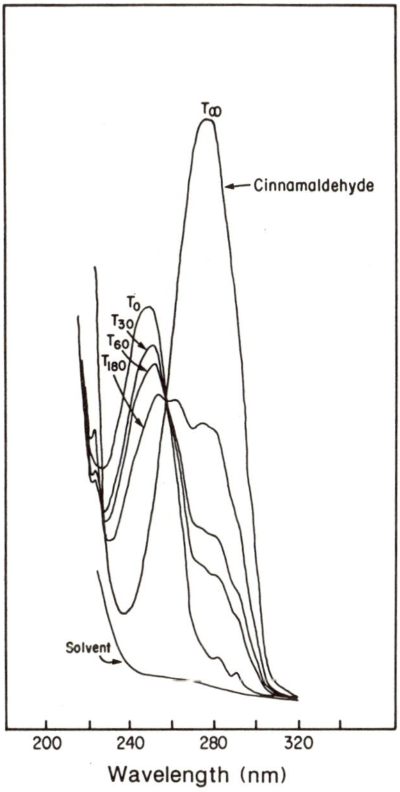

Figure 2. Repetitive scans of the ultraviolet spectrum of a substrate solution containing cinnamyl alcohol and octanal in 99.9% hexane in the presence of immobilized HLADH. The peak appearing at approximately 285 nm corresponds to the appearance of the product cinnamaldehyde.

of an alcohol and aldehyde or ketone, which then are converted into the corresponding carbonyl compound and alcohol. The NAD^+ cofactor must be recycled for the reaction to proceed to an appreciable extent, and because the cofactor is ionically charged and soluble in water but insoluble in virtually all organic solvents, it was of interest to determine whether recycling of NAD^+ could occur in organic media.

The first experiments were carried out as follows: horse liver alcohol dehydrogenase (HLADH) was dissolved in pH 7.0 phosphate buffer containing 1 mM NAD^+, adsorbed onto porous glass, and allowed to air-dry. This catalyst then was added to a hexane solution containing 10 mM cinnamyl alcohol, 10 mM octanal, and 0.1% water (v/v), and the mixture was shaken at approximately 200 rpm on an orbital shaker at a temperature of 20 °C. Aliquots were removed periodically for the spectrophotometric determination of cinnamaldehyde. The uv spectrum was scanned repeatedly, and it showed the simultaneous accumulation of cinnamaldehyde and the disappearance of cinnamyl alcohol (Figure 2).

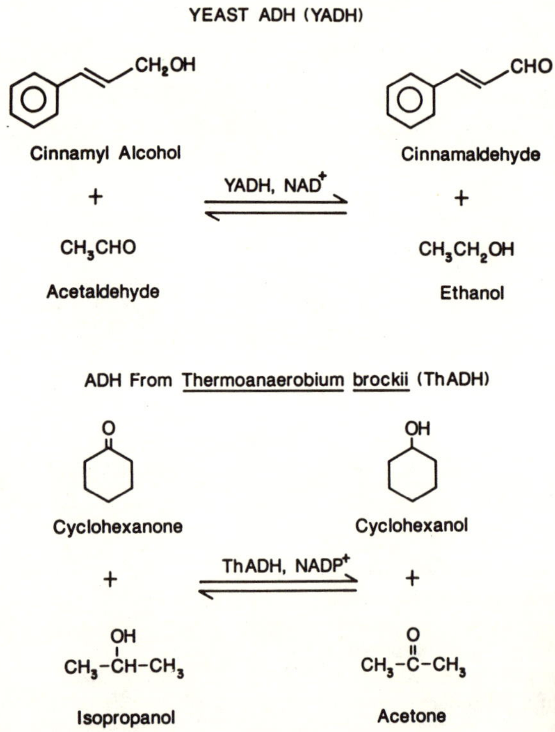

Figure 3. Reactions catalyzed by yeast alcohol dehydrogenase and alcohol dehydrogenase from *Thermoanaerobium brockii* in hexane containing 0.1% added water.

Although this experiment established that HLADH, complete with cofactor recycling, was catalytically active in 99.9% hexane, the applicability of this general technique toward similar reactions catalyzed by other dehydrogenases was also of interest. Therefore, we examined the activity of two other commercially available alcohol dehydrogenases under similar conditions: the enzyme from yeast (YADH) and the enzyme from the thermophilic bacterium *Thermoanaerobium brockii* (ThADH). The latter enzyme requires the cofactor $NADP^+$ rather than NAD^+. The YADH was assayed using benzaldehyde and octanol as substrates, while the ThADH was assayed using isopropanol and cyclohexanone. Both reactions proceeded to equilibrium, thereby demonstrating not only breadth of scope of ADH-catalyzed reactions, but also that the recycling of nicotinamide cofactors in organic solvents containing small amounts of water was not limited to any particular dehydrogenase system (Figure 3). However, at this point, the catalytic activities observed for the enzymes under these conditions were only about 1–2% of those observed for the same reactions catalyzed in aqueous solutions.

A standard protocol was established for preparation of the immobilized alcohol dehydrogenases. A solution of ADH at a concentration of 10 mg/ml was dialyzed against a buffer solution containing 50 mM potassium phosphate, pH 7.0. The cofactor (NAD^+ or $NADP^+$) was added to a final concentration of 1 mM, and the resulting solution was adsorbed into the desired support matrix. Typically, 1 g of support was used per 10 mg of enzyme. The mixture was evaporated to dryness under reduced pressure, and the resulting immobilized ADH was stored at 4 °C until ready for use.

9.3. FACTORS AFFECTING THE CATALYTIC ACTIVITY OF ADH IN ORGANIC SOLVENTS

An investigation of the factors affecting the catalytic activity of ADH in organic solvents was begun by looking at the activity of the enzyme in a range of different solvents. From previously published reports in which lipases were used, a trace of water in the organic solvent had been shown to be an absolute requirement for activity. In order to determine how much water was optimal, the activity of alcohol dehydrogenase was studied as a function of added water in organic solvents with varying dielectric constants. HLADH was chosen as the model enzyme for these studies.

As can be seen from the data in Table 1, HLADH displayed catalytic activity in a wide range of organic solvents. However, the optimal water concentration for enzymatic activity in each solvent varied. In general, the higher the dielectric constant of the solvent, the higher the optimal concen-

Table 1. Effect of [H_2O] on Enzymatic Activity in Various Solvents

Solvents	ε	Percent of H_2O added			
		0.1	0.3	1	5
Hexane	1.9	100	100	20	10.5
Butyl ether	3.1	10	300	90	25
Butyl acetate	5.0	NR	NR	180	10
Acetonitrile	37.5	NR	NR	NR	70

NR = No reaction   Activity in nmoles min^{-1} mg^{-1}

tration of water required for activity. Even in a solvent such as acetonitrile, which is infinitely miscible with water, ADH displayed activity, but only at the highest levels of added water. Upon closer investigation, we found that ADH activity increased as the concentration of water was further increased to 10% and above. Klibanov has suggested that the more hydrophilic the solvent (higher dielectric constant), the greater the tendency of the solvent to strip the essential water of hydration from the enzyme.[7] Laane has made similar observations using a correlation involving the parameter log P, the ratio of the partition coefficient in an octanol/water two-phase system.[8] Our results with ADH support these conclusions. Recently, Clark and co-workers have developed spin-labeling techniques to probe changes in the environment at the active site as a function of added water.[13] These types of experiments promise to provide a much clearer picture of the effect of added water on enzymes in organic media at the molecular level.

The support on which the enzyme is deposited was also found to affect the catalytic activity by altering the microenvironment of the enzyme. ADH immobilized on untreated porous glass displayed an activity that was 1.6% that of the native enzyme. Treatment of the porous glass support with bovine serum albumin prior to immobilization of the alcohol dehydrogenase or the use of glycophase glass (a porous glass, the surface of which has been modified with a glycerol analog, obtained from Pierce) increased the activity of the enzyme to 9.8% and 43% that of the native enzyme activity, respectively. The use of thiolated glycophase glass which produced a hydrophilic and reducing environment led to a catalyst with an activity of 64% that of the soluble enzyme in aqueous solution (Table 2). If Sepharose supports were substituted for glass, the water required to achieve optimal activity under otherwise identical conditions increased. Such differences suggest that the support has an important influence on the microenvironment of the enzyme and affects the interaction of essential water with the enzyme. Optimization of the support in conjunction with

Table 2. Effect of Surface Modification of CP–Glass on Activity of YADH in Organic Media

Support	% Aqueous activity
CPG	1.6
CPG–BSA	9.8
CPG–Glycophase	43
CPG–Thiolated	64

Conditions: Hexane containing 0.1% H_2O
10 mM cinnamyl alcohol, 10 mM octanal
Enzyme loading: 10 mg YADH/gram support
Room temperature, 200 rpm rotary shaking

other parameters such as solvent and the amount of added water is an important component of any strategy to optimize enzyme activity in organic solvents containing limiting concentrations of water.

Another observation, which may be related to the effect of the support matrix, is the relief of inhibition of alcohol dehydrogenases by its substrates when used in organic media. HLADH is severely inhibited by its substrate, for example cinnamyl alcohol, at concentrations above 0.1 mM in aqueous solution. Typical values of K_i for a variety of substrates are in the range of 10–100 mM. This inhibition is substantially relieved using immobilized ADH in organic solvents such as butyl acetate. Figure 4 shows a Lineweaver–Burk plot for the HLADH-catalyzed reaction of cinnamyl alcohol to cinnamaldehyde in butyl acetate containing 1% added water. Inspection of Figure 4 reveals that inhibition by substrate occurs only at concentrations of approximately 50 mM or greater, giving a value for K_i about 3 orders of magnitude higher than that observed in aqueous solution. The partition coefficient for cinnamyl alcohol between water and butylacetate cannot account for this large effort. The reduction of substrate inhibition under substantially nonaqueous conditions would clearly be of benefit in organic synthesis applications.

Another potentially useful property conferred upon ADH in organic solvents containing only trace amounts of water is the saturation of the enzyme at relatively low ratios of NAD^+:enzyme subunit. Experiments were conducted in which HLADH, that had been previously dialyzed against buffer to remove all the bound NAD^+ cofactor, was immobilized by the method described earlier in the presence of varying levels of added cofactor. At amounts of added NAD^+:enzyme subunit of 1:1, the catalytic activity of the HLADH was 50% of maximal. At ratios of cofactor:ADH subunit of 10:1 the enzyme was functioning at maximal catalytic rates; further increases in this ratio did not produce significantly higher rates of

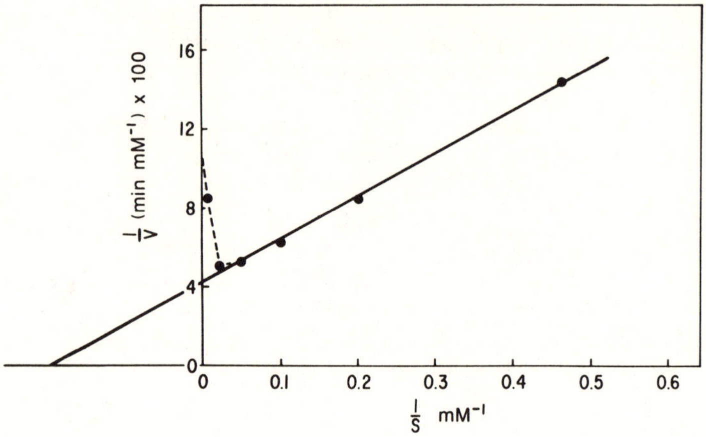

Figure 4. Lineweaver–Burk plot for cinnamyl alcohol as substrate in the presence of octanal as cosubstrate. Solvent was butyl acetate containing 1% added water.

catalysis. Reduction of the amount of cofactor required could have favorable implications on the operating costs in such systems.

Klibanov and co-workers conducted experiments with HLADH in organic media to optimize the number of cycles for the nicotinamide cofactor. Using the isobutyraldehyde/ethanol substrate combination a recycle number of greater than 10^6 was reported, suggesting that the cost contribution of the cofactor could be a very small portion of the overall cost in such biocatalytic systems.[10]

9.4. COFACTOR REGENERATION BETWEEN TWO DIFFERENT ENZYMES

Although the experiments described thus far demonstrate that the cofactor can be recycled efficiently in organic solvents, they do not discriminate between the cofactor which is "fixed" at the NAD-binding site and is regenerated in place and the cofactor which dissociates from the active site of one enzyme and diffuses to the active site of a second enzyme where it is regenerated. In order to determine whether reactions catalyzed by two separate enzymes can be coupled in organic solvents at the lower critical limit of water, we examined the reaction shown in Figure 5 in both butyl acetate (1% v/v H_2O) and hexane (0.1% v/v H_2O). In this scheme, NAD^+ is reduced first to NADH in the YADH-mediated reaction with octanol as the reductant. In turn, reducing equivalents are transferred from NADH to the dye 2,6-dichlorophenolindophenol (DCIP) in the

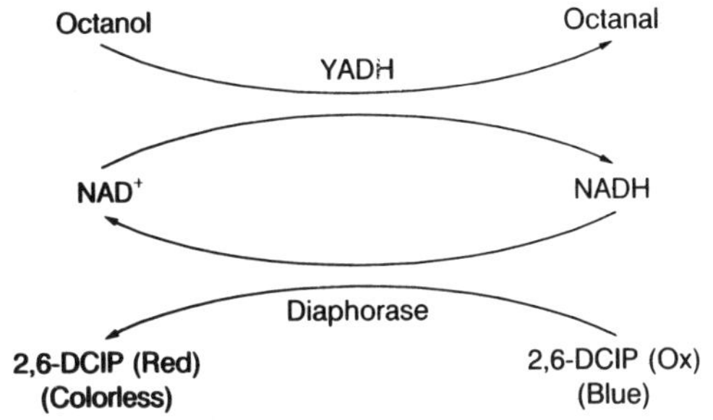

Figure 5. Reaction used to demonstrate recycling of NAD^+ cofactor between the active sites of two distinct enzymes: yeast alcohol dehydrogenase and diaphorase.

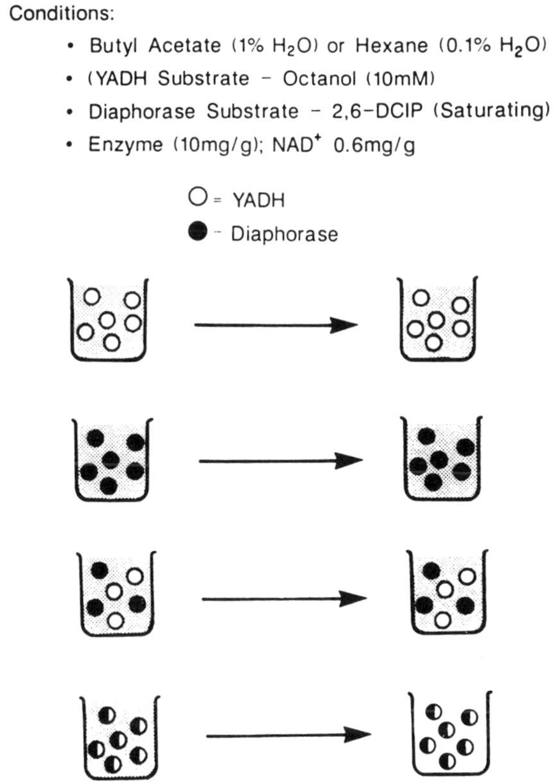

Figure 6. Experiments were carried out in both butyl acetate containing 1% added water and hexane containing 0.1% added water. Only when YADH and diaphorase were coimmobilized on the same support particle, was bleaching of the 2,6-DCIP dye observed.

diaphorase-catalyzed reaction. Because the substrate specificities of both enzymes exclude cross-reactivity, a positive reaction (as evidenced by the bleaching of the oxidized form of DCIP from blue to colorless) can occur only if the cofactor can shuttle between the active sites of the two enzymes.

The reaction was initiated by the addition of catalyst to the solvent containing substrates and the minimum amount of water required for YADH to function, as determined previously. Because only the oxidized form of the cofactor was used in catalyst preparation, reducing equivalents could only originate from octanol via the YADH-dependent half-reaction. As shown in Figure 6, bleaching of DCIP was observed only in the complete reaction mixture that contained "coimmobilized" catalyst, that is, a catalyst with both YADH and diaphorase deposited on the same support particle. In contrast, no reaction was observed when octanol was omitted or when the YADH and diaphorase were immobilized on separate support particles. For this reaction to proceed, the nicotinamide cofactor had to diffuse through a "fluid" microenvironment within the support from the active site of one enzyme to another, even at limiting water concentrations. This demonstration suggests that other multienzyme systems, even those that involve water-soluble intermediates, may be coupled in nonaqueous media.[9, 11]

9.5. CONCLUSIONS

In summary, the potential for the application of enzymatic catalysts in organic catalysis under conditions of water limitation appears to be generally applicable to a wide range of enzymes. Already, applications for the use of alcohol dehydrogenases in organic solvents are beginning to emerge.[12] It is important to note that even those enzymes with cofactor requirements, demonstrated here with oxidoreductases, can be used successfully under nonaqueous conditions. By optimizing the reaction conditions that include the solvent, the amount of added water, and the support, enzymatic rates using alcohol dehydrogenases approaching those observed in aqueous solution have been attained. It has also been shown that the cofactor can diffuse between the active sites of two different enzymes in an environment of low water, suggesting that other multienzyme systems may be used in organic solvents containing low amounts of water.

REFERENCES

1. A. Zaks and A. M. Klibanov, *Proc. Natl. Acad. Sci. USA 82*, 3192 (1985).
2. B. Cambou and A. M. Klibanov, *J. Am. Chem. Soc. 106*, 2687 (1984).

3. R. Kazandjian, J. S. Dordick, and A. M. Klibanov, *Biotechnol. Bioeng. 28*, 417 (1986).
4. A. Zaks and A. M. Klibanov, *Science 224*, 1249 (1984).
5. T. Yoshimoto, K. Takahashi, H. Nishimura, A. Ajima, Y. Tamaura, and Y. Inada, *Biotechnol. Lett. 6(6)*, 337 (1984).
6. K. Takahashi, A. Ajima, T. Yoshimoto, and Y. Inada, *Biochem. Biophys. Res. Commun. 125(2)*, 761 (1984).
7. A. M. Klibanov, *Chemtech. 16*, 354 (1986).
8. C. Laane, S. Boeren, R. Hilhorst, and C. Veeger, (1987); *Stud. Org. Chem. 29*, 65 (1987).
9. J. S. Deetz and J. D. Rozzell, *The World Biotech. Report, Vol. 2, Part 3*, 25 (1986).
10. J. Grunwald, B. Wirz, M. P. Scollar, and A. M. Klibanov, *J. Am. Chem. Soc. 108*, 6732 (1986).
11. J. S. Deetz and J. D. Rozzell, *Trends in Biotechnol. 6*, 15 (1988).
12. T. Kawamoto, A. Aoki, K. Sonomoto, and A. Tanaka, *J. Ferment. Bioeng. 67*, 361 (1989).
13. D. S. Clark, L. Creagh, P. Skerker, M. Guinn, J. Prausnitz, and H. Blanch, *ACS Symp. Ser. 392*, 104 (1989).
14. P. V. Halling and A. R. MaCrae, European Patent Application number 0,064,855 (1982).
15. A. M. Klibanov and G. M. Kirchner, *US Patent 4,601,987* (1986).

10

Lipase Catalysis and Its Applications

VIJAY T. JOHN and GEORGE ABRAHAM

10.1. INTRODUCTION

Lipases have been traditionally defined as enzymes "capable of hydrolyzing esters of oleic acid."[1] The definition of a lipase as a hydrolytic enzyme originated primarily from its physiological function of triglyceride hydrolysis. In more recent years, the recognition that the enzymes are effective catalysts both for ester hydrolysis

$$R_1COOR_2 + H_2O \rightleftharpoons R_1COOH + R_2OH \tag{1}$$

and the reverse reaction of synthesis, has resulted in a tremendous interest in applications to synthetic organic chemistry. Additionally, the realization that the enzyme can also catalyze the transesterification reaction

$$R_1COOR_2 + R_3OH \rightleftharpoons R_1COOR_3 + R_2OH \tag{2}$$

and interesterification reactions (between two acyl substrates – acid + ester, or ester 1 + ester 2) has led to interest in lipase applications to ester modification.

The role of water in lipase catalysis is of much importance. As seen from the reversibility of equation (1), at a given temperature, the equi-

VIJAY T. JOHN • Department of Chemical Engineering, Tulane University, New Orleans, LA 70118. GEORGE ABRAHAM • US Department of Agriculture, Southern Regional Research Center, New Orleans, LA 70124.

librium composition and hence the extent of conversion, depend on the amount of water present in the reaction medium. Thus, hydrolysis is favored in aqueous solutions, while synthesis is favored in reaction systems with minimal water. Comparing equations (1) and (2), it is clear that selectivity to transesterification over hydrolysis also increases in reaction systems with minimal water. The ability of lipases to function in limited water systems under appropriate conditions is thus crucial to synthesis efficiency and to the direction of reaction selectivity.

We have organized this review of lipase catalysis into two parts: one dealing with lipase function in high water-content systems, and the other, with lipase function in essentially nonaqueous media. Much of the fundamental work on lipase activation and function has been done using aqueous systems, while current industrial interest lies in the use of the enzyme for synthesis reactions in nonaqueous systems. The focus of the paper therefore is to provide a review on the physicochemical factors influencing lipase action with the hope that some of these factors may be common to both aqueous and nonaqueous systems and may result in improved biocatalyst design. We cite several industrial applications of lipases in both aqueous and nonaqueous systems, but the list is not all-inclusive, and the reader is often referred to other recent reviews on lipases in industry.

10.2. LIPASES IN AQUEOUS MEDIA

Although we categorize this section as lipase action in aqueous systems, the necessity of an oil–water interface to activate the enzyme is recognized, and we are really referring to lipase behavior in high-water-content systems involving an oil–water interface, such as an oil-in-water emulsion, for example. Lipase catalysis in such aqueous systems primarily concerns ester hydrolysis. Table 1 lists several applications, almost all dealing with the hydrolysis of triglycerides to fatty acids, glycerol, and the mono- and diglycerides. Several review articles dealing with the industrial implications of triglyceride hydrolysis, and the applications of lipases to fats and oils processing are available in the literature.[2–9] A synthesis of the information provided in the literature indicates a consensus that lipase-catalyzed hydrolysis appears most promising for partial hydrolysis to the mono- and diglycerides with applications to the food industry. The mono- and diglycerides serve as biocompatible emulsifiers and food additives. Partial hydrolysis is also cited as leading to flavor enhancement, especially with dairy products.[3, 4, 10] Lipase usage for total hydrolysis in fatty acid production appears uneconomical when compared to conventional alkali-refining methods operating at high pressures (ca. 700 psi) and temperatures

Table 1. Applications of Lipase-Catalyzed Hydrolysis[a]

1.	Blood triglycerides measured through the rate of hydrolysis. Applications in medical diagnostics.
2.	Hydrolysis of milk fats. Flavor improvement of dairy products.
3.	Lipid breakdown and subsequent flavor improvement in meat and fish products.
4.	Partial hydrolysis leading to better bioemulsification. Applications to mayonnaise, dressings, whippings, etc.
5.	Removal of oil and fat stains. Uses in detergent manufacture.
6.	Fermentation acceleration through lipid removal. Applications to alcoholic beverages.
7.	Unusual cleansing applications. Contact lens cleaners. Windshield-washing fluids.
8.	Removal of fats from animal skins; leather products.

[a] Adapted from Iwai and Tsujisaka[28]

(300 °C). The mildness of enzymatic hydrolysis may however be used to advantage in situations involving recovery of highly unsaturated long-chain fatty acids that are thermally unstable.[12]

Selectivity in triglyceride hydrolysis can be achieved by utilizing 1,3-specific lipases.[9, 13] A lipase from *Geotricum candidum* has been reported to be specific to the unsaturated fatty acids, and to oleic and linoleic acids.[14] On the other hand, pancreatic lipase and certain microbial lipases have been found to be very inefficient in hydrolyzing unsaturated fatty acids out of triglycerides.[15, 16] The concept has been used in the interesting application of fish oils processing wherein the dietary component—eicosapentaenoic acid—is recovered.[15] The process involves release of fatty acids from fish oils and simultaneous esterification with ethanol to give the fatty ethyl esters, using sodium methoxide as catalyst. The ethyl ester mixture is then treated with lipase, which hydrolyzes all esters except highly unsaturated ones such as the eicosapentaenoic ethyl ester. Subsequent conversion of the fatty acids to the sodium salts using NaOH solution followed by centrifugation, separates out the unsaturated fatty esters which are recovered. Macrae's 1983 review[9] gives a good classification of microbial lipases in terms of physical properties and substrate specificities.

From an engineering perspective, some interesting reactor schemes have been utilized fairly recently for lipase-catalyzed hydrolysis.[10, 17–19] The fact that reaction occurs at an oil–water interface requires some ingenuity in reactor design. Ishida[10] describes two schemes of commercial applicability. One involves a column reactor containing lipase in aqueous solution. Fat droplets are pumped in through the bottom of the reactor; reaction takes place during their rise to the top. The lipid product is continually removed from the top of the column. In the other scheme, lipase is immobilized on to a hydrophobic membrane.[17] Crossflow over the membrane with lipid on one side and the aqueous phase on the other results in reaction at the interface; the partial glycerides and long-chain fatty acids are recovered from the lipid phase, while glycerol is removed through the aqueous phase. Other configurations involve the use of hollow fiber membranes,[19] and lipase (from *Rhizopus arrhizus*) supported in native mycelia, used in packed bed and stirred tank reactor configurations.[18]

We now turn to some fundamental physicochemical aspects of lipase action. Much of the work has been carried out using pancreatic lipase due to its physiological significance. Although it is likely that microbial lipases will be more widely used in an industrial context due to a greater ease in enzyme production, a knowledge of the action of pancreatic lipase should help in the design of industrial biocatalysts with microbial lipases. The remarkable aspect of lipase catalysis is its activation at the oil–water interface. This was first clarified by Sarda and Desnuelle in a now classical paper.[20] Using triacetin as a substrate, these authors showed that the hydrolysis rate increased sharply when triacetin concentration exceeded its solubility limit in water, thus creating a separate lipid phase. Refinements of this observation have been reported, among them the enhancement of activity when a second hydrophobic phase is added,[21] and the ability of the enzyme to efficiently act on emulsified lipids.[22]

The general picture of lipolysis appears to be the requirement for an interface to allow the enzyme to achieve an optimal conformation. It appears that the catalytic sites are topologically distinct from the amino acid residues that cause binding to the interface.[23,24] Catalysis proceeds through initial formation of an acyl–enzyme intermediate, with deacylation being the rate-determining step; it has been proposed that deacylation is accelerated by an interface.[25] Calcium salts accelerate hydrolysis; the effect has also been observed with a fungal lipase from *Aspergillus niger*.[27] Colipase (a coenzyme of pancreatic lipase) helps anchor pancreatic lipase to the interface and thus helps sustain activity[23,27]; microbial lipases do not require a colipase for activity sustenance.[28] The role of bile salts has been in dispute with both activation and inhibition being claimed.[23] Momsen and Brockman[27] have shown that low concentrations of sodium taurodeoxycholate increase lipase stability, while higher concentrations although still below the critical micelle concentration inhibit the enzyme, perhaps through decreasing lipase adsorption at the interface. The authors postulate that the beneficial aspect of colipase is to allow lipase adsorption to the interface even in the presence of high bile salt concentrations.

Based on these observations, Brockman has come up with a simple "working" model for lipolysis shown schematically in Figure 1. The reaction mechanism involves adsorption of lipase and transport of the lipid substrate to the interface. The velocity of lipolysis is determined not by the

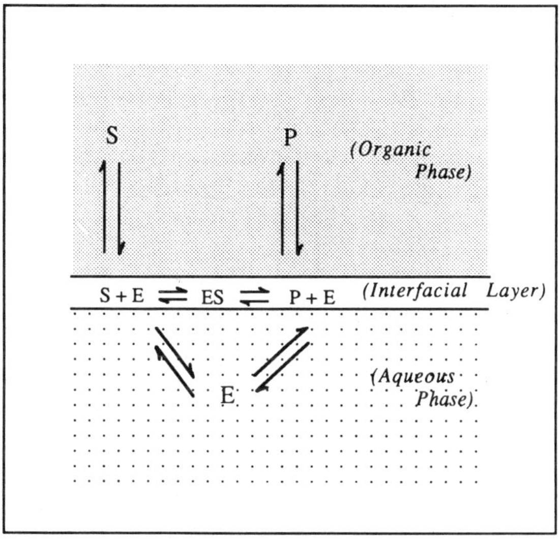

Figure 1. A model for lipase-catalyzed hydrolysis.

overall concentration of the substrate but by the interfacial concentration. Other amphiphilic molecules that compete for interfacial sites reduce observed reaction rates by decreasing enzyme or substrate interfacial concentrations.

Some elegant work with Langmuir-Blodgett films by Brockman and co-workers[30–32] has shown how surface compositions and lipase surface adsorption are affected not only by the nature of the amphiphilic species present, but also by overall species composition. Langmuir–Blodgett film studies are traditionally carried out with amphiphile concentrations below the critical micellar concentration by spreading a monolayer of the amphiphiles on the air–water interface and correlating the decrease in surface tension in the presence of the amphiphile (the surface or spreading pressure) to the molecular surface area. The low amphiphile concentration and the use of the air–water interface may seem inapplicable for emulsion systems where the oil–water interface is of primary concern. However, such studies reveal significant information about interfacial compositions and the role of different amphiphiles in affecting interfacial properties. These authors measured the surface pressure at which a monomolecular surface film collapses into a bilayer (called the transition pressure) as a function of amphiphile composition; the amphiphiles being binary mixtures of 13,16-docosadienoic acid (DA) and 1-palmitoyl-2-oleoyl phosphatidylcholine (POPC), or 1,3-diolein (DO) and POPC. Both DA and DO are substrates for pancreatic lipase while POPC is a nonsubstrate. Figure 2 schematically

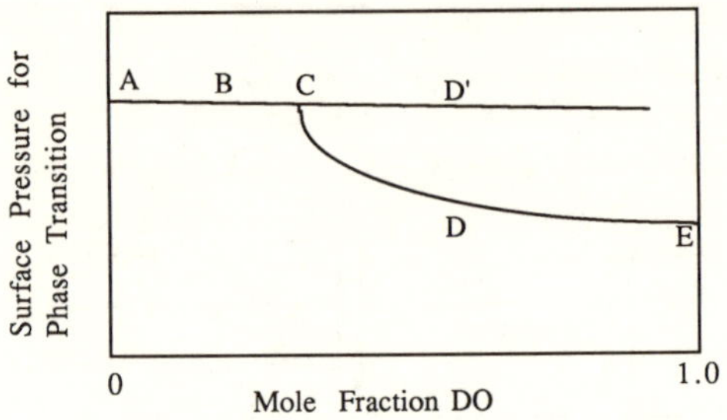

Figure 2. Surface pressures at film collapse for DO + POPC mixtures.

shows the transition pressure vs mixture composition trend for a sample binary, e.g., DO and POPC. At point A, the film collapse or transition pressure is simply governed by the ordering of POPC molecules on the surface. At point B, there is some DO on the surface and the DO–POPC complex has essentially the same collapse pressure as a pure POPC film. From C onwards, the surface packing of molecules tends to reflect the presence of DO. The presence of two collapse pressures at the same overall composition corresponding to D and D′ is interesting. Brockman postulates that the lower pressure is due to collapse of surface regions containing uncomplexed DO; once these uncomplexed molecules are removed from the surface, the surface resembles the state at B, consisting of a film with essentially POPC. The upper collapse pressure thus corresponds very closely to the collapse pressure of a POPC–DO complex. Point E corresponds to collapse of a pure DO film.

The interesting observation from these studies is the binding of pancreatic lipase to the interface. With pure DO, lipase adsorbs strongly to the interface, and adsorption can be modeled by the Langmuir isotherm. With pure POPC, on the other hand, there is little binding to the interface. However, binding of lipase to the interface does not change linearly as a function of overall composition of DO and POPC; Brockman and coworkers found that in DO–POPC complexes, lipase binding was almost essentially eliminated.[31, 32] While the rationale is not completely clear, the authors postulate that perhaps a cluster of interfacial DO molecules is necessary to effect binding. This cluster may be present only at compositions approaching pure DO. At intermediate compositions, the DO–POPC complex is not capable of binding the enzyme and presents an excluded surface area to the large protein molecule.

These observations are of interest not only in understanding the location of the lipases in physiological environments, such as those constituting cell membranes, but also in the industrial context of biocatalyst design. As will be discussed later, lipase encapsulation in reversed micelles made with anionic surfactants such as sodium (bis)2-ethylhexyl sulfosuccinate (AOT) or even with biocompatible surfactants such as phosphatidylcholine (lecithin), presents a viable method of conducting reaction in essentially nonaqueous media. Studies with idealized Langmuir–Blodgett films may offer insight into lipase activity in such systems and a knowledge of lipase–surfactant interactions.

Triglyceride hydrolysis is not the only ester hydrolysis of industrial importance. The ability of lipases to act as highly selective chiral catalysts has been recognized.[33, 34] Hydrolysis of chiral esters, for example, those derived from epoxy alcohols, leads to very high enantioselectivities of the resulting alcohols. The resolution of such chiral alcohols has application to pharmaceutical synthesis.[34]

Although ester hydrolysis is the main mode of lipase action in bulk aqueous solutions, the ability of the enzyme to catalyze the reverse reaction of ester synthesis has also been observed.[35, 36] The role of water in directing lipase activity to synthesis or hydrolysis has been long recognized,[35] but the methods to get total synthesis selectivity by making lipases function in essentially nonaqueous media have been developed only over the last decade. Hoq *et al.*[37] have used the ingenious concept of the membrane bioreactor to effect glyceride synthesis. Again, countercurrent flow of a concentrated glycerol solution containing lipase on one side of a hydrophobic membrane, which contains immobilized lipase, and fatty acids on the other side, leads to glyceride synthesis, with the glycerides being removed through the lipid phase.

The next section of the paper deals with the more recent advances in maintaining lipase activity in systems, where the presence of water leads to low equilibrium conversions and/or undesired side reactions. Ester synthesis, trans- and interesterifications are the primary reactions considered.

10.3. ENZYMES IN ESSENTIALLY NONAQUEOUS MEDIA

It was conventionally thought that water is essential for an enzyme to maintain its conformation and that organic solvents would denature the enzyme. The realization that enzymes can function in limited-water environments when conditioned appropriately, has resulted in tremendous interest in applications of the phenomenon. As Klibanov[38] has stated, water is essential to maintain enzyme conformation, but the real question is the amount of water needed. Two modes of substrate–enzyme contact in limited-water environments have been studied intensively. The first is the concept of providing just sufficient water to hydrate the enzyme and allow it to stay locked in an active conformation, subsequently using the enzyme in apolar organic solvents that do not remove the essential water of hydration. Such hydrated enzymes are used either directly or immobilized on inert supports. This is the concept of enzyme utilization in monophasic organic solvents. The second mode is that of encapsulating the enzyme in water-in-oil microemulsions called reversed micelles. The microaqueous phase of reversed micelles provides a biomimetic environment for the enzyme. While reversed micellar solutions are thermodynamically biphasic, the macroscopic homogeneity and clarity distinguishes these water-in-oil microemulsions from the oil-in-water emulsion systems encountered in traditional studies on lipolysis. The two modes of substrate–enzyme contact are illustrated in Figure 3, and we discuss them separately.

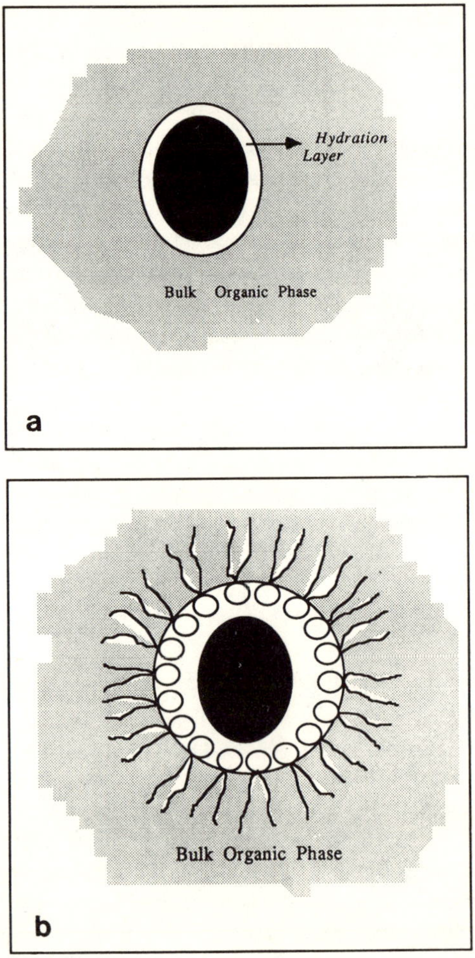

Figure 3. a) Enzymes in organic solvents; b) enzymes in reversed micelles.

10.3.1. Lipases in Organic Solvents

Much of the work involving lipase activation and chemical synthesis in organic solvents was pioneered by Klibanov and co-workers in the US,[39–46] by Tanaka and co-workers in Japan[47–51]; more recently, Triantaphylides and co-workers in France have contributed significantly to the field.[52–56] We describe some of the potentially important reactions conducted by these and other researchers, prior to reviewing physicochemical aspects of the enzyme in organic solvents. All reactions considered involve the synthesis of novel esters, either directly from acids and bases, or through trans- and interesterifications. We classify them purely on the basis of industrial applicability.

10.3.1.1. Lipase-Catalyzed Stereoselective Reactions

Such reactions involve either the synthesis of stereoselective esters, or applications of lipase catalysis to the resolution of racemic alcohol mixtures. A specific example, the reaction of racemic menthol (1) with 5-phenylvaleric acid (2) has been found to take place with conversion of only *l*-menthol to *l*-menthyl-5-phenylvalerate (3), leaving *d*-menthol (4) essentially unreacted. Lipase from *Candida cylindracea* (or *Candida rugosa* as termed recently) has been found to be effective in carrying out the reaction with high enantioselectivity.[47]

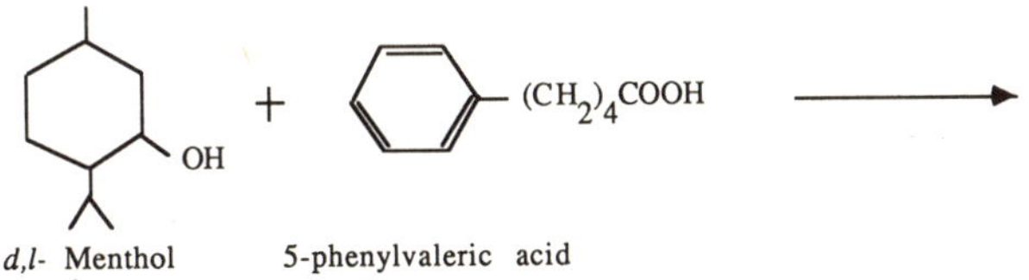

d,l- Menthol
1

5-phenylvaleric acid
2

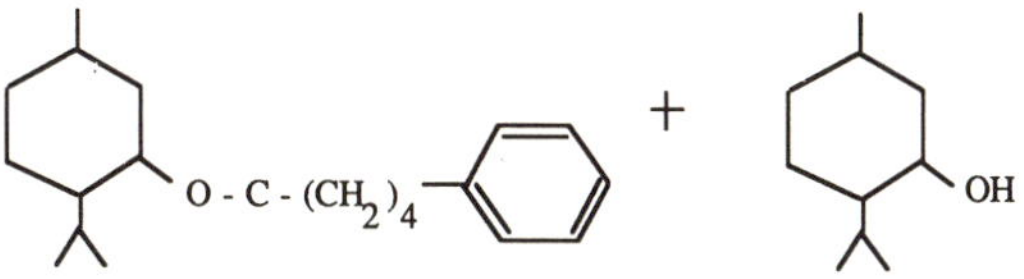

l -menthyl-5-phenylvalerate
3

d -menthol
4

(3)

In addition to obtaining an enantiomeric ester in high selectivity, the principle can be used in the resolution of racemic alcohols.[42, 43, 52, 53, 57] Using the example of racemic menthol, various lipase-catalyzed reactions can be used to obtain *d* and *l* menthol from their racemic mixture or from a racemic mixture of menthyl esters. The schemes are:

a. esterification between (+/−)menthol and an acid: (+/−)menthol + acid → (+)menthol + (−)menthyl ester + water. Separation of (+)menthol from the (−)menthyl ester followed by hydrolysis (chemical or enzymatic) of the (−)menthyl ester to (−)menthol leads to resolution of the optical isomers.
b. transesterification between (+/−)menthol and an ester: (+/−)menthol + ester → (+)menthol + (−)menthyl ester + alcohol, followed by hydrolysis of the (−)menthyl ester.
c. transesterification between racemic menthyl ester and alcohol: (+/−)menthyl ester + alcohol → (−)menthol + (+)menthyl ester + ester, followed by chemical hydrolysis of the (+)menthyl ester.
d. hydrolysis of racemic menthyl esters: (+/−)menthyl ester + water → (−)menthol + (+)menthyl ester + acid, and further chemical hydrolysis of the (+)menthyl ester after separation from (−)menthol.

Chiral terpene alcohols and derived esters have applications in pharmaceuticals and in food flavor chemistry,[55] and in the cosmetics industry as fragrance constituents and enhancers.

An interesting extension of the stereoselective properties of lipases is the work by Margolin *et al.*[46] who carried out polycondensations based on transesterifications between a chiral diester and an achiral diol or between an achiral diester and a chiral diol. A sample reaction is that between bis(2-chloroethyl)( )-2,5-dibromoadipate (A) and 1,6-hexanediol (B) to form the optically active trimers A-B-A and pentamers A-B-A-B-A. As stated by these authors, optically active polymers have applications as chiral adsorbents and in the preparation of liquid crystals.

Using the protease subtilisin Carlsberg, Sakurai *et al.*[58] have shown that enzyme enantioselectivity can be altered by changing the hydrophobicity of the solvent. Similar modifications of lipase enantioselectivity need to be researched.

10.3.1.2. Lipase-Catalyzed Regioselective Reactions

Of interest here, is the ability of lipases to catalyze ester synthesis or transesterifications using only selected positions of polyols. Cesti *et al.*[41] have studied the transesterification of an ethyl carboxylate (e.g., ethyl acetate) with a glycol (e.g., 1,2-butanediol), and have found highly selective acetylation of just the primary hydroxyl group to give the monoacyl glycol. Another reaction is the synthesis of primary monoesters from sugar alcohols and fatty acids[45]; such esters have excellent surfactant properties

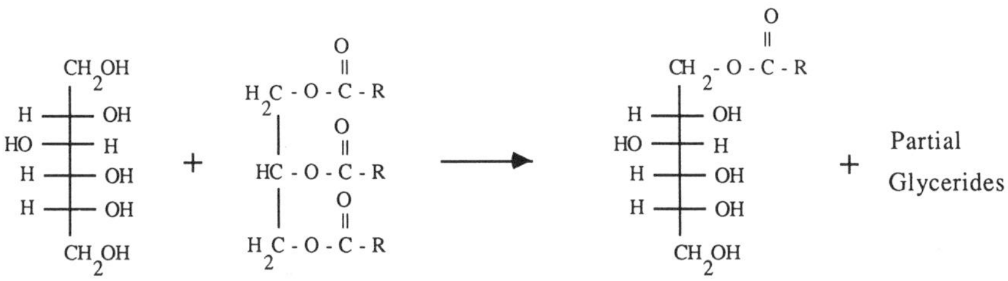

5 6 7 (4)

and are of industrial importance due to their biocompatibility. A typical reaction is that between oleic acid and glucose,[60] or the transesterification between triolein (6) and sorbitol (5).[45]

The long-chain alkyl groups of the sugar monoester (7) add hydrophobicity to the molecule. Therisod and Klibanov[40, 59] have shown that the tendency to acylate primary hydroxyl groups of sugars can be shifted to acylation of secondary hydroxyl groups through blocking the primary hydroxyl group by transesterification with 2,2,2-trichloroethyl butyrate; the resulting 6-*O*-butyrylglucose is subject to acylation at secondary hydroxyl positions. Which of the secondary hydroxyl groups become acylated depends on the lipase used, with lipase from *Chromobacterium viscosum* synthesizing the 3,6-diester with high selectivity and pancreatic lipase synthesizing primarily the 2,6-diester.

10.3.1.3. Reactions Involving Glycerides

This is an artificial category since the reactions involve ester synthesis, trans- and interesterifications, with the relevant esters being the mono-, di- and triglycerides. We discuss the category separately due to the industrial importance of these compounds. The synthesis of glycerides from glycerol and long-chain fatty acids has applications to the production of food emulsifiers. Interesterification reactions are of commercial importance in the modification of vegetable oils to produce higher-value products such as cocoa butter substitutes.[61] For example, the reactions between hydrogenated cottonseed oil, which has a high fraction of stearic acid residues on the triglyceride backbone (C18:0), with olive oil, which contains a high fraction of oleic acid residues (C18:1), give combinations containing stearic and oleic acid residues. Utilization of a 1,3-specific lipase helps retain oleic acid moieties at the 2-position. Triglycerides of the type POP, SOS, and POS (P = palmitic, O = oleic, S = stearic) are common in cocoa butter; such triglycerides contribute to the desired property of a sharp melting point at physiological temperatures. An important concern in interesterification is the amount of water in the system. In the presence of water, lipase selectivity switches drastically to hydrolysis. The whole range of possible interesterification and hydrolysis products for the sample reaction between tristearin and triolein, catalyzed by a nonspecific lipase, is shown below.

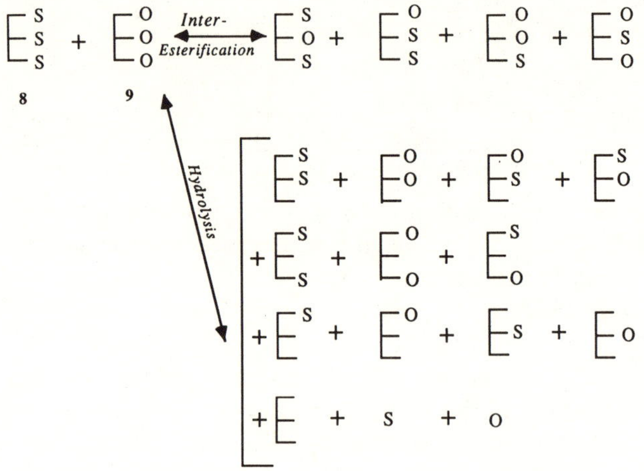

(5)

Exactly how much water is needed to maintain enzyme activity and maximize selectivity to interesterification over hydrolysis remains an unanswered question, although some progress has been made.[62]

The ability of lipases to carry out transesterifications, interesterifications, and glyceride synthesis has been reported by several researchers.[5, 9, 50, 61–67] Progress in high-molecular-weight triglyceride interesterification has been hampered somewhat by the difficulty in analysis; most analytical procedures involve thin-layer chromatography to separate the partial glycerides from the triglycerides, perhaps separation of the triglycerides by HPLC, and conversion of the triglyceride constituents to the fatty acid methyl esters which are then analyzed by gas chromatography. Recent advances in high temperature gas chromatography and in cold on-column injection techniques, have led to procedures to resolve triglycerides using gas chromatography. The figures below show chromatograms illustrating the interesterification between tripalmitin (PPP), and stearic (S) and linoleic (L) acid (Figure 4), and the interesterification of hydrogenated cottonseed oil with olive oil (Figure 5).[68] A Chrompack 15m high-temperature capillary column with a phenylmethylsilicone stationary phase was used in analysis. The enzyme used was immobilized lipase from *Mucor miehei* supplied by Novo (Lipozyme). The lipase is 1,3-specific; as Figure 5 indicates, the initial products reveal the 1,3-specificity of the enzyme, but with extended time, acyl migration of the fatty acid moieties tends to reveal a more uniform distribution of all possible products.

Finally, we cite a few other lipase-catalyzed reactions of importance to chemical synthesis. Chen *et al.*[69] have shown that lipase can catalyze the

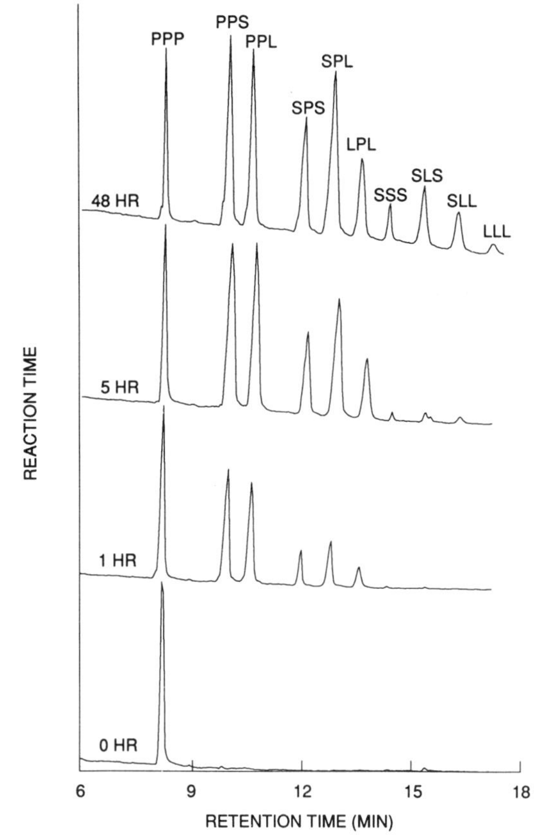

Figure 4. Interesterification between tripalmitin and stearic and linoleic acids. High temperature chromatogram (oven temperature 350–375 °C).

condensation of diacids with diols; intramolecular esterification leads to dilactones. The feasibility of peptide synthesis has been demonstrated by Margolin and Klibanov.[44] Esterification of hydroxy amino acids with fatty acids again to produce biocompatible emulsifiers has been shown to be feasible.[70] Abramowicz and Keese[71] have synthesized polycarbonates through the condensation reaction between diphenyl carbonate (11) and a number of bifunctional alcohols. The reaction between 11 and 4,4′-isopropylidenediphenol (10), for example, results in polycarbonates of commercial value as high-performance plastics (e.g., Lexan).

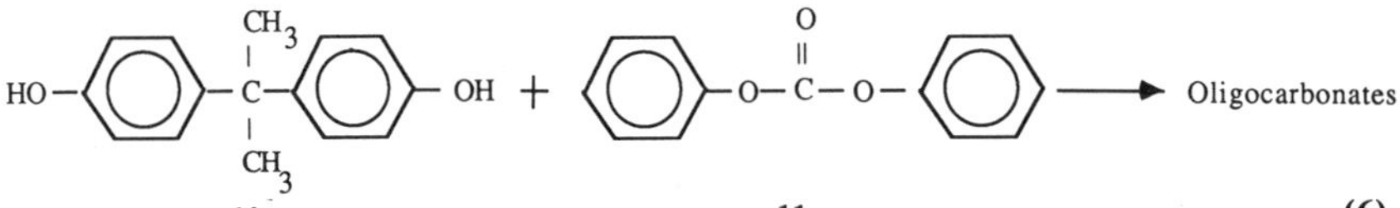

10 **11** **(6)**

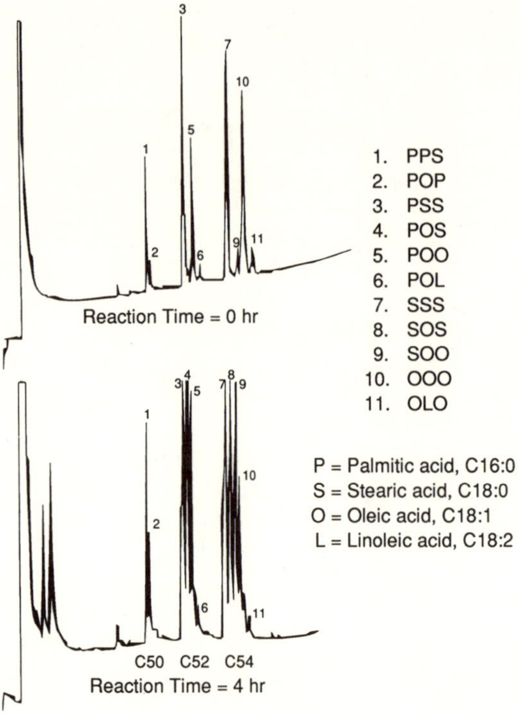

Figure 5. Olive oil/hydrogenated cottonseed oil. High temperature chromatogram.

The ability of lipases to accept not only water, but other nucleophiles such as alcohols, amines, thiols, and so on, implies a vast synthesis potential with such enzymes.

We now turn to a brief review of some of the aspects important to lipase activity in organic solvents. Enzyme activation can be established using several methods, the simplest being to allow the enzyme to attain its optimal conformation in aqueous solution at optimal pH and then to remove excess water through lyophilization, which leaves the essential water of hydration intact. The dependence of enzyme activity on the pH of the aqueous solution from which the enzyme was initially treated has been clearly demonstrated by Zaks and Klibanov.[39] Enzyme immobilization on various supports, such as Celite, helps keep the enzyme well-dispersed; research groups in Japan have pioneered the use of lipase immobilized in hydrophobic or hydrophilic polymer matrices.[47–51, 72] There appears to be no immediate advantage of this procedure over simple immobilization over solid supports. However, it is possible that intra-gel substrate diffusional limitations may lead to substrate concentrations at the enzyme interface being different from those in the bulk. The concept may be used advantageously in competitive reactions between two substrates and in manipulating selectivities in series and parallel reactions perhaps involving multi-enzyme systems.

Zaks and Klibanov[73] have demonstrated the remarkable thermal stability of lipases (up to 100 °C) in organic media. Such thermal stability is a general feature of enzymes in organic solvents and has been attributed to the fact that the minimal water content of the system is insufficient to cause protein denaturation by hydrolysis at aspartic acid residues and breakage of disulfide bonds at higher temperatures.[38] The general nature of biocatalysis in organic solvents has been well discussed by Klibanov[38, 74] and in a comprehensive review by Dordick.[75]

One very interesting aspect of lipase kinetics in organic solvents developed by Triantaphylides and co-workers[54, 56] is worthy of note. These authors studied competitive reaction in reaction systems containing two acylated substrates and one nucleophile, or one acyl substrate with two nucleophiles. The general scheme presented is

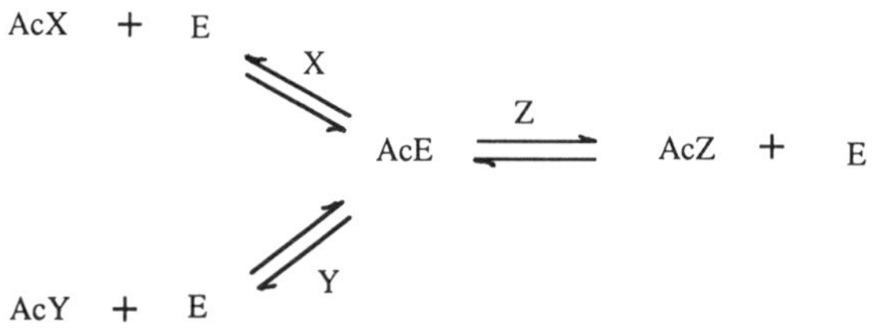

(7)

where AcX, AcY, and AcZ are acyl donors (acids or esters), and X, Y, Z are acceptors (alcohols). The authors found that competitive reaction between two acyl donors AcX and AcY was independent of the acceptor Z used. For example, the ratio of reaction efficiencies for the competitive reaction between cyclohexyl butyrate (AcX) and isopentyl butyrate (AcY) was the same regardless of whether geraniol, lauryl alcohol, or (−)menthol was used as the acyl acceptor Z. The reverse side of the concept is that competitive reaction between two acyl acceptors X and Y (e.g., geraniol and nerol) is independent of the type of acyl donor (AcZ) used (e.g., tributyrin, isopentyl butyrate, and butyric acid). A qualitative rationale for the phenomenon is obtained by considering the enzyme to consist of acylated enzyme and free enzyme. Competition for adsorption sites between two acyl donors determines the reaction efficiency with regard to a single alcohol. On the other hand, competition between two alcohols (X and Y) to accept the acyl group at an acylated site (AcE) determines the relative effectiveness of compound formation from a single acyl donor (AcZ) and the two nucleophiles (X and Y). Both adsorption to the active site and solvolysis at the site are rate-determining steps in the overall sequence. A comprehensive kinetic model based on such considerations has been developed.[56]

The choice of solvents is an important consideration in activation and activity maintenance in organic solvents. Early work by Zaks and Klibanov[39] led to the conclusion that apolar solvents are much preferred over highly polar solvents. The rationale is simply the fact that apolar solvents have no affinity for water and thus do not disturb the hydration layer about the enzyme, whereas polar solvents remove the essential water of hydration, destroying enzyme conformation.[74] Laane and co-workers have defined a factor, P, to serve as a design parameter in the choice of solvents.[76, 77] P is simply the partition coefficient of the solvent in an *n*-octanol-water system. Through analysis of existing data on solvent effects,[39, 78] the authors conclude that $\log P$ values greater than 4 are indicative of a solvent that does not interfere with the enzyme's essential water. The results with solvents with $\log P$ values between 2 and 4 are quite variable; catalytic activity depends on the ability of the individual enzyme species to retain its essential water. Most enzymes are inactive when placed in solvents with $\log P$ less than 2. There are exceptions however, to the classification of solvents through their $\log P$ values. The recent study by Abramowicz and Keese[71] on polycarbonate synthesis using lipase from *C. cylindracea* indicates that synthesis efficiency is greater in methylene chloride than in an apolar solvent such as hexane. In light of the empiricism involved, the $\log P$ correlation should simply be interpreted as stating that apolar solvents help the enzyme retain its essential water, which in most cases enables the enzyme to maintain activity. Fundamental research on enzyme conformation and the dynamics of the active sites in the presence of various solvents is necessary to provide better solvent design parameters.

Knowledge of the exact role of water is as yet incomplete. For example, in ester synthesis, the effect of water is understood through the equilibrium constant

$$\begin{aligned} K &= a_{\text{ester}} a_{\text{w}} / a_{\text{acid}} a_{\text{alc}} \\ &= (x_{\text{ester}} x_{\text{w}} / x_{\text{acid}} x_{\text{alc}})(\gamma_{\text{ester}} \gamma_{\text{w}} / \gamma_{\text{acid}} \gamma_{\text{alc}}) \\ &= K_x K_\gamma \end{aligned} \tag{8}$$

where the activities of the different species, a, can be expressed as the product of the mole fractions x, and the activity coefficients, γ. High yields of the ester can be obtained by reducing the activity of water either by reducing its concentration (x_w), or its activity coefficient (γ_w), i.e., by binding the water such that it is made unavailable for reaction. Reducing water content has its limits since sufficient water must be present to hydrate the enzyme. The situation is even more interesting in the case of competitive reactions such as triglyceride interesterification vs hydrolysis, where the presence of excess water increases the undesired selectivity to hydrolysis. There is no definitive conclusion as to how much water is needed to maintain lipase activity while minimizing the selectivity to hydrolysis. Goderis *et al.*[62] have provided insight to the problem by studying the effects of varying the water activity through changing the relative humidity or the partial pressure of water in equilibrium with the reaction system. Using interesterification between palmitic acid and triolein as the model system, these authors have shown that at low equilibrium partial pressures hydrolysis levels are reduced, but so is the rate of interesterification. The observation lends validity to the hypothesis that the interesterification is really the sequence of hydrolysis (of triolein to diolein) and ester synthesis (of the diolein with palmitic acid). The reduction of water activity has also been discussed by Halling.[79, 80] Operating under reduced pressure or in a closed system where the solvent is continually recycled through molecular sieves to remove saturation moisture levels, are methods used to reduce water activity. Further work needs to be done like perhaps introducing various water adsorbents into the reaction system to bind excess water and determining their effects not only on lipase activity, but also on selectivity.

We next turn to biphasic reversed micellar systems as an alternative mode for substrate enzyme contact.

10.3.2. Lipases in Reversed Micelles

In recent years, extensive studies have been carried out on protein-containing reversed micelles for their biomembrane mimetic properties[81] and for their potential to biocatalysis[82–86] and protein separation processes.[87, 88] While several papers have been written on enzymes in reversed micelles with references 82–86 providing good reviews of such work, relatively few have dealt with lipases. Rhee and co-workers[89–92] have extensively studied triglyceride hydrolysis using lipases in reversed micelles. At first glance it is hard to justify why hydrolysis should be carried out in reversed micellar systems rather than in bulk aqueous systems. Yet the potential advantage of conducting hydrolysis in reversed micelles is the reduction of substrate and product inhibition during reaction. In reversed micelles, the lipid partial glycerides and fatty acids that are formed during hydrolysis diffuse back into the bulk organic solution, in theory leaving the catalytic sites accessible to the triglyceride substrate. Han and Rhee[89] show that increasing the water content of the reversed micelles increases the reaction rate, but no direct comparison of the aqueous system vs the reversed micellar system is reported. The work by Robinson and co-workers[93, 94] is a direct comparison of hydrolysis in reversed micelles and in bulk aqueous solutions, using *p*-nitrophenyl alkanoate esters of varying chain length as substrates. These authors have found that reaction in reversed micelles is inhibited compared to reaction in aqueous systems. The rationale proposed is that while the lipase is as active in reversed micelles as in aqueous or oil-in-water emulsion systems, the partitioning of substrate into the organic phase adversely affects the reaction rate. Thus transport of substrate to the oil–water interface and thence to the catalytic sites may be the rate-determining factor. Further work appears necessary if a conclusion is to be drawn that all hydrolytic reactions are better catalyzed in bulk aqueous solutions than in reversed micelles.

Other fascinating aspects of lipase behavior in reversed micelles are discussed in Rhee's papers.[89–92] Among them is the observation that inhibitors and promotors of lipase activities in physiological environments play a role in regulating lipase activity in reversed micelles. Histidine and glycine help maintain activity, while the cations Cu^{2+}, Hg^{2+}, and Fe^{3+} inhibit activity. An interesting observation is that Ca^{2+} which helps maintain lipase activity in bulk aqueous solutions, has relatively little effect on lipases in reversed micelles.

Some work on ester synthesis using lipases in reversed micelles has been carried out, and the literature reports the feasibility of the reaction.[93–95] In synthesizing mono-, di-, and triglycerides from glycerol and long-chain fatty acids, Morita *et al.*[96] report good yields of the partial glycerides but only a small yield of triglycerides; the observation appears to be corroborated by Fletcher *et al.*[95] Again, the analytical difficulties involved in triglyceride identification made unambiguous determination of synthesis efficiency difficult. Further research using high-temperature gas chromatography may help clarify the ability of lipase to synthesize triglycerides in reversed micellar solutions. A point of interest is the use by Morita *et al.*[96] of phosphatidylcholine (POPC) rather than bis(2-ethylexyl) sodium sulfosuccinate (AOT). A noteworthy point is the use of phosphatidylcholine as the surfactant in the glyceride synthesis studies of Morita *et al.*[96] Standard practice in studies with reversed micelles is to use the anionic surfactant bis(2-ethylhexyl) sodium sulfosuccinate (AOT) purely as a model surfactant to provide conformity among various research groups. The use of biodigestible and biocompatible compounds to form reversed micelles is indicative of the trend to extend research on proteins in reversed micelles to real applications. It is also interesting to note that while lipase may not bind well to the interface in the presence of POPC, as shown by Brockman's surface studies,[30–32] it does exhibit activity in POPC reversed micelles.

An important observation relevant to ester synthesis in reversed micelles is the fact that synthesis can be carried with high yields,[94,95] even though the water content is an order of magnitude greater than that in the alternative system comprising lipase in monophasic organic solvents. The fact that the reaction does not appear to be limited by equilibrium is an indication that the water in reversed micelles is bound and has reduced activity compared to bulk water. Direct comparisons of reversed micellar systems with monophasic organic solvent systems both for lipase activity and for selectivity between hydrolysis and trans/interesterifications would be most useful.

Another useful observation related to several enzymes in reversed micelles is the ability to control activity through adjustment of the water to surfactant molar ratio w_0. Lipases fall under the category of enzymes that have an optimal w_0 for maximum activity.[90] Figure 6 illustrates the activity variance with w_0 for the synthesis of butylpalmitate from palmitic acid and butanol.[68] Kabanov *et al.*[97] have proposed a rationale which has intuitive validity. These authors state that at low w_0, the tightly held enzyme has an inactive conformation. On the other hand, at high w_0, the enzyme approaches behavior in pure water, where natural fluctuations in conformation imply that some substrate–enzyme contact takes place when the enzyme is in an inactive conformation. The authors cite several instances where reaction rate in reversed micelles exceeds that in pure water. Again, a direct comparison of lipase hydrolysis in reversed micelles and in bulk aqueous solutions would be of use.

The importance of the micellar environment in affecting lipase activity was also realized through some recent experiments.[68] It was found that in encapsulating lipase from *C. cylindracea* into reversed micelles from the solid powder, the enzyme was inactive unless initially contacted with the

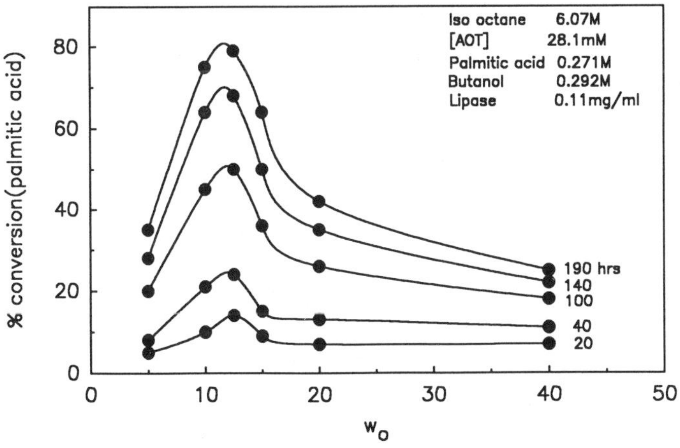

Figure 6. Effect of the [water]/[AOT] ratio (w_0) on lipase activity.

acyl substrate. Figure 7 illustrates the concept for synthesis of octyl palmitate; significant variations in enzyme activity are observed depending on the method of reaction preparation. The (...) symbol in the legend indicates all intermediate steps of reversed micelle preparation (adding isooctane, AOT, water, lipase-stirring, etc.). For example, the plot with the filled squares refers to the preparation wherein palmitic acid was added first followed by isooctane, AOT, water, and lipase. The system was stirred to encapsulate the lipase and form micelles. Excess lipase was removed by centrifugation and the reaction was then started by adding the second substrate, octanol.

It is seen that this preparation method where the acyl substrate is present at the beginning and the alcohol substrate is added at the end, leads to the most active system. Acyl-enzyme coupling prior to and during encapsulation is thus beneficial to maintaining activity; the alcohol tends to inhibit the reaction, perhaps through a cosurfactant effect and competitive adsorption on interfacial sites. We have also found that lipase kept in contact with an acyl substrate maintains its activity for much longer times than the noncoupled lipase.

The question arises as to whether lipases or, for that matter, any enzyme, should be used in the environment of a reversed micelle rather than in monophasic organic solvents for chemical synthesis. The answer is really based on the distinction between homogeneous and heterogeneous catalysis; one is assured of excellent catalyst dispersion in homogeneous systems, while the heterogeneous system is far superior where ease of product separation from catalyst and catalyst recovery are concerned. One

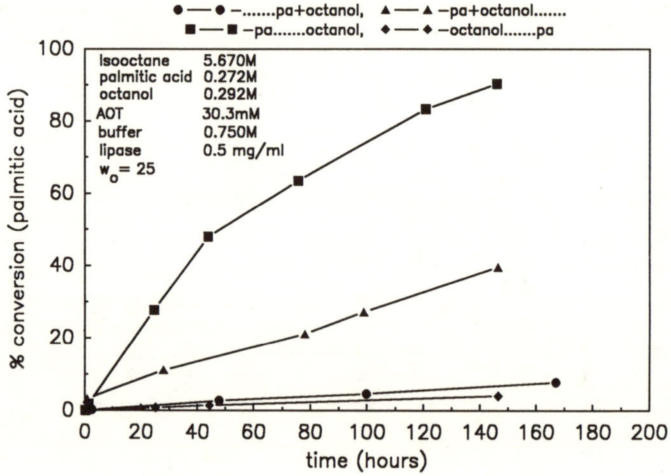

Figure 7. Effect of the order of substrate addition on lipase activity.

point in the reversed micelles favor is the ability to control catalytic activity through manipulation of w_0 values; in the monophasic systems enzyme activity seems to be more of an on–off situation. Whether manipulation of w_0 could lead to control in stereoselective and regioselective synthesis is an interesting question to be considered, and one certainly worth looking at.

10.4. FUTURE DIRECTIONS

It is thus evident that lipases are efficient catalysts for reactions involving the ester bond. The regio- and stereoselectivity characteristics of lipase catalysis make it possible to synthesize a variety of esters. Most research is directed therefore at the synthesis of novel esters and there is an unlimited range of possibilities. For example, can the ω-3 fatty acids of fish oils be incorporated into the triglycerides of vegetable oils without flavor degradation? Can biodegradable surfactants be synthesized economically by lipase catalysis? Such surfactants would be of tremendous importance in large-scale applications of detergents, in enhanced oil recovery, etc., i.e., where there is significant concern over the lingering effects of surfactants on the environment. Other biocompatible surfactants would find applications in the food industry. Werdelmann and Schmid[8] cite unusual bioemulsifiers produced by microorganisms—the trehalose lipids and rhamnolipids, for example. Since the ability of lipases to synthesize sugar esters has been recognized, it may be possible to synthesize such specialty compounds in larger quantities than through fermentation.

Lipase-catalyzed ester hydrolysis, while a more mature area, still does have application possibilities. Among the more recent and unusual ones is the use of lipases in contact lens cleaners and as additives in windshield cleaning fluids. Lipases are finding increased usage in detergent formulations and in partial glyceride preparations for the food industry.

Along with research on novel synthesis, it is important to carry out fundamental work on the physicochemical factors affecting lipase activity as a preliminary step to reactor development and industrial scale-up. In most research reports, lipases are categorized as being active or inactive for particular reactions. Continuing work must attempt to develop a concerted view on the factors that regulate activity. The choice of organic solvents, the thermodynamic conditions imposed, prior conditioning of lipase before use in solvents, are all factors that need to be considered. While lipases are usually active in hydrophobic solvents, such solvents may be inappropriate in applications to the food industry if their complete removal is difficult. In such cases, the use of supercritical solvents may be beneficial. The use of enzymes in supercritical solvents is a nascent field[98] of much promise. While enzyme activity in supercritical solvents has been established, activity maintenance, retention of enzyme conformation during pressure let-down, separation of substrate and products from the enzyme are all factors that need further research.

Thus there is scope for much further research on the factors that affect lipase catalysis. Fundamental correlations between solvent type, water content, and enzyme conformation still need to be established. With reversed micellar systems, we have the additional interesting concept of regulating catalysis by controlling micellar size and aggregation number. Location of lipase in the micelles and conformation at various surfactant to water ratios correlated to activity are important aspects to be addressed. Again, the applicability of findings from the literature on lipases in physiological environments may be important in using the enzyme as an industrial biocatalyst.

Reactor development and scale-up are other factors that need to be addressed to translate novel synthesis to commercial possibilities. Here it is not just activation of lipase that is important, but rather the maintenance of activity for long-time periods to permit continuous operation. Methods of immobilization, the thermodynamic variables of flow, temperature and pressure, are all relevant aspects that need to be understood for optimal operation.

ACKNOWLEDGMENT

Mr. Madhusudhan Rao and Dr. Michael Murray carried out some of the experiments described in this chapter. Support by the National Science Foundation during this period is gratefully acknowledged by VTJ.

REFERENCES

1. H. Brockerhoff and R. G. Jensen, *Lipolytic Enzymes*, Academic Press, New York (1974).
2. C. Ratledge, *Fette, Seifen, Anstrichm. 86(10)*, 379 (1984).
3. A. Kilara, *Process Biochem.*, p. 35, April (1985).
4. E. W. Seitz, *JAOCS 51*, 12 (1974).
5. A. R. Macrae, in *Proc. World Conference on Emerging Technologies in the Fats and Oils Industry* (A. R. Baldwin, ed.), p. 7, Amer. Oil Chem. Soc., Champaign, Illinois (1986).
6. T. Nielsen, *Fette, Seifen, Anstrichm. 87(1)*, 15 (1985).
7. T. Yamane, *JAOCS 64(10)*, 1657 (1987).
8. B. W. Werdelmann and R. D. Schmid, *Fette, Seifen, Anstrichm. 84(11)*, 436 (1982).
9. A. R. Macrae, in: *Microbial Enzymes and Technology* (W. M. Fogarty, ed.), p. 225, Elsevier, Amsterdam (1983).
10. S. Ishida, in: *Proc. World Conference on Emerging Technologies in the Fats and Oils Industry* (A. R. Baldwin, ed.), p. 359, Amer. Oil Chem. Soc., Champaign, Illinois (1986).

11. R. H. Potts and V. J. Mukerheide, in: *Fatty Acids and Their Industrial Applications* (E. S. Pattison, ed.), p. 28, Marcel Dekker, New York (1968).
12. W. M. Linfield, D. J. O'Brien, S. Serota, and R. A. Barauskas *JAOCS 61(6)*, 1067 (1984).
13. L. H. Posorske, *JAOCS 61(11)*, 1758 (1984).
14. R. G. Jensen, S. A. Gerrior, M. M. Hagerty, and K. E. McMahon, *JAOCS 55*, 422 (1978).
15. Nippon Oil and Fats Co. Ltd., Japanese Patent 59-14793 (1984).
16. H. Brockerhoff, *Arch. Biochem. Biophys. 110*, 586 (1965).
17. M. M. Hoq, T. Yamane, S. Shimizu, T. Funada, and S. Ishida, *JAOCS 62(6)*, 1016 (1985).
18. G. Bell, J. R. Todd, J. A. Blain, J. D. E. Patterson, and C. E. L. Shaw, *Biotech. Bioeng. 23*, 1703 (1981).
19. M. M. Hoq, M. Koike, T. Yamane, and S. Shimizu, *Agric. Biol. Chem. 49(11)*, 3171 (1985).
20. L. Sarda and P. Desnuelle, *Biochim. Biophys. Acta 30*, 513 (1958).
21. B. Entressangles and P. Desnuelle, *Biochim. Biophys. Acta 159*, 285 (1968).
22. H. L. Brockman, J. H. Law, and F. J. Kezdy, *J. Biol. Chem. 248(14)*, 4965 (1973).
23. R. Verger, in: *Lipases* (B. Borgstrom and H. L. Brockman, eds.), Elsevier, New York (1984).
24. H. Brockerhoff, *Chem. Phys. Lipids 10*, 215 (1973).
25. C. Chapus, M. Semeriva, C. Bovier-Lapierre, and P. Desnuelle, *Biochemistry 15(23)*, 4980 (1976).
26. M. Iwai, Y. Tsujisaka, and J. Fukumoto, *J. Gen. Appl. Microbiol. 10*, 87 (1964).
27. W. E. Momsen and H. L. Brockman, *J. Biol. Chem. 25(2)*, 378 (1976).
28. M. Iwai and Y. Tsujisaka, in: *Lipases* (B. Borgstrom and H. L. Brockman, eds.), Elsevier, New York (1984).
29. H. L. Brockman, in: *Lipases* (B. Borgstrom and H. L. Brockman, eds.), Elsevier, New York (1984).
30. T. Tsujita and H. L. Brockman, *Biochemistry 26*, 8423 (1987).
31. T. Tsujita, J. M. Smaby, and H. L. Brockman, *Biochemistry 26*, 8430 (1987).
32. H. L. Brockman, in: *Proc. Symp. on the Biology, Biochemistry and Technology of Lipases* (S. Stegink and G. Abraham, eds.), p. 1, Amer. Oil Chem. Soc., Champaign, Illinois (1987).
33. W. E. Ladner and G. M. Whitesides, *J. Am. Chem. Soc. 106*, 7250 (1984).
34. A. Akiyama, M. Bednarski, M.-J. Kim, E. S. Simon, H. Waldmann, and G. M. Whitesides, *Chemistry in Britain*, p. 647, July (1987).
35. M. Iwai, Y. Tsujisaka, and J. Fukumoto, *J. Gen. Appl. Microbiol. 10*, 13 (1964).
36. S. Okumura, M. Iwai, and Y. Tsujisaka, *Biochim. Biophys. Acta 575*, 156 (1979).
37. M. M. Hoq, H. Tagami, T. Yamane, and S. Shimizu, *Agric. Biol. Chem. 49(2)*, 335 (1985).
38. A. M. Klibanov, in: *Protein Engineering. Applications in Science, Medicine, and Industry* (M. Inouye and R. Sarma, eds.), p. 341, Academic Press (1986).
39. A. Zaks and A. M. Klibanov, *Proc. Natl. Acad. Sci. USA 82*, 3192 (1985).
40. J. Chopineau, F. D. McCafferty, M. Therisod, and A. M. Klibanov, *Biotech. Bioeng. 31*, 208 (1988).
41. P. Cesti, A. Zaks, and A. M. Klibanov, *Appl. Biochem. Biotech. 11*, 401 (1985).
42. B. Cambou and A. M. Klibanov, *J. Am. Chem. Soc. 106*, 2687 (1984).
43. G. Kirchner, M. P. Scollar, and A. M. Klibanov, *J. Am. Chem. Soc. 107*, 7072 (1985).
44. A. L. Margolin and A. M. Klibanov, *J. Am. Chem. Soc. 109*, 3802 (1987).
45. M. Therisod and A. M. Klibanov, *J. Am. Chem. Soc. 109*, 3977 (1987).
46. A. L. Margolin, J.-Y. Crenne, and A. M. Klibanov, *Tetrahedron Lett. 28(15)*, 1607 (1987).
47. S. Koshiro, K. Sonomoto, A. Tanaka, and S. Fukui, *J. Biotech. 2*, 47 (1985).
48. M. Kawase and A. Tanaka, *Biotech. Lett. 10(6)*, 393 (1988).

49. A. Tanaka and S. Fukui, in: *Enzymes and Immobilized Cells Biotechnology. Biotech. Ser. 5*, 149 (1985).
50. K. Yokozeki, S. Yamanaka, K. Takinami, Y. Hirose, A. Tanaka, and K. Sonomoto, *Eur. J. Appl. Microbiol. Biotechnol. 14*, 1 (1982).
51. T. Kawamoto, K. Sonomoto, and A. Tanaka, *Biocatalysis 1*, 137 (1987).
52. G. Langrand, J. Baratti, G. Buono, and C. Triantaphylides, *Tetrahedron Lett. 27(1)*, 29 (1986).
53. J. Baratti, G. Buono, H. Deleuze, G. Langrand, M. Secchi, and C. Triantaphylides, in: *Proc. World Conference on Emerging Technologies in the Fats and Oils Industry* (A. R. Baldwin, ed.), p. 355, Am. Oil Chem. Soc., Champaign, Illinois (1985).
54. H. Deleuze, G. Langrand, H. Millet, J. Baratti, G. Buono, and G. Triantaphylides, *Biochim. Biophys. Acta 911*, 117 (1987).
55. G. Langrand, C. Triantaphylides, and J. Baratti, *Biotech. Lett. 10(8)*, 549 (1988).
56. G. Langrand, J. Baratti, G. Buono, and C. Triantaphylides, *Biocatalysis 1*, 231 (1988).
57. A. M. Klibanov and B. Cambou, *Methods in Enzymology 136*, 117 (1987).
58. T. Sakurai, A. L. Margolin, A. J. Russel, and A. M. Klibanov, *J. Am. Chem. Soc. 110*, 7236 (1988).
59. M. Therisod and A. M. Klibanov, *J. Am. Chem. Soc. 108*, 5638 (1986).
60. H. Seino, T. Uchibori, T. Nishitani, and S. Inamasu, *JAOCS 61(11)*, 1761 (1984).
61. A. R. Macrae, *JAOCS 60(2)*, 291 (1983).
62. H. L. Goderis, G. Ampe, M. P. Feyten, B. L. Fouwe, W. M. Guffens, S. M. Van Cauwenbergh, and P. P. Tobback, *Biotech. Bioeng. 30*, 258 (1987).
63. R. Schuch and K. D. Mukherjee, *J. Agric. Food Chem. 35*, 1005 (1987).
64. T. T. Hansen and P. Eigtved, *Proc. World Conference on Emerging Technologies in the Fats and Oils Industry* (A. R. Baldwin, ed.), p. 414, Am. Oil Chem. Soc., Champaign, Illinois (1985).
65. R. W. Stevenson, F. E. Luddy, and H. L. Rothbart, *JAOCS 56*, 676 (1979).
66. D. K. Bhattacharya, S. Majumdar, and S. Khatoon, *Proc. World Conference on Emerging Technologies in the Fats and Oils Industry* (A. R. Baldwin, ed.), p. 414, Am. Oil Chem. Soc., Champaign, Illinois (1985).
66a. T. Tanaka, E. Ono, M. Ishihara, S. Yamanaka, and K. Takinami, *Agric. Biol. Chem. 45*, 2387 (1981).
67. G. Abraham, M. A. Murray, and V. T. John, *Biotech. Lett. 10(8)*, 555 (1988).
68. M. Rao, M. A. Murray, G. Abraham, and V. T. John, in press (1991).
69. C. S. Chen, S.-H. Wu, G. Girdaukas, and C. J. Sih, *J. Am. Chem. Soc. 109*, 2812 (1987).
70. A. Nagao and M. Kito, *JAOCS 66(5)*, 710 (1989).
71. D. A. Abramowicz and C. R. Keese, *Biotech. Bioeng. 33*, 149 (1989).
72. Y. Inada, H. Nishimura, K. Takasaki, T. Yoshimoto, A. R. Saha, and Y. Saito, *Biochem. Biophys. Res. Commun. 122(2)*, 845 (1984).
73. A. Zaks and A. M. Klibanov, *Science 224*, 1249 (1984).
74. A. M. Klibanov, *Chemtech.*, p. 354, June (1986).
75. J. S. Dordick, *Enzyme Microb. Technol. 11*, 194 (1989).
76. C. Laane, S. Boeren, and K. Vos, *Trends in Biotech. 3(10)*, 251 (1985).
77. C. Laane, S. Boeren, K. Vos, and C. Veeger, *Biotechnol. Bioeng. 30*, 81 (1987).
78. L. E. S. Brink and J. Tramper, *Biotechnol. Bioeng. 27*, 1258 (1985).
79. P. J. Halling, in: *Biocatalysis in Organic Media* (C. Laane, J. Tramper, and M. D. Lilly, eds.), p. 125, Elsevier, Amsterdam (1986).
80. P. J. Halling, *Biocatalysis 1*, 109 (1987).
81. J. Wirz and J. P. Rosenbusch, in: *Reverse Micelles* (P. L. Luisi and B. E. Straub, eds.), Plenum Press, New York (1984).
82. M. E. Leser, G. Wei, P. Lüthi, G. Haering, A. Hochkoeppler, E. Blöchliger, and P. L. Luisi, *Journal de chimie physique 84(9)*, 1113 (1987).

83. K. Martinek, I. V. Berezin, Yu. L. Khmelnitski, N. L. Klyachko, and A. V. Levashov, *Biocatalysis 1*, 9 (1987).
84. L. Magid, P. Walde, G. Zampieri, E. Battistel, Q. Peng, E. Trotta, M. Maestro, and P. L. Luisi, *Colloids and Surfaces 30*, 193 (1988).
85. P. L. Luisi, M. Giomini, M. P. Pileni, and B. H. Robinson, *Biochim. Biophys. Acta 947*, 204 (1988).
86. J. W. Shield, H. D. Ferguson, A. S. Bommarius, and T. A. Hatton, *Ind. Eng. Chem. Fundam. 25(4)*, 605 (1986).
87. M. E. Leser, G. Wei, P. L. Luisi, and M. Maestro, *Biochem. Biophys. Commun. 135*, 629 (1986).
88. R. S. Rahman, J. Y. Chee, J. M. S. Cabral, and T. A. Hatton, *Biotech. Prog. 4*, 218 (1988).
89. D. Han and J. S. Rhee, *Biotech. Lett. 7(9)*, 651 (1985).
90. D. Han and J. S. Rhee, *Biotechnol. Bioeng. 28*, 1250 (1986).
91. D. Han, J. S. Rhee, and S. B. Lee, *Biotechnol. Bioeng. 30*, 381 (1987).
92. D. Han, D. Y. Kwon, and J. S. Rhee, *Agric. Biol. Chem. 51(2)*, 615 (1987).
93. P. D. I. Fletcher, B. H. Robinson, R. B. Freedman, and C. Oldfield, *J. Chem. Soc., Faraday Trans. 1, 81*, 2667 (1985).
94. C. Oldfield, G. D. Rees, B. H. Robinson, and R. B. Freedman, in: *Biocatalysis in Organic Media* (C. Laane, J. Tramper, and M. D. Lilly, eds.), p. 119, Elsevier, Amsterdam (1987).
95. P. D. I. Fletcher, R. B. Freedman, B. H. Robinson, G. D. Rees, and R. Schomäcker, *Biochim. Biophys. Acta 912*, 278 (1987).
96. S. Morita, H. Narita, T. Matoba, and M. Kito, *JAOCS 61(10)*, 1571 (1984).
97. A. V. Kabanov, A. V. Levashov, N. L. Klyachko, A. V. Pshezhetesky, and K. Martinek, *J. Theor. Biol. 133*, 327 (1988).
98. T. W. Randolph, D. S. Clark, H. W. Blanch, and J. M. Prausnitz, *Science 238*, 387 (1988).

11

Biocatalysis in Supercritical Fluids

THEODORE W. RANDOLPH, HARVEY W. BLANCH, and DOUGLAS S. CLARK

11.1. INTRODUCTION

General industrial application of biocatalysis has long been hindered by the nearly exclusive use of water as a solvent, although it has been known for some time that biocatalysis can occur in nonaqueous environments. Recent interest in nonaqueous systems for biocatalysis has been kindled by the work of Klibanov and co-workers,[1–3] who showed that enzymatic catalysis can take place even in nearly anhydrous organics. Since the appearance of that work, many different solvent systems have been investigated as media for biocatalysis, and general rules for solvent selection have been proposed.[4–7]

One result of the search for suitable solvent systems for biocatalysis is the possibility of biocatalysis in supercritical fluids. Supercritical fluids are those fluids which exist at a temperature above the fluid's critical temperature (T_c) and at a pressure above the fluid's critical pressure (P_c).

While supercritical fluids have been known for over 100 years,[8,9] the patents of Zosel[10–13] at the Max Planck Institut für Kohlenforschung in the early 1970s sparked a renewed wave of interest in supercritical fluid pro-

THEODORE W. RANDOLPH • Department of Chemical Engineering, Yale University, New Haven, Connecticut 06520. HARVEY W. BLANCH and DOUGLAS S. CLARK • Department of Chemical Engineering, University of California, Berkeley, Berkeley, California 94720.

cessing, especially for biological materials. Supercritical fluid processes have been developed for the extraction of many biochemicals, including caffeine from coffee,[13] odor compounds and fatty acids from plant oils,[11] hop-oils, alpha-acids and beta-acids from hops,[14] nicotine from tobacco,[15] and essential oils and flavors (piperine, eugenol, aceteugenol, vanillin, and cinnamaldehyde) from spices.[16]

While most attention in the literature on supercritical fluid processing has been devoted to the advantages of supercritical fluid solvents over liquid solvents for extraction purposes (reviews on the subject include those by Randall,[17] Paulaitis *et al.*,[18] Brunner and Peter,[19] and McHugh and Krukonis[20]), reactive systems that utilize supercritical fluid solvents offer the same advantages, plus certain advantages that pertain only to reactive systems. These advantages are described in the following sections.

11.2. TRANSPORT PROPERTIES OF SUPERCRITICAL FLUIDS

The transport properties of supercritical fluids are, in general, intermediate between those of liquids and gases, and vary strongly with pressure and temperature, especially near the critical point. These properties have been exploited for applications involving extraction, but can also be advantageous for reactive systems.

11.2.1. Low Viscosity

One physical property which is favorable for reactions in the supercritical state is the viscosity. Supercritical fluid viscosities are about an order of magnitude less than liquid viscosities, resulting in reduced pressure drops and faster settling time for precipitates.[21] Near the critical point, viscosities increase rapidly with pressure. Supercritical fluid viscosities, unlike gas-phase viscosities, decrease with increasing temperature.

11.2.2. High Diffusivity

As expected from the low viscosities, diffusivities typically found in supercritical fluids are substantially larger than those found in liquids. Generally, an increase of one to two orders of magnitude in diffusion coefficients is observed between the liquid state and the supercritical fluid state. For example, diffusion coefficients of naphthalene and benzene in supercritical ethane or carbon dioxide (at reduced pressures of 1–5 and reduced temperatures of 1.0–1.2) are about $10^{-4}\ cm^2\ s^{-1}$, compared with diffusion coefficients in the liquid state of about $10^{-5}\,cm^2\,s^{-1}$.[18] Such high diffusivities may reduce or eliminate mass transport limitations on reactions in the supercritical state.

11.2.3. Liquid-like Density

The density of supercritical fluids is very sensitive to pressure, changing from gas-like densities ($1\,kg\,m^{-3}$) in the near-critical gaseous region to nearly liquid-like densities ($600\,kg\,m^{-3}$) in the supercritical region. It should be noted that the combination of low viscosity and liquid-like density results in kinematic viscosities even lower than those exhibited by liquid metals.

11.3. OTHER IMPORTANT SUPERCRITICAL FLUID PROPERTIES

There are several other potentially advantageous features of supercritical fluids. To date, little information exists in the literature concerning the exploitation of these features for purposes other than extraction, but it is apparent that they apply equally to reactive systems.

11.3.1. High Sensitivity of Solubilities to Pressure and Temperature

Probably the most important feature of supercritical fluids is the remarkable sensitivity of solubilities to small changes in pressure and tem-

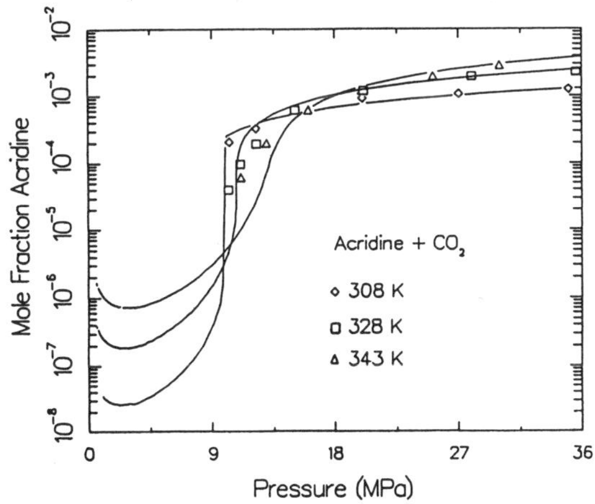

Figure 1. Experimental and calculated solubility of acridine in carbon dioxide. Reprinted with permission from ref. 22.

perature. Solubility changes of orders of magnitude may occur as a result of small (10–50 bar) changes in pressure. For example, Figure 1 shows the solubility of acridine in carbon dioxide as a function of pressure at several different temperatures.[22]

The high sensitivity of solubility to pressure may be exploited in separations steps in order to reduce energy costs, to produce small crystals of a solute with minimal solvent entrainment, or to change crystal morphology.[23–25]

11.3.2. Low Surface Tension

The surface tension of a supercritical fluid is nearly zero. Thus, a supercritical fluid solvent is able to penetrate smaller pores than those accessible to liquids. Such an ability to wet small pores, already used for drying of ceramics and impregnation of solid matrices, may prove to be of use in heterogeneous catalysis.

11.3.3. Low-Temperature Processing

A major advantage of supercritical fluid processing (especially for thermally unstable biomolecules) is the ability to operate at temperatures near room temperature with a variety of supercritical fluids. Because most processes operate at reduced temperatures of 1.0–1.2, the temperatures for separations steps may be substantially lower than those for typical liquid distillation. This feature of supercritical fluid processing is now in commercial use in supercritical fluid chromatography/mass spectroscopy (SFC/MS), where large, low-vapor-pressure molecules may be solubilized, separated, and analyzed without thermal degradation. Several potential supercritical solvents with temperatures near room temperature are listed in Table 1.

Table 1. Supercritical Fluids with Critical Temperatures near Room Temperature

Solvent	Critical temperature, °C	Critical pressure, atm
Carbon dioxide	31.1	73
Ethane	32.3	48
Ethylene	9.5	49
Nitrous oxide	36.5	70
Trifluoromethane	25.9	46

11.4. REACTIONS IN SUPERCRITICAL FLUIDS

The use of supercritical fluids as solvents for biological reactions is new, and relatively few data exist in the literature. It is expected, however, that information garnered from studies of nonbiological reactions in supercritical fluids will prove useful for biological catalysis as well. Chemical reactions in supercritical fluids have been reviewed by Subramaniam and McHugh.[26]

11.4.1. Effect of Pressure on Reaction Rate

Reactions in supercritical fluids may be controlled by changing either the temperature or the pressure. The effect of pressure on reaction rates is quite complex, but may be generally divided into three categories: pressure effects on concentrations of reactants and products in solution, pressure effects on mass transfer, and pressure effects on the reaction-rate constant itself.

Pressure effects on reaction rates caused by changes in concentration are well known and have been exploited for years in gas-phase reactions to control rates. The large changes in solubility with small pressure changes may result in corresponding changes in the reaction rate. Likewise, if the pressure dependence of density, viscosity, and diffusivity is known, the effect of pressure on mass-transfer limitations can be estimated.

On the other hand, the effect of pressure on the reaction-rate constant is difficult to predict, especially for biocatalytic reactions. While the theory of pressure effects on rate constants for biocatalytic reactions is not well developed, an approximation may be made by extrapolation from the existing theories for pressure effects on homogeneous reactions in ideal gases. Such a nonrigorous approach ignores many nonideal effects (e.g., solvent caging) and no longer retains its statistical-mechanical meaning, but it provides a useful basis for understanding reaction-rate constants in terms of "effective" activation volumes.

11.4.1.1. Eyring Transition-State Theory

The Eyring rate theory[27] shows that the rate constant k may be written as

$$k = r(k_B T/h) K^{\#} \tag{1}$$

where k_B is the Boltzmann constant, $K^{\#}$ is the equilibrium constant for reactant and the transition state, h is Plank's constant, and r is a pressure- and temperature-independent constant. The equation:

$$\Delta G^{\#} = \Delta H^{\#} - T\Delta S^{\#} = -RT \ln K^{\#} \tag{2}$$

can then be used to relate the transition-state-equilibrium constant (and thus the reaction-rate constant) to the free energy of activation, $\Delta G^{\#}$. Taking a derivative with pressure yields

$$-RT\left(\frac{\partial \ln k_x}{\partial P}\right) = \Delta\bar{v}^{\#} = \bar{v}_x^{\#} - \bar{v}_A - \bar{v}_B \tag{3}$$

where k_x is the rate constant in pressure-independent units, such as molality or mole-fraction units, $\bar{v}_x^{\#}$ is the partial molar volume of the activated complex, and $\bar{v}_A$ and $\bar{v}_B$ are the partial molar volumes of the two reactants. In concentration units,

$$k_c = k_x V_0^{-\Sigma_i \nu_i} \tag{4}$$

where V_0 is the molar volume of pure solvent if k_x is in mole-fraction units, or the specific volume of pure solvent if k_x is in molal units, and ν_i is the stoichiometric coefficient of the ith component (negative for reactants). Therefore,

$$-RT\left(\frac{\partial \ln k_c}{\partial P}\right) = -RT\left(\left(\frac{\partial \ln k_x}{\partial P}\right) - \sum_i \nu_i\left(\frac{\partial \ln V_0}{\partial P}\right)\right)$$

$$= \Delta\bar{v}^{\#} - RTB_0 \sum_i \nu_i \tag{5}$$

where B_0 is the solvent isothermal compressibility. Supercritical fluids near their critical points are highly compressible, and the compressibility term may be appreciable for all but unimolecular reactions where $\sum_i \nu_i = 0$. It must be noted that the activation volume $\Delta\bar{v}^{\#}$ is *not* independent of pressure. Thus, equations (3) and (5) cannot be directly integrated to give the effect of pressure on the reaction rate, unlike the corresponding derivation for temperature effects that leads to the Arrhenius activation energy, which is approximately independent of temperature.

To date, no activation volumes for biocatalytic reactions in supercritical fluids have been reported. It is expected, however, that the volumes may be large. Kim and Johnston[28] report an activation volume which reaches $-6000\ \text{cm}^3/\text{mol}$ for the unimolecular decomposition of α-chlorobenzyl methyl ether in 1,1-difluoroethane. This value is an order of magnitude larger than the largest previously reported value for a homogeneous reaction. Values for activation volumes in liquids are more typically $\pm 50\ \text{cm}^3/\text{mol}$.

11.4.2. Pressure Effects on Enzymatic Reactions

Interpretation of activation volumes for enzyme-catalyzed reactions is more complicated than for noncatalyzed reactions, and the effect of pressure on an enzymatic reaction is difficult to predict *a priori*. Care must always be exercised in interpreting reaction rate data in terms of activation volumes in order to avoid masking effects caused by mass-transfer limitations.

As an example of the possible complexities introduced in an enzymatic reaction system as a function of pressure, a simple one-substrate, one-product reaction has been considered (Morild[29]). The reaction may be written as

$$E+S \underset{k_{-1}}{\overset{k_1}{\longleftrightarrow}} ES \underset{k_{-2}}{\overset{k_2}{\longleftrightarrow}} EP \underset{k_{-3}}{\overset{k_3}{\longleftrightarrow}} E+P \tag{6}$$

If k_{-2} and k_{-3} are small enough to be neglected, the rate equation may be expressed as

$$v = \frac{k_2[\mathrm{E}]_0\,[\mathrm{S}]}{((k_{-1}+k_2)/k_1)+((k_2+k_3)/k_3)[\mathrm{S}]} \tag{7}$$

Taking appropriate pressure derivatives and assuming k's are written in mole-fraction or molality units, the observed activation volume for the overall reaction may be shown to be

$$\Delta\bar{v}_0^{\#} = \frac{k_3\,\Delta\bar{v}_2^{\#} + k_2\,\Delta\bar{v}_3^{\#}}{k_2+k_3} \tag{8}$$

If k's are written in pressure-independent units and the reaction step ES to EP is rate-limiting, the reaction rate as a function of pressure may be written as

$$v(p) = k_2(0)[\mathrm{E}]_0 \exp(-p\,\Delta\bar{v}_2^{\#}/RT) \tag{9}$$

where $k_2(0)$ is the rate constant at reference of atmospheric conditions. A similar equation may be written if the dissociation step represented by k_3 is rate-limiting. The dissociation represented by k_3 will generally have a positive activation volume, because bonds are being stretched. Thus, the rate constant for dissociation will generally decrease with increasing pressure. The reaction step, on the other hand, may be associated with either a positive or negative activation volume. In either case, the effect of pressure on the two rate constants k_2 and k_3 will be different, and by appropriate adjustment of pressure the rate-limiting step may actually be changed.

More complicated rate equations and pressure effects may be envisioned for reactions involving multiple substrates and/or products. Because reaction-rate constants which are small or negligible at atmospheric conditions may increase with pressure and thereby influence the observed rates, and conversely, very fast reaction steps may slow to become rate-limiting, caution must be exercised in interpreting reaction-rate data as a function of pressure. For example, Fukuda and Kunugi[30] have studied pressure effects on peptide hydrolysis catalyzed by carboxypeptidase Y. Apparently, the rates of the acylation and deacylation processes were comparable at normal pressure; however, peptide hydrolysis became more acylation-controlled as the pressure increased up to 1000 bar.

11.4.3. Conformational Changes in Enzymes with Pressure

Penniston[31] has found that a major effect on enzymatic activity in aqueous solutions is the dissociation of multimeric enzyme units with increasing pressure. Generally, he found that the activity of monomeric enzymes increased with increasing pressure, while multimeric-enzyme activity decreased at higher pressures. Changes in enzymatic activity were small until the pressure was increased to about 1000 bar, and irreversible denaturation occurred at about 4000 bar.

Randolph *et al.* used high-pressure electron paramagnetic resonance spectroscopy to probe possible conformational changes in cholesterol oxidase from *Gloecocysticum chrysocreas* in supercritical carbon dioxide and supercritical carbon dioxide/cosolvent mixtures.[32] As shown in Figure 2, little change in the EPR spectra of spin-labeled enzyme can be detected between 1 and 112 bar, suggesting that conformational changes were minimal.

Enzyme denaturation during depressurization in supercritical carbon dioxide was studied by Kasche *et al.*[33] Slow depressurization (~5–10 min)

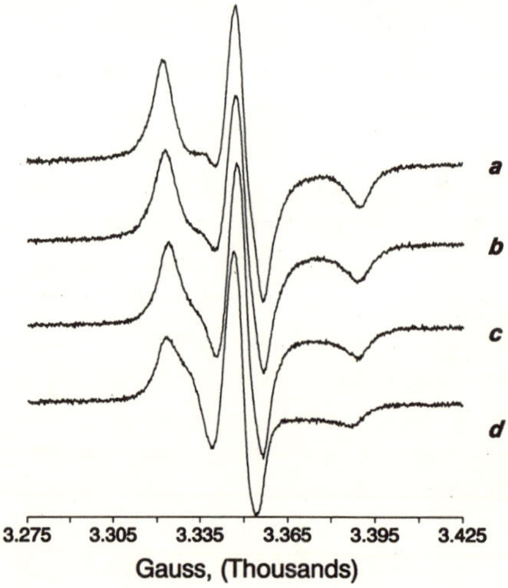

Figure 2. EPR spectra of spin-labeled cholesterol oxidase from *Gl. chrysocreas* under atmospheric and supercritical conditions. Spectra: (a) atmospheric pressure (308 K); (b) in carbon dioxide at 104 bars (308 K); (c) in carbon dioxide and 3% (vol/vol) 2-methyl-2-propanol at 104 bars (308 K); (d) in carbon dioxide and 3% (vol/vol) methanol at 104 bars (308 K). Reprinted with permission from ref. 48.

from 100 bar resulted in partial inactivation of chymotrypsin and trypsin (both of which contain disulfide bridges), and the extent of inactivation increased with the number of pressurization–depressurization steps. Moreover, the degree of inactivation was larger in humid carbon dioxide than in dry carbon dioxide. Interestingly, penicillin amidase, which does not contain cysteine, was the least stable during depressurization. In this case stability was not increased in dry carbon dioxide. Shifts in the maximum fluorescence emission wavelengths suggested that inactivation of each enzyme was caused by partial unfolding of the protein. Electrophoretic analysis showed that neither peptide bond cleavage nor oligomerization accompanied denaturation.

11.4.4. Observed Enzymatic Activation Volumes

Activation volumes for a variety of enzymatic reactions in aqueous solutions have been measured. These enzyme systems include acetylcholinesterase,[34] alcohol dehydrogenase,[35] alkaline phosphatase,[32] chymotrypsinogen,[36] lysozyme,[37] peroxidase,[31] pyruvate kinase,[38] trypsin,[39] and xanthine oxidase[29] among others. Reported activation volumes for these enzyme systems are of order $\pm 50\ \mathrm{cm^3/mol}$. The numbers are probably valid only near atmospheric conditions, and, as pointed out by Morild,[29] may even change sign at higher pressures. It must be noted again that the measured activation volume of an enzyme at a given pressure is an "*effective*" activation volume representative of the rate-limiting step in the reaction; this measured overall activation volume may be representative of different reaction steps as a function of pressure.

11.5. EXAMPLES OF ENZYMATIC CATALYSIS UNDER SUPERCRITICAL CONDITIONS

Supercritical fluids offer several advantages as solvents for enzymatic catalysis. Because enzymes are not solubilized by most of the "room-temperature" supercritical fluids, post-reaction recovery and separation of the catalyst is simplified. Immobilization techniques may be as simple as lyophilizing the enzyme on the surface of glass beads. For hydrophobic substrates like lipids or alkaloids, solubilities may be greater than those found in aqueous solution. Finally, the separation of solvent and product downstream of a reaction is easily accomplished by decompressing the fluid.

11.5.1. Alkaline Phosphatase

Enzyme activity in a supercritical fluid was first demonstrated by Randolph *et al.*, using the enzyme alkaline phosphatase (EC 3.1.3.1) to catalyze the hydrolysis of disodium *p*-nitrophenyl phosphate to *p*-nitrophenol.[40] Preliminary data showed that the enzyme retained full activity after 24 h of exposure to supercritical carbon dioxide at 100 bar, 35 °C. Subsequent batch reaction experiments demonstrated that the enzyme was catalytically functional under supercritical conditions, although the rate of hydrolysis was much slower than in aqueous solution.

Analysis of the alkaline phosphatase system was complicated by the requirement that water be present in the system as a reactant. The observed reduction in activity under supercritical conditions might be expected, considering the acidity of high-pressure water–carbon dioxide mixtures. Also, as noted by Randolph[41] and by Krukonis and Hammond,[42] the reactant, disodium *p*-nitrophenyl phosphate, is negligibly soluble in supercritical carbon dioxide, and the observed activity under supercritical conditions was probably due to substrate dissolving in a water film surrounding the enzyme.

11.5.2. Polyphenol Oxidase

Hammond *et al.*[43] used mushroom polyphenol oxidase to catalyze the oxidation of *p*-cresol with oxygen to methyl-substituted catechol and *o*-quinone. Oxygen and *p*-cresol were dissolved in water-saturated supercritical carbon dioxide, then passed over a packed-bed reactor containing enzyme immobilized by drying onto glass beads. At the end of each experiment the enzyme was quite damp, apparently having absorbed water from the saturated gas stream. It was suggested that the enzymatic activity, as in the alkaline phosphatase example, took place in this "water film," rather than in the carbon dioxide continuous phase.[42] Enzyme activity for polyphenol oxidase was also demonstrated using supercritical fluoroform as a solvent.

11.5.3. Cholesterol Oxidase

Cholesterol oxidase (EC 1.1.3.6) catalyzes the oxidation of a variety of 3-β-ol-5-ene steroids.[44] In aqueous solutions, the oxidation of cholesterol is quite slow due to the low solubility of cholesterol in water (4.7 m*M*).[45] In carbon dioxide at 123 bar, cholesterol is about 50 times more soluble.[46] This solubility may be increased by another order of magnitude by the addition of small amounts of cosolvents such as methanol or ethanol.

Randolph *et al.* have studied cholesterol oxidase from *Norcardia* sp.,

Pseudomonas sp., *Streptomyces* sp., and *Gl. chrysocreas* in supercritical carbon dioxide and carbon dioxide/cosolvent mixtures.[27, 39, 47, 48] Enzymatic activity was observed with enzymes from all species, although stability varied dramatically from species to species. Cholesterol oxidase from *Streptomyces* sp. had a half-life of only about an hour in carbon dioxide at 40 °C and 100 bar, while no loss of activity could be detected after three days of operation in a packed-bed reactor containing enzyme from *Gl. chrysocreas.*

The effect of water content on cholesterol oxidase activity in supercritical carbon dioxide was examined by first operating a packed-bed reactor using water-saturated carbon dioxide until a steady-state production was achieved, then switching to bone-dry carbon dioxide. A rapid 10-fold decrease in productivity was observed as the enzyme was dried. The loss of activity was reversible; switching back to water-saturated carbon dioxide resulted in a regaining of the initial activity.

Cosolvents were added to supercritical carbon dioxide to enhance the solubility of cholesterol. Steady-state productivities (expressed as a turnover number, moles product per mole enzyme per second) were measured at 35 °C, 100 bar, and compared with turnover numbers measured in aqueous solutions. As shown in Table 2, an increase in turnover number of over 200-fold was seen in solutions of carbon dioxide and *iso-* or *tert*-butanol as compared to a buffered aqueous system with isopropanol added as a solubilizing agent.

Electron paramagnetic resonance (EPR) spectroscopy under supercritical conditions revealed a possible explanation for the strong cosolvent effect on cholesterol oxidation rates under supercritical conditions. Those solvents which produced the highest turnover numbers (ethanol, *iso*-butanol, and *tert*-butanol) also promoted the aggregation of cholesterol,

Table 2. Turnover Numbers for the Oxidation of Cholesterol with Cholesterol Oxidase from *Gloeocysticum chrysocreas* at 35 °C. All Turnover Numbers Measured at Saturation with Cholesterol and under Conditions where Reaction is Zero-Order in Oxygen Concentration[47]

Solvent	Turnover number, s^{-1}
Water	Not measurable
Water (5 m*M* phosphate, pH 7.0) + 5 % v/v isopropanol	1
Carbon dioxide, 100 atm	75
Carbon dioxide + 2.0 % v/v methanol	62
Carbon dioxide + 2.0 % v/v ethanol	165
Carbon dioxide + 2.0 % v/v *tert*-butanol	274
Carbon dioxide + 2.0 % v/v *iso*-butanol	238
Carbon dioxide + 2.0 % v/v acetone	86
Carbon dioxide + 2.0 % v/v *n*-butanol (extrapolated)	100

while those solvents which gave smaller increases (*n*-butanol, methanol, and acetone) did not promote aggregation. It was surmised that cholesterol oxidase is more active towards cholesterol aggregates than cholesterol monomers.

EPR investigations on the role of water in cholesterol binding by cholesterol oxidase in supercritical carbon dioxide showed that when damp, the enzyme was able to "bind" more than one cholesterol molecule at a time (perhaps by promoting adsorption to the enzyme's surface). When the enzyme was dried, however, it lost the ability to bind multiple substrate molecules.

11.5.4. Lipase

Lipases have been used in supercritical carbon dioxide by several groups to catalyze interesterification, transesterification, and hydrolysis reactions. Nakamura and Yano,[49] and Chi, Nakamura, Yamada, and Tano[50, 51] studied the transesterification and interesterification of triolein and stearic acid, catalyzed by lipases isolated from *Rhizopus delemar*, *Rhizopus japonicus*, *Alcaligenes* sp., and *Mucor miehei.* In order to solubilize the sparingly soluble stearic acid, they chose a temperature of 50 °C and a pressure of 249 bar.

Study of enzyme-catalyzed interesterification reactions is complicated by the role of water. Water is a reactant in the main competing side reaction, hydrolysis of the triglycerides to the corresponding acids and alcohols.

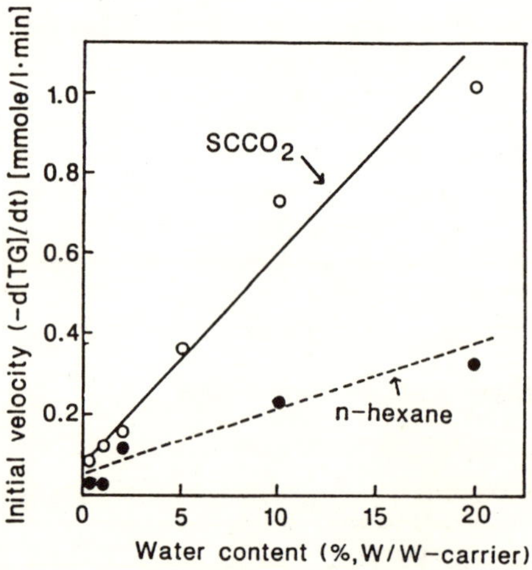

Figure 3. Effect of water content on the initial rate of hydrolysis catalyzed by *M. meihei* lipase. Reprinted with permission from ref. 51.

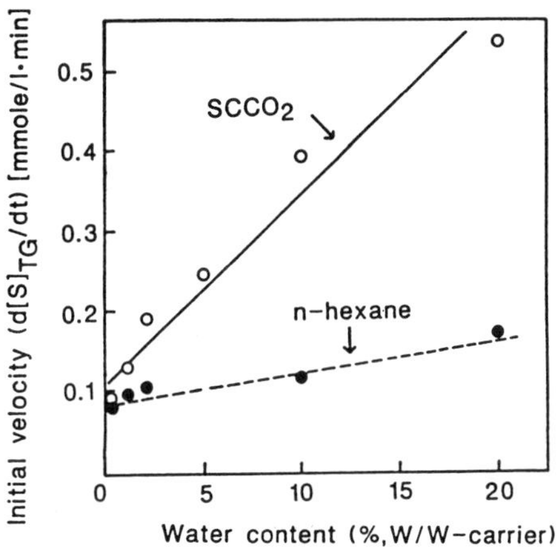

Figure 4. Effect of water content on the initial rate of interesterification catalyzed by *M. meihei* lipase. Reprinted with permission from ref. 51.

However, water is also needed as an intermediate reactant in trans- and interesterification reactions, which require an initial hydrolysis. In order to avoid the hydrolysis reaction favored by equilibrium considerations in aqueous solutions, several workers have investigated the use of nonpolar organic solvents.

The initial rates of hydrolysis and of interesterification found by Chi *et al.* in both *n*-hexane and supercritical carbon dioxide varied with the water content of the medium. For lipase from *M. meihei* immobilized on Duolite at low water content (less than 5% w/w-carrier), interesterification and hydrolysis rates in hexane and carbon dioxide were approximately equal. However, as shown in Figures 3 and 4, initial rates at higher water content (20% w/w-carrier) for both reactions were about four times higher in carbon dioxide. The lipase from *M. meihei*, unlike that from *Rhizopus* sp., was active at water contents below 1% w/w-carrier.

Chi *et al.* also reported that the equilibrium extent of interesterification and the equilibrium amount of residual triglycerides in carbon dioxide were shifted by changing the system's water content. The shift was different for each lipase. However, since each lipase was immobilized on a different type of support, a given percent water per weight carrier probably corresponded to a different water activity in each case.

The lipase-catalyzed interesterification of triglycerides in supercritical carbon dioxide has also been studied by Miller *et al.*[52] The interesterification of trilaurin and myristic acid was conducted in a continuous-flow packed-bed reactor containing lipase immobilized by covalent attachment to porous glass beads. The authors studied the effects of superficial velocity and pressure on the observed interesterification rate in water-saturated carbon dioxide; the results are shown in Figure 5. The interesterification rate did not depend on superficial velocity over the range of flows studied; therefore, no external mass-transfer limitations were present. The rate of interesterification increased as pressure increased between 83 and 111 bar. This increase may have been due to enhanced substrate solubilities at higher pressures.

In another example of lipase-catalyzed reactions under supercritical conditions, van Eijs *et al.* used a commercial immobilized enzyme, Lipozyme TM (Novo Industry, Denmark) to catalyze the transesterification of ethyl acetate and isoamyl alcohol to isoamyl acetate and ethanol.[53, 54] In contrast to the work of Chi *et al.*, a flow-through reactor was used, facilitating analysis of water content on reaction rates. At a fixed pressure and temperature (100 bar, 60 °C), they observed a maximum in the productivity of about 120 g isoamyl acetate per kg enzyme per hour at a water content of 0.2% in carbon dioxide. Under these conditions, the lipase preparation contained about 6% w/w water at equilibrium.

Van Eijs *et al.*[53, 54] varied the superficial fluid velocity in their packed column and found a linear relationship between isoamyl acetate production rate and fluid velocity. No loss of activity was seen after several hours under supercritical conditions.

A more valuable ester, nonyl acetate, was synthesized in the same

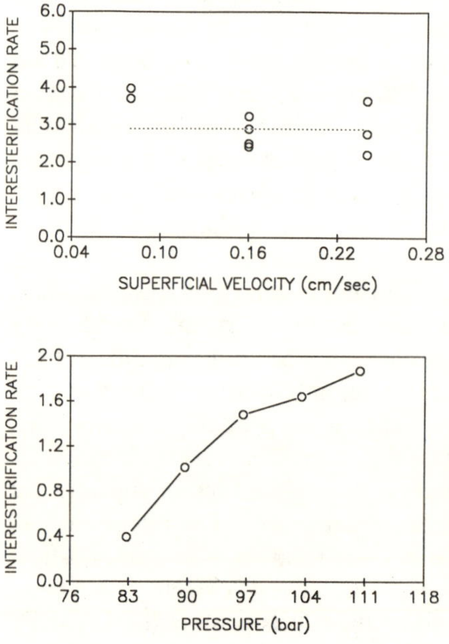

Figure 5. (Top) Effect of superficial velocity on observed rate of interesterification of trilaurin and myristic acid (MA) at constant residence time (42 s). $P = 96$ bar; $T = 35$ °C. (Bottom) Effect of pressure on interesterification rate. $T = 35$ °C; residence time = 28 s. The interesterification rate is expressed in (mols MA incorporated $L^{-1}s^{-1}$) × 10^6.

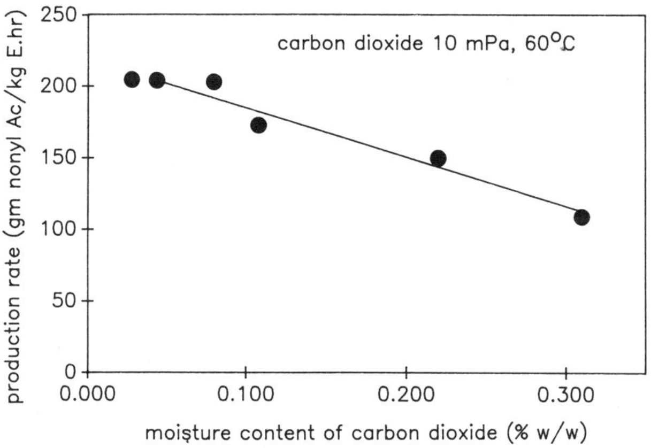

Figure 6. Lipase-catalyzed production of nonyl acetate as a function of moisture content of supercritical carbon dioxide. Data taken from ref. 54.

reactor using the transesterification of nonyl alcohol and ethyl acetate. Unlike the results obtained with isoamyl alcohol and ethyl acetate, productivity decreased continuously as the water content of the carbon dioxide was increased above 0.03% w/w. Figure 6 shows the productivity of nonyl acetate as a function of water content in the supercritical carbon dioxide.

11.6. BIOCATALYSIS BY WHOLE CELLS IN SUPERCRITICAL FLUIDS

Supercritical fluids may be appealing for use as solvents to extract compounds such as ethanol[55] or carboxylic acids[56] from fermentation broths, but there is little information on the possibility of carrying out fermentations under supercritical conditions or on the toxicity of supercritical fluids toward cells. Several conflicting reports are now in the literature; a Japanese patent for ethanol production by immobilized yeasts in supercritical carbon dioxide[57] contrasts sharply with the work of Kamihira,[58] who suggests that the same conditions may be used to sterilize broths containing yeasts.

11.6.1. *Saccharomyces rouxii*, *Saccharomyces cerevisiae*, and Commercial Baker's Yeast

L'Italien *et al.* have studied the effects of near-critical carbon dioxide (30 °C, 70 bar) on ethanol productivity for three different yeasts.[59] Two of the yeasts (*S. rouxii* and *S. cerevisiae* ATCC 24702) are known osmotolerants, and the third was isolated from a commercial baker's yeast preparation. Under hyperbaric conditions, severe inhibition was observed, and volumetric productivity dropped from the 29.7 $gL^{-1}h^{-1}$ observed under atmospheric conditions to 2.4 $gL^{-1}h^{-1}$. No advantage was seen in using osmotolerant yeast strains. Figure 7 shows ethanol concentrations found during fermentations with baker's yeast under ambient and hyperbaric conditions.

The inhibition observed by L'Italien *et al.* was reversible. After short periods of exposure to near-critical conditions (3–6 h), normal activity resumed immediately when the culture was returned to atmospheric conditions. After longer periods of exposure, lags of up to 36 h were observed.

As a result of their investigations, l'Italien *et al.* proposed that a cyclic fermentation might be advantageous, alternating between ambient pressure for maximum ethanol production and near-critical conditions for efficient removal of ethanol from the broth. Preliminary tests of this idea seemed favorable; 10 pressurization–depressurization cycles did not alter cell viability, and a constant specific rate of ethanol synthesis was maintained.

11.6.2. *Lactobacillus plantarum*, *E. coli*, and *Candida albicans*

Van Eijs *et al.* studied the effect of supercritical carbon dioxide on *L. plantarum*, *E. coli* and *C. albicans.*[60, 61] They considered at least four possible mechanisms for cell damage due to supercritical conditions:

1. Extraction of essential compounds from the cells.
2. Acidification of the medium.
3. Cell bursting.
4. Toxicity of carbon dioxide.

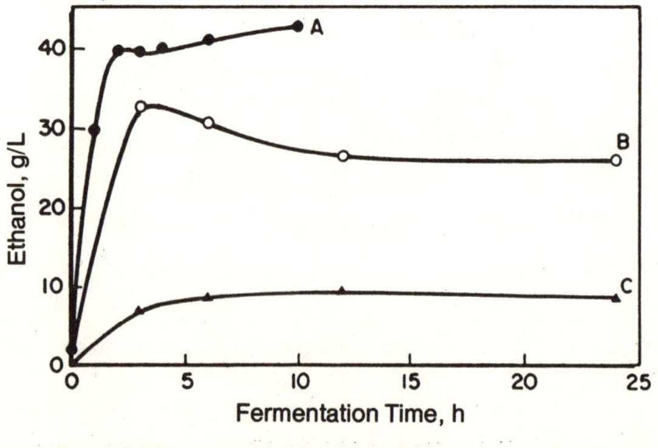

Figure 7. Ethanol fermentation using high concentrations of baker's yeast ($S_0 = 95\ gL^{-1}$): (A) 0.1 MPa, $C_0 = 87.4\ gL^{-1}$; (B) 7.0 MPa, $C_0 = 113\ gL^{-1}$; (C) 7.0 MPa, $C_0 = 8.1\ gL^{-1}$. Reprinted with permission from ref. 59.

By controlling compression and expansion rates, van Eijs *et al.* determined that compression/decompression rates are indeed important, and that the decompression step in particular should be done slowly (< 60 bar/h) to avoid viability losses. Acidification as a damage mechanism was also confirmed, as *Lactobacillus* survival rates after 3 h of exposure to supercritical conditions (150 bar, 35 °C) jumped from only 1% in an unbuffered system to 60% in a buffered system. Toxicity of carbon dioxide *per se* was not implicated; carbon dioxide solubility in aqueous media is higher at low temperatures than at high temperatures, yet survival rates were higher below the critical temperature of carbon dioxide. The decrease in viability above the critical temperature suggests that the extraction of essential compounds (e.g., membrane components) may be responsible for long-term cell damage.

11.6.3. *Candida utilis*

The results of van Eijs *et al.* were qualitatively confirmed by Gigon *et al.*,[62] who measured cell survival rates for *C. utilis*, ATCC 9950 as a function of pressure and temperature in near- and supercritical carbon dioxide. In general, higher fractional cell survival rates (measured as the ratio of colony forming units before and after high-pressure treatment) were observed when the broth was buffered. Survival rates dropped when the temperature was raised from 27 °C (subcritical) to 32 °C (supercritical), but the viability was fairly insensitive to pressure.

Gigon *et al.* also studied the survival rates of *C. utilis* in carbon dioxide (27 °C, pressure 80–160 bar) as a function of culture age for cells previously stored at 4 °C for varying lengths of time. A maximum in the survival rate was seen after 40–60 h storage at 4 °C. A possible explanation for the increased resistance to high-pressure carbon dioxide with increasing culture age is the production of an agent which acts to protect cell membranes from damage.

After exposure to high-pressure carbon dioxide, cells cultured on agar plates showed varying lag times (indicated by widely varying colony sizes). This result suggests that *C. utilis* cells may be more sensitive to environmental shock during certain growth phases.

11.7. CONCLUDING REMARKS

Many biocatalytic reactions have now been run in supercritical fluids on a laboratory scale. In addition to the appealing properties of organic solvents for certain enzymatic transformations, supercritical solvents offer the general advantages of high diffusivities and low viscosities. Moreover, the solvent properties (and product solubility) are easily varied near the critical temperature and pressure. Supercritical fluids can also be useful for studying enzymatic mechanisms and for examining the fundamental behavior of enzymes in hydrophobic environments. Disadvantages, at least with respect to industrial processes, include the need for high-pressure equipment, poor solubilities of many substrates, and the toxicity of some solvents. Nonetheless, as advances are made in solubilizing a wider range of substrates, the use of supercritical solvents in both research and bioprocessing is likely to expand well beyond the relatively few applications reported to date.

REFERENCES

1. A. Zaks and A. M. Klibanov, *Proc. Natl. Acad. Sci. USA 82*, 3192 (1985).
2. A. Zaks and A. M. Klibanov, *Science 224*, 1249 (1984).
3. R. Z. Kazandjian and A. M. Klibanov, *J. Am. Chem. Soc. 107*, 5448 (1985).
4. C. Laane, S. Boeren, and K. Vos, *Trends Biotechnol. 3(10)*, 251 (1985).
5. C. Laane, S. Boeren, K. Vos, and C. Veeger, *Biotechnol. Bioeng. 30*, 81 (1987).
6. C. Laane, *Biocatalysis 1*, 17 (1987).
7. C. Laane, R. Hilhorst, R. Spruijit, and C. Veeger, presented in the European Conference on Biotechnology, September (1984).
8. J. B. Hannay and J. Hogarth, *Proc. Roy. Soc. (London) 29*, 324 (1979).
9. J. B. Hannay, *Proc. Roy. Soc. (London) 30*, 484 (1980).
10. K. Zosel, US Patent 3,597,464 [Cl260, 448A; CO7f] (1971)].
11. K. Zosel, Ger. Offen. 2,332,038 [Cl. C11b (1974)].
12. K. Zosel, Ger. Offen. 2,363,418 [Cl. C11b (1974)].
13. K. Zosel, Ger. Offen. 2,441,152 [Cl. C11c (1975)].
14. O. Vitzthum, P. Hubert, and W. Sirtl, US Patent 4,104,409 (1978).
15. W. Roselius, O. Vitzthum, and P. Herbert, Ger. Offen. 2,043,537 [Cl. a23f (1975)].
16. O. Vitzthum and P. Herbert, Patentschrift 2,127,611 (1974).
17. L. G. Randall, *Sep. Sci. Tech. 17(1)*, 1 (1982).
18. M. E. Paulaitis, V. J. Krukonis, R. T. Kurnik, and R. C. Reid, *Rev. Chem. Eng. 1*, 179 (1982).
19. G. Brunner and S. Peter, *Chem. Ing. Tech. 53*, 529 (1981).
20. M. A. McHugh and V. J. Krukonis, *Supercritical Fluid Extraction*, Butterworths, Boston (1986).
21. G. Brunner, D. Stuetzer, and S. Peter, *AIChE Symp. Ser.*, in press.
22. P. G. Debenedetti and S. K. Kumar, *AIChE Symp. Ser. 34(4)*, 646 (1988).
23. R. S. Mohamed, P. G. Debenedetti, and R. K. Prud'homme, *AIChE J. 35(2)*, 325 (1989).
24. D. W. Matson, R. C. Petersen, and R. D. Smith, *Advanced Ceramic Materials 1*, 242 (1986).
25. K. A. Larson and M. L. King, *Biotechnol. Progress 2(2)* (1986).
26. B. Subramaniam and M. A. McHugh, *Ind. Eng. Process Des. Dev. 25*, 1 (1986).
27. S. Glasstone, K. J. Laidler, and H. Eyring, *The Theory of Rate Processes*, McGraw-Hill, New York (1941).
28. S. Kim and K. P. Johnston, *Chem. Eng. Commun. 63*, 49 (1988).
29. E. Morild, *Adv. in Protein Chem. 34*, 93 (1981).
30. M. Fukuda and S. Kunugi, *Eur. J. Biochem. 149*, 657 (1985).

31. J. T. Penniston, *Arch. Biochem. Biophys. 142*, 322 (1971).
32. T. W. Randolph, D. S. Clark, H. W. Blanch, and J. M. Prausnitz, *Science 238*, 387 (1988).
33. V. Kasche, R. Schlothauer, and G. Brunner, *Biotechnol. Lett. 10*, 569 (1988).
34. P. W. Hochachka, K. B. Storey, and J. Baldwin, *J. Comp. Biochem. Physiol. B 52*, 43 (1975).
35. E. Morild, *Biophys. Chem. 6*, 351 (1977).
36. A. L. Curl and E. F. Jansen, *J. Biol. Chem. 185*, 713 (1950).
37. E. D. Holyoke and F. H. Johnson, *Arch. Biochem. Biophys. 31*, 41 (1951).
38. P. S. Low and G. N. Somero, *Comp. Biochem. Physiol. B 52*, 67 (1975).
39. D. Fraser and F. H. Johnson, *J. Biol. Chem. 190*, 2706 (1951).
40. T. W. Randolph, H. W. Blanch, J. M. Prausnitz, and C. R. Wilke, *Biotechnol. Lett. 7, 5*, 325 (1985).
41. T. W. Randolph, Ph.D. Thesis, University of California, Berkeley (1987).
42. V. J. Krukonis and D. A. Hammond, *Biotechnol. Lett. 10, 11*, 837 (1988).
43. D. A. Hammond, M. Karel, A. M. Klibanov, and V. J. Krukonis, *Appl. Biochem. Biotechnol. 11*, 393 (1985).
44. A. G. Smith and C. J. W. Brooks, *Biochem. J. 167*, 127 (1977).
45. M. F. Haberland and J. A. Reynolds, *Proc. Natl. Acad. Sci. USA 70(8)*, 2313 (1973).
46. J. M. Wong and K. P. Johnston, *Biotechnol. Progress 2(2)* (1986).
47. T. W. Randolph, H. W. Blanch, and J. M. Prausnitz, *AIChE 34(8)*, 1354 (1988).
48. T. W. Randolph, D. S. Clark, H. W. Blanch, and J. M. Prausnitz, *Proc. Natl. Acad. Sci. USA 85*, 2979 (1988).
49. K. Nakamura and T. Yano, Japanese Patent Application 59 143138 (1984) and Jpn. Kokai Tokyo 61 21098 (1986).
50. K. Nakamura, Y. J. Chi, Y. Yamada, and T. Yano, *Chem. Eng. Commun. 45*, 207 (1986).
51. Y. M. Chi, K. Nakamura, and T. Yano, *Agric. Biol. Chem. 52(6)*, 1541 (1988).
52. D. A. Miller, H. W. Blanch, and J. M. Prausnitz, in: *Ann. N.Y. Acad. Sci.*, to appear.
53. A. M. M. van Eijs, J. P. J. de Jong, and H. H. M. Oostrom, poster presentation at the Netherlands Congress on Biotechnology, Amsterdam, October 20-21 (1988).
54. A. M. M. van Eijs, J. P. J. de Jong, H. J. Doddema, and D. R. Lindeboom, *Proceedings of the International Symposium on Supercritical Fluids*, Nice, October 17-18 (1988).
55. R. P. De Phillipi and J. M. Moses, *Biotechnol. Bioeng. Symp. 12*, 205 (1982).
56. E. J. Shimshick, *Chemtech.* 374 (1983); E. J. Shimshick, US Patent 4,250,331 (1983).
57. Hitachi, Ltd., Jpn. Kokai Tokyo Koho Japanese Patent 60 94 089.
58. M. Kamihira, M. Taniguchi, and T. Kobayashi, *Agri. Biol. Chem. 51*, 407 (1987).
59. Y. L'Italien, J. Thibault, and A. LeDuy, *Biotechnol. Bioeng. 33*, 471 (1989).
60. A. M. M. van Eijs, J. M. P. Wokke, and B. ten Brink, in: *Preconcentration and Drying of Food Materials* (S. Bruin, ed.), Elsevier, Amsterdam (1988).
61. A. M. M. van Eijs, J. M. P. Wokke, B. ten Brink, and K. A. Dekker, *Proceedings of the International Symposium on Supercritical Fluids*, Nice, October 17-18 (1988).
62. M. Gigon, T. W. Randolph, I. W. Marison, and U. von Stockar, unpublished results.

IV

FUTURE DIRECTIONS OF BIOCATALYSTS

12

Protein Stabilization

SUSAN J. TOMAZIC

12.1. INTRODUCTION

For many industrial processes, biocatalysts provide a desirable specificity not possible with either the analogous uncatalyzed chemical reaction or a catalyzed nonenzymatic counterpart. For example, in the conversion of starch to high-fructose corn syrup, α-amylase consistantly yields a product containing 3–5% more dextrose than simple acid hydrolysis, with less impurities and by-products.[1] Similarly, rennin (chymosin) is used in the making of cheese because it will specifically cleave only the κ-casein component in milk thus initiating the clotting process.[2] The use of biocatalysts however remains limited, because while many applications require harsh conditions (such as temperatures greater than 50 °C) to ensure high productivity, high solubility of substrates, and reduced microbial contamination, these conditions often result in an irreversible inactivation of the enzyme.[3] Hence repetitive, time-consuming additions of the costly catalyst are necessary to maintain the process. This dilemma has compelled scientists to search for means to stabilize biocatalysts against irreversible inactivation.

Before addressing the problem of *how* to make enzymes more stable, it is necessary to first understand *why* the protein is labile, i.e., one must elucidate its molecular mechanism(s) of irreversible inactivation. Enzymes may be inactivated by a variety of conditions such as temperature, pH, ionic strength, denaturing agents, pressure, or mechanical forces.[4] Inactiva-

SUSAN J. TOMAZIC • Abbott Laboratories, Department 90U, Building AP-8B, Abbott Park, Illinois 60064.

tion brought about by high temperature, because of its practical importance, has been extensively studied, hence it will be stressed here. The multitude of processes shown in the literature to contribute to the thermoinactivation of enzymes can be divided into two main classes: conformational processes and covalent processes.[5] Conformational processes are those which do not alter the primary structure of the protein and include aggregation,[6–9] irreversibly denatured, incorrectly folded (scrambled) structures,[10–13] disassociation of an oligomeric protein into subunits,[14, 15] disulfide bond interchange,[16, 17] and disassociation of a coenzyme from the active site of a molecule.[18] Alternatively, covalent processes change the primary structure of the enzyme. Such processes include hydrolysis of the peptide bond (especially at sites adjacent to aspartic acid residues),[17, 19] autolysis/proteolysis,[20] deamidation of amide residues,[13, 17, 19] oxidation of cysteine residues,[13, 21] racemization of amino acids,[22] and β-elimination of cystine residues followed by the formation of new amino acids.[17, 23]

The feature common to all the covalent and chemical processes listed above is that they occur after the protein has undergone a *reversible* unfolding.[24] In other words, this unfolding is a universal and general step in all enzyme inactivations. It is reversible, since upon return to ambient conditions, a return of enzymatic activity will result. Reversible unfolding, a well-studied and understood phenomenon, occurs when the noncovalent forces responsible for the three-dimensional structure of the native enzyme are disrupted.[25–28] These interactions include hydrogen bonds, ionic and van der Waals interactions, and hydrophobic forces.[29] Subsequent to the disruption of such interactions, the irreversible processes occur. Figure 1 schematically depicts the relationship between reversible and irreversible inactivation. It follows, therefore, that a general approach to protein stabilization should be focused upon preventing or slowing down the first step of the two-step inactivation process, the reversible unfolding of the molecule. By strengthening the noncovalent forces, one hopes to "rigidify" the molecule and reduce the propensity for unfolding; stabilization against

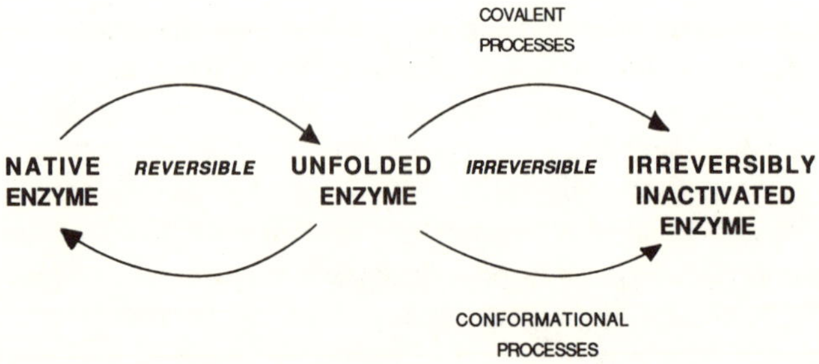

Figure 1. Relationship between reversible unfolding and irreversible inactivation of an enzyme.

irreversible inactivation can follow provided the working temperature is near the enzyme's T_m (i.e., temperature where [native enzyme] = [unfolded enzyme]).[24, 30]

The many approaches taken to obtain stable biocatalysts can be divided into four general categories: (i) the empirical addition of additives to the enzyme's milieu, (ii) chemical modification of the biocatalyst, (iii) genetic engineering, and (iv) the isolation of stable catalysts from thermophiles—organisms found growing at temperatures from 55 °C to the boiling point.[31] Each of these approaches will be discussed in the subsequent sections.

12.2. ADDITIVES

Since the nineteenth century, it has been recognized that stability may be induced by nonprotein additives. Salts were among the first stabilizers. They can be divided into two groups.[4] The first group consists of neutral salts at high concentrations which affect the protein's conformation either by binding to charged groups or dipoles ("salting in"), or by reducing the solubility of hydrophobic groups by increasing the ionic strength of the medium ("salting out").[3] The stabilizing potential of different ions often follows the Hoffmeister lyotropic series.[4] In this series, cations and anions such as NH_4^+, Na^+, K^+, SO_4^{2-}, Cl^-, and Br^- afford greater stability than Mg^{2+}, Ca^{2+}, Ba^{2+}, NO_3^-, and SCN^-. These effects are relatively independent of the protein source. This is, therefore, a general means to limit reversible unfolding (inactivation) by inducing a "rigidity" into the molecule. The rigidity may result from an enhancement of hydrophobic interactions, neutralization of charge repulsion, or the formation of new salt bridges.[33, 34]

The second group encompasses divalent ions, such as Ca^{2+}, Mn^{2+}, Zn^{2+}, and Mg^{2+}, which specifically interact with some enzymes.[3] These ions, used at low concentration, can stabilize the tertiary structure of a protein by providing critical cross-links. An example of this is thermolysin from *Bacillus stearothermophilus*, where two calcium ions cross-link the subunits to provide a rigid polypeptide backbone.[35] It has also been proposed by Fontana *et al.*, that Ca^{2+} in this case may also serve to counteract a potentially destabilizing accumulation of negative charges.[36] In contrast to the general stabilizing effects afforded by high concentrations of neutral salts, this second class is specific to the enzyme's requirement of metals as well as to the actual metal itself.[4, 37]

Polyhydric alcohols and sugars such as glycerol and sucrose are frequently used to enhance the stability, both temperature and storage, of proteins. Ribonuclease, β-glucosidase, and α-amylase are but a few examples where increased thermal stability occurs in the presence of these compounds.[38–40] The exact mechanism(s) of stabilization is still not fully understood, but it is generally attributed to an increase in the structuring of the solvent due to the formation of tight hydrogen bonds between the polyol and solvent.[41–43] The protein is less able to unfold in this altered environment than in the solvent alone. Controversy continues as to whether conformational changes such as a tightening of the protein's hydrophobic core result from this solvent structuring.[44]

The addition of substrates or other ligands has long been recognized as a means to stabilize many enzymes.[3,4,43] α-Amylase from *Bacillus amyloliquefaciens*, for example, is stabilized against irreversible thermoinactivation by a factor of 70 in the presence of 10% starch, its natural substrate.[13] This phenomenon can be explained by the shift, upon ligand binding, in the equilibrium between native and unfolded enzyme away from the unfolded form known to undergo irreversible thermoinactivation (see Figure 1).[24,45] Ligand stabilization has been shown to be synergistic for some enzymes. For an apo-glutamine synthetase from *Bacillus caldolyticus*, the total stabilizing effect induced by the L-glutamate or MnATP addition was less than that observed when both were added together.[46]

Native enzymes can also be stabilized by organic solvents such as ethanol, methanol, or acetone at concentrations up to 20%.[4] Furthermore, it has been shown that lipases and lysozyme are but two examples of many enzymes which exhibit several orders of magnitude greater thermostability at 100 °C when in nearly anhydrous organic solvents.[5,47] By drastically reducing the water content, the many processes which involve water as a reactant and contribute to irreversible thermoinactivation are restricted.[5,17,19,48]

Finally, it is known that sulfur groups in proteins can be extremely labile, participating in inactivating reactions such as oxidation or disulfide interchange.[21] The addition of antioxidants such as vitamin E and ascorbic acid or Cu^{2+} to eliminate thiols which contribute to disulfide interchange is a simple measure, shown to stabilize enzyme solutions.[17,49]

12.3. CHEMICAL MODIFICATION

A second general approach toward enzyme stabilization is that of chemical modification. This topic encompasses alterations which range from the modification of surface amino acid residues with monofunctional reagents to the covalent linkage of an enzyme to a solid supporting polymer.[3,4,41] The resulting stabilization, if it occurs, is frequently attributed to a rigidification of the protein's conformation such that reversible unfolding is slowed (or prevented). This will, as described in Section

12.1, similarly alter the irreversible inactivating processes. Exceptions to this include inactivation via autolysis or aggregation and disassociation of an oligomeric protein. In such cases, it is clear that immobilization to a solid support to physically restrict or slow intermolecular contact for the former cases and the creation of intramolecular subunit cross-links to prevent disassociation for the latter are the likely reasons for their stabilization.[3, 4]

Ionic interactions are important for maintaining the three-dimensional structure of a protein and have been shown to play a crucial role in the thermostability of proteins such as tyrosinase, D-glyceraldehyde-3-phosphate dehydrogenase, and α-amylase.[50–54, 30] In fact, one or two additional ionic bonds provide the decrease in ΔG of 5–7 kcal/mol which can account for a 20–30 °C difference in the thermostability between enzymes isolated from mesophilic and thermophilic organisms.[26, 55] Moreover, it has been observed that in comparing some highly stable proteins with their mesophilic counterparts, a replacement of histidine (His) and lysine (Lys) residues by arginine (Arg) residues frequently occurs.[50, 51] A plausible explanation for this phenomenon is as follows: the pK's for the side chains of His and Lys (6.0 and 10.8, respectively) are much lower than for Arg (p$K = 12.5$), hence upon an increase of the temperature at neutral and alkaline pH's, the ionogenic groups of His and Lys will become increasingly neutral, resulting in a disruption of a stabilizing salt bridge. Salt bridges with Arg, on the other hand, should be stronger than those previously mentioned resulting in a more thermostable molecule. It follows, therefore, that stabilization may be induced by the introduction of new or additional electrostatic interactions, preferably with Arg residues as the positively charged participant into the macromolecule. This can be accomplished through the chemical modification of lysine residues with O-methylisourea or methylacetimidate to create Arg-like side chains.[56, 57] This hypothesis was verified with lactate dehydrogenase, where after treatment with O-methylisourea, the enzyme retained 65 % of its activity after 30 min at 65 °C, while the unmodified native enzyme was completely inactivated.[56] Acetamidination has also been used to stabilize this enzyme against irreversible thermoinactivation.[57]

Hydrophobic interactions are often regarded as important contributors to the overall stability of a protein.[58–60] They are the only non-covalent interactions which actually increase with temperature up to about 60–85 °C.[61] The presence of hydrophobic residues on the surface of a protein, in contact with the aqueous medium, creates a thermodynamically unfavorable situation. Therefore, to lower the free energy of the system, the hydrophobic residues/regions associate resulting in conformational changes which may irreversibly denature the biocatalyst.[61, 62] These two facts have led to an exciting use of chemical modification in protein stabilization. In a recent publication, Mozhaev *et al.* postulated that nonpolar clusters on the protein's surface represent potentially unstable regions, which upon heating, may be driven to self-associate and result in new, irreversibly denatured structures.[63] Stabilization may occur if these hydrophobic regions were "masked" via chemical modification. To test this, they used specific chemical reagents to hydrophilize some nonpolar regions on both trypsin and α-chymotrypsin. For the former, surface tyrosine residues were converted to amino-tyrosines, while in the latter, amino groups were acylated with anhydrides or chloroanhydrides of aromatic carboxylic acids. Upon heating, it was observed that these modified enzymes were 100 to 1000 times more resistant to irreversible thermoinactivation than the native enzyme. In fact, their stability is nearly equal to that of proteolytic enzymes from extreme thermophiles, the most stable proteinases known to date. This high degree of stabilization resulting from a chemical modification with low molecular weight compounds has never before been demonstrated and certainly warrants further consideration and study.

Immobilization of some biocatalysts to a support has proven to be an effective means to reduce the conformational flexibility of an enzyme and thus stabilize it. Many reviews are available which describe the varying procedures.[3, 64–66] These diverse techniques include the immobilization of biocatalysts to both soluble and insoluble supports, covalent and non-covalent attachments and entrapments, and even encapsulation in hollow fibers. Purified enzymes as well as whole cells have been successfully immobilized and used under harsh conditions where the native, unmodified molecule would be very unstable.

Immobilization via multipoint attachments has clear advantages over single-point attachment for stabilization, because more of the molecule is rigidified in the former case. Recently, Germain *et al.* discovered that the thermal stability of a cellobiase–dextran conjugate increased with the number of covalent linkages introduced.[68] Implementation of the principle of multipoint attachment can at times be difficult, since the support and the macromolecule do not have complementary surfaces. This can be circumvented by using chemical agents which vary in length so as to accommodate the three-dimensional structure of the enzyme and link up different parts of the molecule to the support. Copolymerization—the entrapment of an enzyme molecule within a polymer matrix—is yet another way to achieve multipoint attachments.[3, 69] This method is extremely useful, since the polymer conforms to the surface of the macromolecule and allows for many attachments without the use of numerous chemical agents. Chymotrypsin, for example, was copolymerized in polymethacrylate and found to be stabilized against irreversible thermoinactivation by a factor of 1000 at 60 °C.[70]

Chemical reagents which covalently cross-link amino acid residues within a molecule and eliminate the use of a support (carrier) are often used to reduce the enzyme's conformational flexibility and confer stability. β-Glucosidase from *Aspergillus niger* was cross-linked with a glutaraldehyde/sodium borohydride treatment and found to have a half-life 80 times longer than the native enzyme at 65 °C.[71] Use of this modified enzyme could dramatically increase the efficiency of the cellulose digestion process which is currently limited by the instability of the native enzyme. There are many bifunctional reagents available to accomplish this cross-linking ranging from glutaraldehyde, diimidoesters and dithiols to diamines and dicarboxylic acids (used after an appropriate amino acid activation).[41, 72–75] Many of the bifunctional cross-linking agents are variable in length which allows for linkage between different parts of the native molecule to occur. Intramolecular cross-linking can also be used to prevent the disassociation of an oligomeric protein, a process known to inactivate some enzymes.[76] Glyceraldehyde-3-phosphate dehydrogenase, treated with diacids of varying length following carbodiimide activation, lactate dehydrogenase, cross-linked with gluteraldehyde, and yeast acid phosphatase, cross-linked through its carbohydrate chains, are three examples where stabilization following chemical modification was attributed to the prevention of subunit disassociation.[14, 77, 78]

In summary, chemical modification *can* be used to stabilize proteins against irreversible inactivation, but as with any approach, it is not a panacea. Its drawbacks include the lack of complete specificity of the reagents, the limited number of amino acid residues that can be introduced (i.e., there are no suitable reagents for the modification of amino acid residues such as isoleucine, valine, glycine, phenylalanine, or leucine), the drastic changes in the enzyme's conformation which frequently accompany chemical modification (e.g., modification often results in an increase in the side chain bulk which can result in large conformational perturbations), and the lack of site-directed modifications.[79] (Most of these concerns can be overcome through genetic engineering as discussed in the subsequent section.) It is obvious also, from the preceding discussion, that knowledge of the inactivating processes can greatly increase the likelihood that the modification will lead to stabilization, although even then it will remain a highly empirical process.

12.4. GENETIC ENGINEERING

The developments in molecular biology over the past twenty years have provided scientists with a powerful tool to address fundamental biochemical questions regarding protein structure, function, folding and unfolding patterns, and stability. From the practical viewpoint, it provides the means to tailor-make enzymes which potentially possess the stability or selectivity necessary to overcome the native enzyme's limitations. The genetic manipulations vary from the insertion of DNA-coding for protein X from a mesophile into a thermophilic host, to site-directed mutagenesis where preselected amino acid residues are replaced by other natural or even nonnatural amino acid residues.[80–84] Many of these genetic engineering techniques have been used with various proteins to increase their stability.

A rapid approach for obtaining thermostable enzymes via molecular manipulations is to transform an essential gene from a mesophile into a thermophilic host grown at high temperatures and selected for increased stability.[85] Liao *et al.* demonstrated the effectiveness of this technique by selecting a mutant kanamycin nucleotidyl transferase found to tolerate 63 °C for prolonged periods of time as compared to the native enzyme which inactivated rapidly at 55 °C.[80] Subsequent sequencing revealed different undesigned point mutations in each of the mutants. This technique of employing a thermophilic cell to produce stable enzymes, should it prove general, could be extremely useful in view of its rapidness. Random mutations can also be rapidly introduced into a molecule by chemically mutagenizing its gene with agents such as hydrazine, formic acid, or sodium nitrite.[86] Makino *et al.* used this approach to introduce changes into glucose dehydrogenase from *Bacillus megaterium.*[87] Using chemical mutagenesis followed by a rapid temperature screen, twelve mutants were found to retain from 1% to 97% of their original activity after 20 min at 60 °C—conditions under which the native enzyme was completely inactivated. Further analysis revealed the exact sites of amino acid substitutions, although the mechanism of stabilization remains unclear and will require further study. Chemical mutagenesis was also employed in the thermostabilization of the *Bacillus* serine protease, subtilisin BPN′.[88] These two general approaches are useful for situations where little is known about a molecule's three-dimensional structure making it impossible to rationally preselect sites to be altered. Moreover, both are rapid procedures which employ simple selection schemes.

In contrast to the preceding random approaches to genetic engineering, site-directed mutagenesis, as the name suggests, actually introduces point mutations at specific locations in the molecule. These sites are usually selected after careful study of the primary structure of the protein (and other homologous enzymes from different species, e.g., thermophiles), if available, its tertiary structure from X-ray diffraction analysis, and any other biochemical studies which indicate a likely mechanism of inactivation.[89] The changes are then introduced, the new gene expressed, and the mutant protein analyzed. The outcomes, despite the careful, rational planning, are often surprising and require reinterpretation of one's hypothesis. This undoubtedly reflects our limited, but fast-growing understanding of structure–function relationships in proteins. Nonetheless, there are increasing reports of recombinant enzymes containing point mutations which are more resistant to inactivation than the parent enzyme.[90–102]

Site-directed mutagenesis has been used to implement the general strategy of stabilizing an enzyme against irreversible inactivation by reducing its tendency to undergo reversible thermounfolding (see Section 12.1). For example, the hydrophobic interactions in the α-subunit of tryptophan synthetase from *E. coli* were strengthened by substituting a methionine residue for glutamic acid residue at site 49.[90] The observed 1.8-fold stabilization against irreversible inactivation was attributed to a rigidification of the molecule which slowed reversible unfolding. In the case of both RNase A and lambda repressor, stabilization was achieved by strengthening a portion of their secondary structure, the α-helix.[91, 92] An increase in T_m of 6 °C was measured for some mutants in both cases, directly demonstrating a new, inflexible structure. Disulfide bonds can be used to stabilize proteins against reversible thermal unfolding by reducing the entropy of the unfolded state, and, in fact, many proteins have had disulfide bonds introduced via genetic engineering.[93–97] Subtilisin BPN′, engineered to have a disulfide bridge between residues 22 and 87, has a T_m 3.1 °C higher than the native enzyme.[96] Disulfide bonds can also be introduced into proteins and not result in a thermostabilization as seen with dihydrofolate reductase.[97] In fact, their introduction may be a detriment for enzymes that are to work at high temperatures for long periods of time because of their tendency to undergo cleavage (β-elimination) and/or shuffling (disulfide interchange), processes known to lead to the irreversible inactivation of many enzymes.[17, 19] Perry and Wetzel found that the mutant lysozyme molecules that had a disulfide bond introduced and a free cysteine residue replaced via genetic engineering were more stable than either the native enzyme or mutant containing the free thiol.[99] Once the destabilizing thiol–disulfide exchange process was eliminated, the stabilizing effect of the disulfide could be observed. Metallo-enzymes offer yet another alternative for stabilization against reversible unfolding: engineering increased stability through an enhanced binding affinity at the metal binding site.[100] Pantoliano *et al.* chose subtilisin BPN′ which possesses a weak calcium binding site to test the feasibility of this hypothesis.[100] They replaced two uncharged amino acid residues in the vicinity of the bound Ca^{2+} with negatively charged aspartic acid residues and found the binding affinity for Ca^{2+} to be increased over sixfold. The stability against irreversible thermoinactivation of this mutant enzyme was also greater than that of the parent enzyme, and additional increases are predicted with the introduction of other mutations.

Knowledge about the exact mechanism of irreversible inactivation when coupled with site-directed mutagenesis can be extremely useful for designing stable molecules. For example, biochemical studies of subtilisin revealed that when the active-site methionine residue was oxidized, subtilisin was irreversibly inactivated.[101] Replacement of this sensitive residue with alanine, leucine, or serine residues resulted in a dramatic resistance towards oxidative inactivation. Similarly, Ahern *et al.* found that deamidation of amide residues resulted in the irreversible thermoinactivation of dimeric yeast triose phosphate isomerase.[102] The subsequent replacement of two asparagine residues at the intersubunit interface by threonine and isoleucine residues resulted in a thermostable mutant exhibiting a half-life at 100 °C nearly two times greater than the native enzyme.

12.5. ENZYMES ISOLATED FROM THERMOPHILES

Organisms found growing at temperatures from 50 to 100 °C are considered thermophiles.[31] They and their proteins have long played a significant industrial role, initially causing the spoilage of many canned foods.[103] This negative image, however, has diminished over the years, being replaced with the positive anticipation that they may prove valuable to biotechnology in processes requiring harsh conditions or long storage times.

It is logical to expect that proteins isolated from organisms growing in extreme conditions will be more resilient than their counterparts found in moderate environments (e.g., mesophiles cultured at 25–45 °C[31]). This expectation has indeed been born out: most purified proteins isolated from thermophiles possess an intrinsic thermostability greater than the homologous mesophilic enzyme.[104] For example, in examining lactate dehydrogenases, alanine dehydrogenases, and neutral proteinases, it was found that the thermophilic enzymes retained 100% of their activity after 30 min at 70 °C to 90 °C, whereas the mesophilic proteins exhibited the same at only 40 °C to 60 °C.[51] In addition to tolerating high temperatures better than the enzymes from mesophiles, proteins from thermophiles also exhibit better stability at moderate temperature, which could lead to increased shelf life of commercial products.

The remarkable stability exhibited by thermophilic proteins is generally attributed to a rigidification of its structure (see Section 12.1). [To date however, the exact mechanism used to bring about this rigidification remains unclear and is the subject of many studies[51, 67, 85, 105]]. Another consequence of this inflexibility is that for many enzymes the catalytic activity at temperatures well below the optima is drastically reduced.[106–108] For these enzymes high temperatures are required to bring about the degree of structural flexibility that is required for enzyme catalysis.[109–111] This relationship of increased stability–reduced activity holds for many enzymes, of which enolase, kanamycin nucleotidyltransferase, and aldolase are examples.[112–114] It is, however, not without exception and should be tested for the enzyme of interest.[104] The homologous enzymes from mesophiles and thermophiles do exhibit similar specific activities at their respective temperature optima, suggesting that the use of such thermophilic enzymes will be more effective at high than at low temperatures.[106]

Despite the obvious advantages offered by enzymes isolated from thermophiles, there are only a few processes where they currently play a critical role. These areas include the starch and cellulose-processing industry and the detergent industry.[115] The majority of industrial thermostable enzymes are produced by mesophiles which, after perhaps years of screening and strain selection, possess some degree of stability. This contradiction can be explained by the fact that the study of thermophiles is in its infancy. The plethora of organisms found in geothermal areas or deep-sea thermal vents are just beginning to be studied, and simple questions of how to culture and ferment these organisms, or what equipment is needed, are finally being addressed. Hence it appears that, in time, a host of stable biocatalysts isolated from thermophiles will be available for use in biotechnology. Investment in this growing field has a high chance of success since the odds of finding the enzyme of interest in a thermophile are very good considering that the metabolic activities appear to be the same in mesophiles and thermophiles.[106]

In summary, it is safe to assume that the interest in and demands for highly thermostable enzymes will only increase with time. Already enzymes from thermophiles are impacting new fields. For example, NADH oxidase from *Thermus aquaticus* shows great potential for use in biosensors, and *T. aquaticus* DNA polymerase, which greatly facilitates DNA analysis, offers vast possibilities in medicine, forensic science, and fundamental research.[116, 117] This approach for obtaining stable biocatalysts will surely grow in importance and undoubtedly warrants the investment of resources for its development.

12.6. SUMMARY

The preceeding discussions describe some of the wealth of information available to the scientist assigned the task of stabilizing a biocatalyst for a particular application. Generally, knowledge about the mechanisms of inactivation for the enzyme of interest and the fundamentals of protein structure aid in the design of rational approaches to its stabilization. To this end, there is a wide variety of options available particularly if the approaches are coupled (see Table 1). For example, one can use additives and chemical modification to identify amino acid residues for subsequent

Table 1. Approaches Available for Stabilizing Biocatalysts

1. Use of Additives	
	Metals, salts, substrates, sugars, antioxidants
2. Chemical Modification of Existing Enzyme	
	Amino acid modification, cross-linking, immobilization
3. Genetic Manipulation	
	Site-directed mutagenesis, chemical mutagenesis, DNA transformation, and subsequent selection
4. Enzymes Isolated from Thermophiles	
5. Combination of Approaches 1–4	
	Genetic manipulation of thermophile's DNA, chemical modification coupled with genetic engineering

site-directed mutagenesis. Similarly, one can join the techniques of molecular biology with the study of thermophilic organisms to manipulate the DNA of thermophiles in a mesophilic host. These approaches are relatively straightforward, although the consequences are not. The reality of the situation is that despite growing insights into protein structure and stability and the increasingly strong arsenal of stabilization methods, the task of preparing a stable biocatalyst for some applications remains largely empirical. There are many reasons for this, including an incomplete fundamental understanding of most proteins' structure–function relationship as well as the complexity of the milieu the biocatalyst must act in during the application. For example, in many immunoassays employing enzymes, interferences from patient samples are commonly encountered thus reducing the necessary specificity and selectivity of the assay. Extraneous proteins, such as digested gelatin or serum proteins, must be included in the diluent containing the enzyme–antibody conjugate to reduce the interfering nonspecific interactions but may destabilize the biocatalyst.[118] This complexity cannot be eliminated if the test is to be successful, yet it often creates a system too complicated to ascertain what is needed to stabilize the protein.

The need for increasingly stable biocatalysts will continue to grow. Undoubtedly, new and increasingly powerful tools will aid in their design, thus opening the door to many biotechnological applications.

REFERENCES

1. P. H. Blanchard and E. O. Geiger, *Sugar Technol. Rev. 11*, 1 (1984).
2. V. S. Malik, in: *Applied Microbiology* (S. L. Neidleman, ed.), Vol. 34, p. 263, Academic Press, New York (1989).

3. A. M. Klibanov, *Adv. Appl. Microbiol. 29*, 1 (1983).
4. R. D. Schmid, in: *Advances in Biochemical Engineering* (T. K. Ghose, A. Fechter, and N. Blakebragh, eds.), Vol. 12, p. 41, Springer-Verlag, Berlin (1979).
5. T. J. Ahern and A. M. Klibanov, in: *Protein Engineering* (D. L. Oxender and C. F. Fox, eds.), p. 213, Alan R. Liss, New York (1987).
6. H. Neurath, J. P. Putnam, F. W. Putnam, and J. O. Erikson, *Chem. Rev. 34*, 157 (1944).
7. M. L. Anson, *Adv. Prot. Chem. 2*, 361 (1945).
8. W. Kauzmann, *Adv. Prot. Chem. 14*, 1 (1959).
9. R. Lumry and H. Eyring, *J. Phys. Chem. 58*, 110 (1954).
10. L. F. McCoy, Jr. and K. P. Wong, *Biochem. 20*, 3062 (1981).
11. A. M. Klibanov and V. V. Mozhaev, *Biochem. Biophys. Res. Commun. 83*, 1012 (1978).
12. K. Martinek, V. V. Mozhaev, and I. V. Berezin, *Biochim. Biophys. Acta 615*, 426 (1980).
13. S. J. Tomazic and A. M. Klibanov, *J. Biol. Chem. 263*, 3086 (1988).
14. N. Gottchalk and R. Jaenicke, *Biotech. Appl. Biochem. 9*, 389 (1987).
15. H. Gorisch and F. Lingens, *J. Bacteriol. 114*, 645 (1973).
16. F. S. Steven, M. M. Griffin, and R. H. Smith, *Eur. J. Biochim. 119*, 75 (1981).
17. S. E. Zale and A. M. Klibanov, *Biochem. 25*, 5432 (1986).
18. D. M. Kuhn, B. Ruskin, and W. Lovenberg, *J. Biol. Chem. 255*, 4137 (1980).
19. T. J. Ahern and A. M. Klibanov, *Science 228*, 1280 (1985).
20. P. Johnson and T. L. Whateley, *Biochem. J. 193*, 285 (1981).
21. Y. M. Torchinski, *Sulfhydryl and Disulfides in Proteins*, Consultants Bureau, New York (1974).
22. G. E. Pollock and L. H. Frommhagen, *Anal. Biochem. 24*, 18 (1968).
23. Z. Bohak, *J. Biol. Chem. 239*, 2878 (1964).
24. S. E. Zale and A. M. Klibanov, *Biotech. and Bioeng. 25*, 2221 (1983).
25. C. Tanford, *Adv. Protein Chem. 23*, 121 (1968).
26. W. Kauzmann, *Adv. Protein Chem. 14*, 1 (1959).
27. S. Lapanje, in: *Physiochemical Aspects of Protein Denaturation*, Wiley, New York (1978).
28. R. Jaenicke, *Ann. Rev. Biophys. Bioeng. 10*, 1 (1981).
29. G. E. Schultz and R. H. Schirmer, in: *Principles of Protein Structure*, Ch. 3, Springer-Verlag, New York (1979).
30. S. J. Tomazic and A. M. Klibanov, *J. Biol. Chem. 263*, 3092 (1988).
31. T. D. Brock, *Science 230*, 132 (1985).
32. P. H. Von Hippel and K. Y. Wong, *Science 145*, 577 (1964).
33. H. B. Bull and K. Breese, *Arch. Biochem. Biophys. 202*, 116 (1980).
34. S. Damodaran and J. E. Kinsella, *J. Biol. Chem. 256*, 3394 (1981).
35. F. W. Dahlquist, L. W. Long, and W. L. Bigbee, *Biochem. 15*, 1103 (1976).
36. A. Fontana, E. Boccu, and F. M. Veronese, in: *Enzymes and Proteins from Thermophilic Microorganisms* (H. Zuber, ed.), p. 55, Birkhauser Verlag, Basel (1976).
37. W. Heinen and A. M. Lauwers, in: *Enzymes and Proteins from Thermophilic Microorganisms* (H. Zuber, ed.), p. 77, Birkhauser Verlag, Basel (1976).
38. S. Y. Gerlsma, *J. Biol. Chem. 243*, 957 (1968).
39. R. F. Bernier and F. J. Stutzenberger, *J. Biotech. 7*, 293 (1983).
40. K. Yasumatsu, M. Ohno, C. Matsumuro, and H. Shimazono, *Agr. Biol. Chem. 29*, 665 (1965).
41. C. O'Fagain, H. Sheehan, R. O'Kennedy, and C. Kilty, *Proc. Biochem. 23*, 199 (1988).
42. J. C. Lee and S. N. Timasheff, *J. Biol. Chem. 256*, 7193 (1981).
43. C. O'Sullivan and F. W. Thompson, *J. Chem. Soc. 57*, 834 (1890).
43. K. Gekko and S. N. Timasheff, *Biochem. 20*, 4667 (1981).
44. J. F. Back, D. Oakenfull, and M. B. Smith, *Biochem. 18*, 5191 (1979).
45. R. Lumry and H. Eyring, *J. Phys. Chem. 58*, 110 (1954).

46. D. J. Merkler, K. Strikumar, and F. C. Wedler, *Biochem. 26*, 7805 (1987).
47. A. Zaks and A. M. Klibanov, *Science 224*, 1249 (1984).
48. A. Zaks and A. J. Russell, *J. Biotech. 8*, 259 (1988).
49. A. L. Tappel, *CA 75*, 1068007 (1971).
50. M. F. Perutz, *Science 201*, 1187 (1978).
51. H. Zuber, in: *Biochemistry of Thermophily* (S. M. Friedman, ed.), p. 267, Academic Press, New York (1978).
52. S. C. Fujita, T. Oshima, and K. Imahori, *Eur. J. Biochem. 64*, 57 (1976).
53. C. Ruegg, D. Amer, and K. Lerch, *J. Biol. Chem. 257*, 6420 (1982).
54. J. E. Walker, A. J. Wonacott, and I. J. Harris, *Eur. J. Biochem. 108*, 581 (1980).
55. J. L. Finney, B. J. Gettatly, I. C. Golton, and J. Goodfellow, *Biophys. J. 32*, 17 (1980).
56. N. Minotani, T. Sekiguchi, J. G. Bautista, and Y. Nosoh, *Biochim. Biophys. Acta 581*, 334 (1979).
57. F. Tuengler and G. Pfleiderer, *Biochim. Biophys. Acta 284*, 1 (1977).
58. R. Singleton and R. E. Amelunxen, *Bacteriol. Rev. 37*, 320 (1973).
59. P. Argos, M. G. Rossman, U. M. Girau, G. Zuber, G. Frank, and J. D. Tratschin, *Biochem. 18*, 5698 (1979).
60. A. R. Ferst, in: *Enzyme Structure and Mechanism*, W. H. Freeman, England (1977).
61. H. A. Scheraga, in: *The Proteins* (H. Neurath, ed.), Vol. 1, p. 477, Academic Press, New York (1963).
62. T. E. Creighton, in: *Proteins, Structure and Molecular Properties*, W. H. Freeman and Company, New York (1984).
63. V. V. Mozhaev, V. A. Siksnis, N. S. Melik-Nubarov, N. Z. Galkantaite, G. J. Denis, E. P. Butkus, B. Y. Zaslavsky, N. M. Mestechkina, and K. Martinek, *Eur. J. Biochem. 173*, 147 (1976).
64. M. Mosbach, *Methods Enzymol. 44*, 453 (1976).
65. G. P. Royer, *Catal. Rev. 22*, 29 (1980).
66. A. M. Klibanov, *Anal. Biochem. 93*, 1 (1979).
67. V. V. Mozhaev and K. Martinek, *Enzyme Microb. Technol. 6*, 50 (1984).
68. P. Germain, T. Stagmolen, and R. R. Chichton, *Biotech. Bioeng. 33*, 563 (1989).
69. A. Pollak, H. Blumefeld, M. Wax, R. L. Baughn, and G. M. Whitesides, *J. Am. Chem. Soc. 102*, 6324 (1980).
70. K. Martinek, A. M. Klibanov, V. S. Goldmacher, and I. V. Berezin, *Biochim. Biophys. Acta 485*, 1 (1977).
71. J. O. Baker, K. K. Oh, K. Grohman, and M. E. Himmel, *Biotech. Lett. 10*, 325 (1988).
72. O. R. Zaborski, in: *Enzyme Engineering* (E. K. Pye and L. B. Wingard, eds.), Vol. 2, p. 115, Plenum Press, New York (1973).
73. R. Reiner, H. U. Siebeneick, I. Christensen, and H. Doring, *J. Mol. Catal. 2*, 119 (1977).
74. G. E. Means and R. E. Feeney, in: *Chemical Modification of Proteins*, Holden-Day, San Francisco (1971).
75. H. Fasold, J. Klappenberb, C. Meyer, and H. Remold, *Angew. Chem. Int. Ed. Eng. 10*, 854 (1971).
76. E. Dziemer-Gryszkiewz, A. V. Maksimenko, V. P. Torchilin, and W. S. Ostrowski, *Biochem. Int. 6*, 627 (1983).
77. V. P. Torchilin, V. S. Trubetskow, V. G. Omel'yanenko, and K. Martinek, *J. Mol. Cat. 19*, 291 (1983).
78. B. Kuzulic, S. Barbaric, B. Ries, and P. Mildner, *Biochem. Biophys. Res. Commun. 122*, 1083 (1984).
79. A. T. Profy and P. Schimmel, *Prog. Nucleic Acid Res. Mol. Biol. 35*, 1 (1988).
80. H. Liao, T. McKenzie, and R. Hageman, *Proc. Natl. Acad. Sci. USA 83*, 576 (1986).
81. K. M. Ulmer, *Science 219*, 666 (1983).

82. G. Winter, A. R. Fersht, A. Y. Wilkinson, M. Zoller, and M. Smith, *Nature 299*, 756 (1982).
83. J. R. Roesser, M. S. Chorghade, and S. M. Hecht, *Biochem. 25*, 6361 (1986).
84. C. J. Noren, S. J. Anthony-Cahill, M. C. Griffith, and P. G. Schultz, *Science 244*, 182 (1989).
85. J. R. Knowles, *Science 236*, 1252 (1987).
86. B. D. Davie, R. Dulbecco, H. N. Eisen, G. S. Ginsberg, and W. B. Wood, Jr., in: *Microbiology* (M. McCarty, ed.), p. 254, Harper and Row, New York (1973).
87. Y. Makino, S. Negoro, I. Urabe, and H. Okada, *J. Biol. Chem. 264*, 6381 (1989).
88. A. L. Rollence, D. Filpula, M. W. Pantoliano, and P. N. Bryan, *CRC Crit. Rev. Biotech. 8*, 217 (1988).
89. V. V. Mozhaev, I. V. Berezin, and K. Martinek, *CRC Crit. Rev. Biochem. 23*, 235 (1988).
90. K. Yutani, K. Ogasahara, Y. Sugino, and A. Matsushiro, *Nature 267*, 274 (1977).
91. C. Mitchinson and R. L. Baldwin, *Prot. Struct. Func. and Genet. 1*, 23 (1986).
92. M. H. Hecht, T. M. Sturtvant, and R. T. Sauer, *Prot. Struct. Func. and Genet. 1*, 43 (1986).
93. C. G. Anfinsen and H. A. Scheraga, *Adv. Prot. Chem. 29*, 205 (1975).
94. J. E. Villafranca, E. E. Howell, O. H. Voet, M. S. Strobel, R. C. Ogden, J. N. Abelson, and J. Kraut, *Science 222*, 782 (1983).
95. L. J. Perry and R. Wetzel, *Science 226*, 555 (1984).
96. M. W. Pantoliano, R. C. Ladner, P. N. Bryan, M. L. Rollence, J. F. Wood, and T. L. Poulos, *Biochem. 26*, 2077 (1987).
97. J. E. Villafranca, E. H. Howell, S. J. Oatley, N. Xuong, and J. Kraut, *Biochem. 26*, 2182 (1987).
98. D. B. Volkin and A. M. Klibanov, *J. Biol. Chem. 262*, 2945 (1987).
99. L. J. Perry and R. Wetzel, *Biochem. 25*, 733 (1986).
100. M. W. Pantoliano, M. Whitlow, J. F. Wood, M. L. Rollence, B. C. Finzel, G. L. Gilliland, T. L. Poulos, and P. N. Bryan, *Biochem. 27*, 8311 (1988).
101. D. A. Estell, T. P. Graycar, and J. A. Wells, *J. Biol. Chem. 260*, 6518 (1985).
102. T. J. Ahern, J. I. Casal, G. A. Petsko, and A. M. Klibanov, *Proc. Natl. Acad. Sci. USA 84*, 675 (1987).
103. E. R. L. Gaughran, *Bacteriol. Rev. 11*, 189 (1947).
104. R. Singleton, Jr. and R. E. Ameluxen, *Bacteriol. Rev. 37*, 320 (1973).
105. P. Argos, M. G. Rossman, U. M. Girau, H. Zuber, G. Frank, and J. D. Tratschin, *Biochem. 18*, 5698 (1979).
106. A. Fonatana, *Biophys. Chem. 29*, 181 (1988).
107. R. Jaenicke, *Annu. Rev. Biophys. Bioeng. 10*, 1 (1981).
108. H. Zuber, in: *Structural and Functional Aspects of Enzyme Catalysis* (H. Eggerer and R. Huber, eds.), p. 114, Springer-Verlag, Heidelberg (1981).
109. D. Ringe and G. A. Petsko, *Prog. Biophys. Mol. Biol. 45*, 197 (1985).
110. G. Careri, P. Fasella, and E. Gratton, *CRC Crit. Rev. Biochem. 3*, 141 (1975).
111. A. Cooper, *Sci. Prog. Oxford 66*, 473 (1981).
112. E. Stillwagen and L. D. Barnes, in: *Enzymes and Proteins from Thermophilic Microorganisms* (H. Zuber, ed.), Exp. Suppl. 26, p. 223, Birkhauser Verlag, Basel (1976).
113. M. Matsumura, S. Yasumura, and S. Aiba, *Nature 323*, 356 (1986).
114. H. Freeze and T. D. Brock, *J. Bacteriol. 101*, 541 (1970).
115. T. K. Ng and W. R. Kenealy, in: *General, Molecular and Applied Microbiology* (T. D. Brock, ed.), p. 197, Wiley, New York (1986).
116. C. J. McNeil and J. A. Spoors, *Anal. Chem. 61*, 25 (1989).
117. J. L. Marx, *Science 240*, 1408 (1988).
118. C. Blake and B. J. Gould, *Analyst 109*, 108 (1984).

13

Protein Engineering of Subtilisin

THOMAS P. GRAYCAR

13.1. INTRODUCTION

The use of enzymes as catalysts in organic synthesis reactions has expanded rapidly in recent years.[1–5] In contrast to conventional organic reactions enzyme-catalyzed reactions offer the potential of highly stereoselective or regioselective transformations. The value of enzymes in organic synthesis can be further attributed to the high reactivity and mild conditions characteristic of enzyme-catalyzed reactions. The increased availability and lower cost of enzymes resulting from the development of recombinant DNA technology has also stimulated interest in the use of enzymes for the production of fine organic chemicals. An apparent impediment to the widespread utilization of enzymes as commercial biocatalysts, however, is the difficulty encountered in optimizing their use in industrial processes. Although enzymes are capable of reacting with compounds having structures similar to that of their natural substrates, the catalytic efficiency of reactions involving nonnatural substrates is often suboptimal. In addition, the inherent instability of proteins and the sensitivity of enzyme activity to alterations in pH further limit their utility in industry. Hence, despite the known advantages of enzymes their use as biocatalysts is currently limited to less than 5% of the total industrial enzyme market.[6] In the past, the only methods available to optimize the performance of an

THOMAS P. GRAYCAR • Department of Enzymology, Genencor International, Inc., 180 Kimball Way, South San Francisco, CA 94080.

Table 1. Functional Properties of Enzymes which Can Be Modified by Protein Engineering

Stability	Specificity
Temperature	Substrate
Denaturants	Nucleophile
pH	Catalytic rate
Organic solvents	pH Activity profile
Oxidation	Allosteric regulation
Proteolysis	Antigenicity

enzyme involved chemical modification of residues on the surface of the enzyme or else random mutagenesis of the corresponding gene. Now, however, the recent development of protein engineering has made it possible to redesign the structure of an enzyme and tailor its functional properties for a particular application, thereby greatly enhancing the potential to create novel industrial biocatalysts.

The field of protein engineering developed in the early 1980s as advances in recombinant DNA techniques made it possible to specifically modify the coding sequence of a cloned gene. This powerful technology enables one not only to determine the functional importance of a particular residue within a protein but also to change systematically the structure and corresponding properties of the protein. The functional properties of an enzyme which can be modified by protein engineering are listed in Table 1. Since the advent of protein engineering, numerous examples of such modifications in different enzymes have been reviewed.[7–10] In particular, the *Bacillus* serine protease subtilisin has proven to be an excellent model system for protein engineering.[11] Indeed, modifications in most of the properties listed in Table 1 have been made for this enzyme. The purpose of this chapter is to review the progress made in the engineering of subtilisin and to describe the various approaches used in the process. As we shall see, an extensive data base has been generated from the analysis of engineered subtilisin mutants. In addition to yielding valuable insight into protein structure–function relationships, this will enable us to make some projections for the future uses of protein engineering for industrial biocatalysts.

13.2. APPROACHES TO THE ENGINEERING OF SUBTILISIN

Although the technology exists to create new proteins or to specifically alter the structure of existing ones, the theoretical methodology for using this technology to create useful proteins remains to be developed. Despite recent advances in the *de novo* design of proteins, our understanding of how the amino acid sequence of a polypeptide chain directs its folding into a stable three-dimensional structure is still largely incomplete.[12–13] Furthermore, due to the complex nature of electrostatic interactions within proteins,[14–15] the structural modification of proteins is largely carried out on an empirical basis. This dilemma in itself is not so serious as to preclude the development of engineered industrial biocatalysts. It does, however, require careful consideration in selecting the enzyme to be redesigned by protein engineering. For many reasons, technical as well as commercial, subtilisin is an excellent protein for studying protein engineering. Serine proteases comprise one of the best understood classes of enzymes in terms of mechanism of catalysis.[16–17] Extensive kinetic measurements have been reported for subtilisin using a variety of synthetic substrates.[18] A wealth of structural information on subtilisin has been obtained from chemical modification[19] as well as X-ray crystallography studies.[20–25] Subtilisin is also readily available. The subtilisin BPN′ gene from *Bacillus amyloliquefaciens* has been cloned[26] and expressed in a protease deficient strain of *Bacillus subtilis*.[27] Subtilisin is secreted extracellularly in large quantities and is easily purified.[28] In addition, commercially sold as an additive to household laundry detergents, subtilisin represents the largest share of the worldwide enzyme market with sales estimated for 1989 at $150 million.[29]

Many methods, both chemical and genetic, are available for modifying proteins (Table 2). As we will see, each of these methods has been utilized in the engineering of subtilisin. The need for a variety of mutagenesis methods arises from our limited understanding of protein structure–function relationships. The specific structural alteration which would best achieve a desired functional change is often not clear. As a result, the

Table 2. Protein Modification Methods

Chemical methods
Total synthesis
Semisynthesis
Side-chain modification
Site-specific mutagenesis
Oligonucleotide-directed
Cassette
Localized random mutagenesis
Chemical
Enzymatic
Modified PCR
Hybrid formation

mutagenesis method of choice depends both on the type of modification required as well as the information available regarding the property to be altered. Potential sites for mutagenesis can often be discerned from sequence comparisons with homologous proteins or by examination of the protein X-ray crystal structure. Mutations can then be introduced at the selected sites by site-specific mutagenesis. However, even given a highly refined crystal structure, random mutagenesis may be required to identify functionally critical residues within a protein. For example, later on we will see how increased thermal stability via the introduction of disulfide bridges proved a difficult property to design into subtilisin. Instead, the selection of random mutants exhibiting increased stability provided a more successful approach. Hence, a brief review of the strengths and limitations of the various methods of protein modification at this point would be useful.

13.2.1. Chemical Methods

There are three different classifications for the chemical modification of proteins: total synthesis, semisynthesis, and side-chain modification.[30] The principal advantage of these methods lies in the ability to make substitutions using noncoded amino acids or analogs. Total synthesis is the method of choice for producing peptides and may soon rival the efficiency of recombinant methods for the production of small proteins. Solid-phase peptide synthesis has advanced rapidly with the development of automated instrumentation. Recently, human transforming growth factor-α, a relatively small protein consisting of 50 amino acid residues, has been chemically synthesized in less than a week.[31] However, because the coupling efficiency of chemical synthesis is not 100%, the production of large proteins is best achieved with recombinant gene expression systems.

Semi-synthesis involves the fragmentation of a protein, either chemically using cyanogen bromide or enzymatically using specific protease cleavages, followed by fragment separation and modification of the residues at the cleavage site. Modified fragments are recombined under conditions which allow for the reformation of a functional protein. Because semisynthesis involves the use of naturally occurring proteins, it is possible to introduce noncoded amino acids into much larger proteins than would be possible via total synthesis. The use of semisynthesis for protein engineering, however, suffers the limitation that modifications are possible only at residues adjacent to the available cleavage sites. Also, fragment condensation can at times be difficult. A novel approach to fragment condensation, however, has been demonstrated for the coupling of tryptic fragments of cytochrome *c*.[32] The fragment comprising residues 1–38 is activated by the enzymatic addition of an amino acid ester. The activated fragment can then facilitate peptide bond formation as the fragments are recombined.

Side-chain modification of proteins involves the chemical modification of surface residues having side chains capable of reacting with various reagents. Prior to the development of site-specific mutagenesis, the chemical modification of side chains was the principal technique used in the identification of active-site residues as well as the alteration of enzyme activity. The chemical conversion of the catalytic serine to cysteine in subtilisin BPN′ is an early example of the utility of chemical methods for generating an enzyme having altered catalytic activity.[33–34] However, the versatility of the chemical modification of side chains is limited in that only residues having reactive side chains may be modified and it is often difficult to modify a single residue. Still, a unique advantage of chemical modification is the ability to introduce compounds such as coenzymes. Kaiser and Lawrence used this approach to convert the thiolprotease activity of papain to an oxidoreductase activity.[35] In addition, side-chain modification can also be combined with site-specific mutagenesis to introduce a variety of structural modifications at a functionally important position in an enzyme. For example, a methionine residue in the substrate binding region of carboxypeptidase Y has been replaced via site-specific mutagenesis with cysteine.[36] Subsequent treatment of the cysteine mutant enzyme with several chemical reagents significantly altered the enzyme's substrate specificity. Thus, it seems clear that chemical methods of protein modification will find applications in the engineering of novel biocatalysts.

13.2.2. Site-Specific Mutagenesis

As its name implies, this technique allows for the specific substitution, insertion, or deletion of any coded amino acid. A commonly used *in vitro* technique for site-specific mutagenesis is oligonucleotide-directed mutagenesis.[37] Briefly described, a mutagenic oligonucleotide is synthesized, which is complementary to a specific region in a cloned gene except for a mismatch at the target codon which directs the desired mutation. The mutagenic oligonucleotide is then annealed to a single-stranded phage DNA vector into which the gene for the protein had previously been cloned. The oligonucleotide can then act as a primer for the Klenow fragment of polymerase I, which, using the single-stranded DNA as a template, replicates the entire vector DNA. Following ligation of the two ends of the newly synthesized DNA by T4-ligase, the heteroduplex DNA is used to transform a host bacterial strain. Single-stranded phage DNA is isolated from plaques and probed for the desired mutant gene by hybridization with a radiolabeled preparation of the mutagenic oligonucleotide. Several improvements in this simple method have been reported which increase the yield of mutant DNA.[38] Although oligonucleotide-directed mutagenesis is a general method for site-specific mutagenesis, it is a laborious method for introducing multiple amino acid substitutions at a single site. This limitation can be troublesome in cases where it is impossible to anticipate the optimal amino acid substitution for the alteration of a particular property. To facilitate the insertion of multiple mutations at a target codon a technique called cassette mutagenesis has been developed.[39] The strategy for cassette mutagenesis involves first the use of oligonucleotide-directed mutagenesis to introduce unique restriction sites into the cloned gene such that they closely flank the codon of interest. The mutagenized gene is then ligated into a shuttle vector plasmid and digested using the restriction enzymes specific for the flanking restriction sites. Pools of mutagenic duplex DNA cassettes (10–25 base pairs in length) are inserted into the gap in the gene, ligated, and the DNA transformed into bacteria. The mutagenic cassettes are designed to restore the coding sequence except at the target codon as well as to eliminate one or both restriction sites. Elimination of a restriction site simplifies selection of transformants containing the mutagenic cassette. The power and utility of these techniques lies in the specificity of mutagenesis achieved with both oligonucleotide-directed and cassette mutagenesis. However, this same specificity can also be a limiting factor in the utilization of site-specific mutagenesis since its application requires precise knowledge of where in the protein to incorporate an amino acid change.

13.2.3. Random Mutagenesis

The combination of random *in vitro* mutagenesis with a suitable phenotypic screening procedure has proved to be a powerful approach for identifying residues in a protein associated with a particular functional property.[40] Recombinant DNA techniques have expanded the types of mutagenesis methods available as well as allowed for mutations to be localized to a specific region of interest in the cloned gene. Various chemical and enzymatic methods are available for generating random point mutations and have been extensively reviewed.[41–43] In addition, a new method which utilizes a modified polymerase chain reaction has been developed and is reported to be both simple and efficient.[44] Although these techniques are capable of generating large libraries of mutant proteins, they all share a common problem in that not all amino acid substitutions can be made at a single site. This limitation arises from the fact that some amino acid substitutions require two or three base changes in a codon, whereas the predominant mutation is of a single base change only. Hence, following the sequence identification of a selected mutant, it is prudent to not only confirm the observed effect of the mutation by site-directed mutagenesis, but it may be also necessary to make additional substitutions to generate sufficient information regarding the structure–function relationship for the site.

Depending on the length of DNA being mutagenized, a further limitation on using random mutagenesis can be the efficiency of the screening procedure used to select mutants exhibiting the phenotype of interest. A rapid and efficient assay is essential for testing a large number of mutants and attaining a high percentage of true positives. The integration of automated pipetting devices, microtiter plate readers, and robotics for sample handling, is a new development in culture screening and enzyme assays which can greatly increase the potential utility of random mutagenesis for protein engineering.[45]

13.2.4. Hybrid Formation

Closely related proteins can often have significantly different functional properties despite a high level of amino acid sequence homology. Identification of the individual residues or segments responsible for the functional difference would lend valuable insight into the relationship between protein structure and function. One approach to mapping the functional properties of related proteins has been devised using the formation of hybrid proteins.[46] Hybrid proteins are generated by *in vivo* recombination of homologous DNA coding sequences. A plasmid is constructed containing the gene for one protein joined near the 3′ end of the gene for the second protein. A poly linker having one or more unique restriction sites is placed at the junction site to facilitate the enrichment of recombinant genes. Gene crossovers occur at regions of high DNA homology and the range and frequency of crossovers is directly related to the degree of homology. Radiolabeled oligonucleotide probes which correspond to regions of low homology are used to localize crossover sites. The exact location of the crossover site can then be easily obtained by direct sequencing of that region of DNA. The functional properties of the protein can be mapped by comparison of the location of crossover sites with the functional properties of the corresponding hybrid proteins. Although this method of analysis is limited to families of homologous proteins it does provide a means for correlating functional properties to the amino acid sequence of a protein. This type of information can be quite valuable for the engineering of proteins especially when a three-dimensional X-ray crystal structure is unavailable. Once sites of functional importance have been identified through the analysis of hybrid proteins, modifications at these sites can be introduced via site-specific mutagenesis to confirm the functional role of the residue or segment and to gain additional structure–function information.

13.3. PROTEIN ENGINEERING OF SUBTILISIN

The chemical or genetic modification of protein structure is only one facet of protein engineering. The analysis of modified proteins to elucidate the relationship between protein structure and function is where most of the effort involved in protein engineering is directed. Unfortunately, mutant proteins can be generated far faster than the three-dimensional structure can be determined or the effect of each mutation on a particular function can be adequately studied. On the other hand, because the chemical forces which direct protein folding and functionality are the same for all enzymes, information gained through the engineering of a particular property in one protein may be applicable to the engineering of that same property or even a separate property in a different protein. Hence, a review of the extensive protein engineering experiments performed with subtilisin should provide insight regarding the potential engineering of functional properties in other enzymes and the approaches which might prove most successful.

13.3.1. Engineering Substrate Specificity

An important goal in the engineering of enzymes is to increase the catalytic efficiency of the enzyme for a chosen substrate. The chemical forces involved in substrate binding and catalysis include a combination of electrostatic (ion-pairing, hydrogen bonding) as well as hydrophobic and steric interactions.[47] The addition or removal of residues in the active site of enzymes via site-specific mutagenesis has proven to be a valuable method for determining the contribution of these different chemical forces toward catalysis and specificity.[8, 10] Measurement of changes in the kinetic parameters k_{cat} and K_M of mutant enzymes for a defined substrate provides a sensitive assay for characterizing the functional consequence of the mutation. The reaction of free enzyme plus substrate to form products follows an apparent second-order rate constant equal to the ratio of k_{cat}/K_M. This ratio is a useful measurement of the catalytic efficiency of an enzyme for a particular substrate as the logarithm of the k_{cat}/K_M ratio is proportional to the activation energy of the transition state for the reaction.[47] Changes in substrate specificity are determined by analysis of the changes in catalytic efficiency.

In the case of subtilisin BPN′, site-specific mutagenesis was utilized to alter substrate specificity by redesigning the structure of the active site such that electrostatic, hydrophobic, and steric interactions with a bound substrate molecule would be modified. The active site of subtilisin consists of an extended trough in the surface of the protein along which numerous pockets called subsites interact with the bound substrate molecule. The nomenclature used in referring to the binding of a polypeptide substrate to the active site of proteases follows a numbering system initially proposed by Schechter and Berger.[48] The substrate amino acids providing the carbonyl carbon and the amino group of the scissile peptide bond are called the P1 and P1′ amino acids respectively. The side chain of the P1 residue binds in the S1 subsite while the side chain of the P1′ residue binds in the S1′ subsite. Substrate residues leading away from the site of bond cleavage toward the N-terminus of the bound substrate are termed the P1, P2,..., P_n residues. Substrate residues leading in the opposite direction toward the C-terminus of the bound substrate are termed the P1′, P2′,..., $P_n{}'$ residues. Figure 1 gives a stereoview of a model tetrapeptide substrate molecule bound in the active site of subtilisin BPN′. The S1 subsite in subtilisin is a large open cleft and is the primary determinant of substrate specificity. Although the substrate specificity for subtilisin BPN′ is quite broad, for a particular set of substrate molecules it has been shown to correlate directly with the hydrophobicity of the P1 amino acid.[49] Catalytic efficiency is greatest for substrates having large hydrophobic P1 amino acids such as tyrosine whereas it is decreased by four orders of magnitude with glutamic acid at P1. A glycine at position 166 is located at the far end of the open cleft of the S1 subsite. Computer modeling suggests that the side chains of amino acid substitutions at position 166 may extend down into the binding site. Hence, it appears possible to alter the substrate specificity of the enzyme by altering the size and hydrophobicity of the S1 subsite via substitutions at position 166. To fully characterize the effect of substitutions at position 166 on substrate specificity, cassette mutagenesis was employed to generate all nineteen possible substitutions.[49]

The analysis of the position 166 mutants for alterations in substrate specificity was completed by determining the catalytic efficiency of each mutant for various synthetic tetrapeptide substrate molecules identical in structure except for the P1 amino acid. The optimal substitutions at position 166 for substrates having different P1 amino acids are listed in Table 3 along with k_{cat}/K_M values and combined side chain volumes. It is significant to note that not only have enzymes been engineered which are more specific for various P1 amino acids but also the catalytic efficiency of these mutant enzymes is of the order of the best substrate of this type for the wild-type enzyme. Also, from this collection of data it appears that for optimal activity the average combined side-chain volume should not exceed 164 ± 30 Å^3.[49] In general, enzyme–substrate pairs which surpassed this combined volume limit exhibited significantly decreased catalytic efficiency. For example, the k_{cat}/K_M for the tryptophan 166 mutant on the substrate having tyrosine at P1 (combined side-chain volume = 308.4 Å^3) is decreased by over three orders of magnitude compared to when glycine is at 166. The tryptophan side chain occupies the previously open S1 subsite and sterically excludes the optimal binding of the substrate tyrosine side

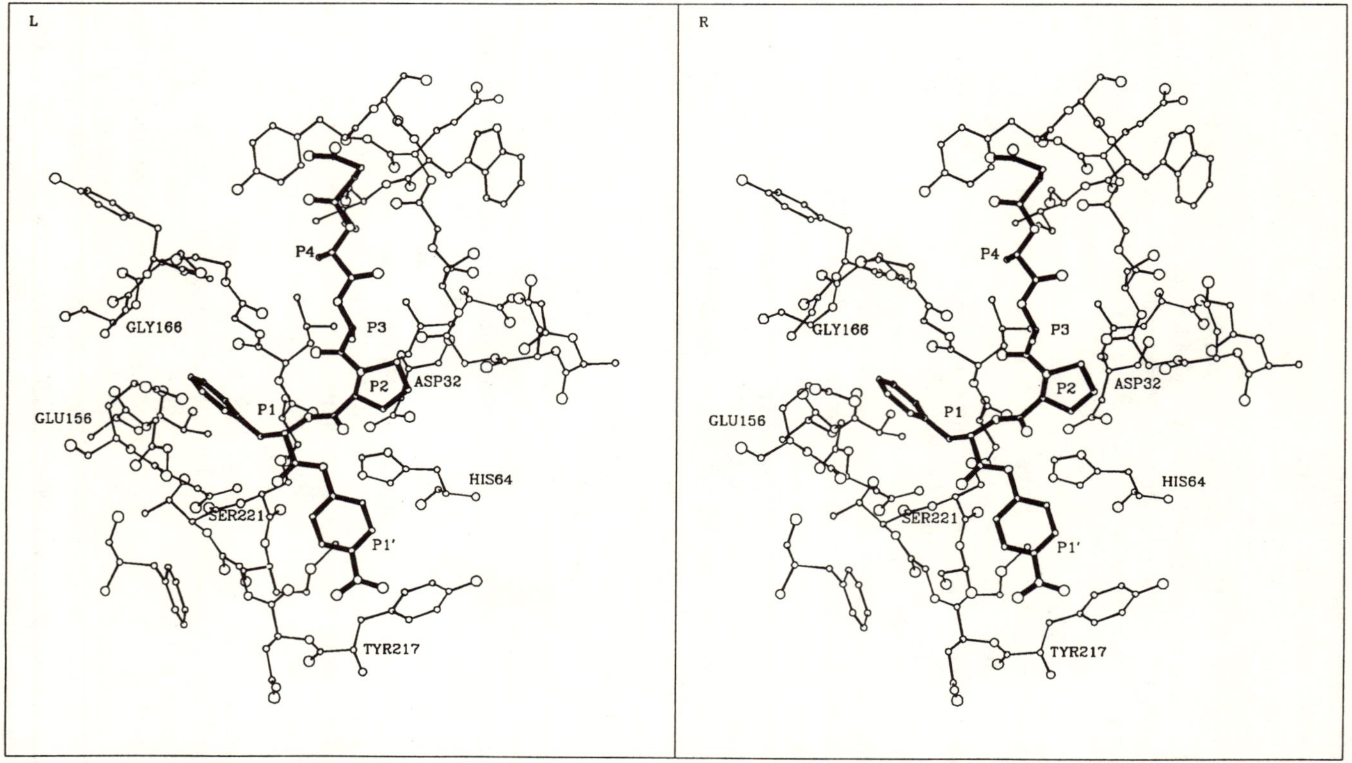

Figure 1. Stereographic view of a model substrate, *N*-succinyl-L-Ala-L-Ala-L-Pro-L-Phe-*p*-nitroanilide (shown with thicker bonds), bound to the active site of subtilisin BPN′. The catalytic residues aspartate 32, histidine 64, and serine 221, as well as other relevant residues, are labeled. Substrate residues are also labeled according to the nomenclature of Schechter and Berger.[48] The graphic was produced by Dr. R. Bott.

Table 3. Catalytic Efficiency (k_{cat}/K_M) and Combined Side-Chain Volume of Optimal Amino Acid Substitutions at Poisition 166 for Different Substrate P1 Amino Acids[a]

Optimal amino acid at position 166	Substrate P1 amino acid	k_{cat}/K_M ($s^{-1}\,M^{-1}$)	Combined side-chain volume ($Å^3$)
Glycine	Tyrosine	$1.1 \cdot 10^6$	138
Alanine	Phenylalanine	$9.5 \cdot 10^5$	163
Valine	Methionine	$1.2 \cdot 10^6$	181
Leucine	Methionine	$1.3 \cdot 10^6$	207
Isoleucine	Alanine	$1.4 \cdot 10^5$	129
			Average = 164 ± 30

[a] All substrates have the form *N*-succinyl-L-Ala-L-Ala-L-Pro-L-[X]-*p*-nitroanilide, where X is the P1 amino acid. Kinetics data and combined side-chain volumes are from ref. 49.

chain. Still, the combined side-chain volumes for the position 166 substitution and the substrate P1 amino acid fail to explain all of the substrate specificity alterations observed for the mutant enzymes.

Orientation of the side chain for the amino acid substituted at position 166 is also important in determining catalytic efficiency. For example, activity for the substrate having phenylalanine at P1 is two orders of magnitude greater for the leucine 166 mutant than it is for the isoleucine 166 mutant.[49] Combined side-chain volumes cannot account for this difference in activity since the side-chain volumes of leucine and isoleucine are nearly identical. High resolution X-ray crystal structures of the mutants having leucine and isoleucine substituted at position 166 have been determined and reveal a significant structural difference for the two enzymes.[50] The electron density map of the leucine 166 structure shows the γ-branched aliphatic side chain of the leucine to be oriented up towards the surface of the enzyme. By contrast, the aliphatic side chain of isoleucine has a β-branch and the electron density map for the isoleucine 166 structure reveals that the larger branch of the side chain is directed down into the S1 subsite. Hence, the side chain of the isoleucine substitution sterically blocks the binding of phenylalanine whereas the leucine side chain does not.

Substitutions at position 166 alter the substrate specificity of subtilisin BPN′ not only by sterically decreasing the size of the S1 subsite but also by increasing the hydrophobicity of the binding site.[49] Substituting nonpolar amino acids at position 166 increased the catalytic efficiency for substrates having small nonpolar P1 amino acids. For example, the isoleucine 166 mutant exhibits over ten times the activity of the glycine 166 (wild-type) enzyme for the substrates having alanine and valine at P1. These results are best explained by the greater hydrophobicity of the isoleucine side chain compared to that of glycine. Interestingly, the difference in substrate specificity profiles between wild-type subtilisin BPN′ and the isoleucine 166 mutant is similar to that between the mammalian serine proteases chymotrypsin and elastase. Both chymotrypsin and subtilisin are more specific for large hydrophobic P1 amino acids whereas both elastase and the isoleucine 166 mutant exhibit greater activity with small non-charged amino acids at P1.

In addition to steric and hydrophobic interactions, electrostatic interactions are also directly involved in determining the catalytic efficiency of enzymes. For the purpose of evaluating electrostatic interactions in the S1 subsite of subtilisin BPN′, mutations were introduced at two separate sites, glutamate 156 and glycine 166.[51] The side chain of glutamate 156 forms a portion of one side of the S1 subsite. Computer modeling of charged P1 amino acids bound into the S1 subsite indicates that ion pair interactions could be formed with substitutions at either position 156[52] or position 166.[53] In general, favorable ion pairing between charged substrates and complementary charged amino acid substitutions at the two sites lowers the activation energy of the transition state. For example, increasing the net positive charge of the binding site by substituting glutamine for glutamate at 156 and lysine for glycine at 166 resulted in significantly increased catalytic efficiency for negatively charged substrates whereas for positively charged substrates catalytic efficiency decreased.[51] Indeed, the catalytic efficiency of the double mutant Gln156/Lys166 for the glutamic acid substrate is increased 1900-fold over that of the native enzyme. This substantial increase in k_{cat}/K_M is achieved by the synergistic effect of the two mutations on lowering the activation energy for the reaction (Table 4). These results suggest that the engineering of electrostatic interactions can improve the catalytic efficiency of an enzyme for a charged substrate. Furthermore, compared to modifications in steric and hydrophobic inter-

Table 4. Catalytic Efficiency of Substitutions at Residue Positions 156 and 166 on a Substrate Having Glutamate at P1[a]

Residue position 156	Residue position 166	k_{cat}/K_M (s^{-1} M^{-1})
Glutamate	Glycine (wild-type)	$3.5 \cdot 10^1$
Glutamine	Glycine	$6.2 \cdot 10^2$
Glutamate	Lysine	$1.2 \cdot 10^4$
Glutamine	Lysine	$6.6 \cdot 10^4$

[a] Kinetic measurements were performed at 25 °C in 0.1 *M* Tris, pH8.6, using the substrate *N*-succinyl-L-Ala-L-Ala-L-Pro-L-Glu-*p*-nitroanilide.[51]

actions, it is apparent that significantly larger effects on specificity can be obtained with alterations in electrostatic interactions.

The success achieved in modifying the substrate specificity of the S1 subsite in subtilisin BPN′ suggests that substitution of residues making direct contact with the bound substrate molecule is a useful approach to the design of enzyme specificity. To test further the use of this approach in protein engineering an attempt was made to convert the substrate specificity of subtilisin BPN′ to that of a related subtilisin from *B. licheniformis* called subtilisin Carlsberg.[54] The activity for different synthetic substrates of these two subtilisins varies by up to 60-fold (Figure 2). Although the primary amino acid sequence of these two enzymes differs by 31%, the alpha carbon backbones given by the X-ray crystal structures can be superimposed with a deviation of about 0.5 Å.[55] Examination of a model substrate bound in the active site of subtilisin BPN′ reveals that of the 86 sequence differences between the two enzymes only two residues, 156 and 217, are within 4 Å of the bound substrate molecule and a third residue, 169, lies within 7 Å.[54] The amino acids at positions 156 and 217 make contact with the substrate P1 and P1′ amino acids respectively. By making contact with the residue at position 152, the amino acid at position 169 can interact indirectly with the substrate P1 amino acid. Substitution at these three sites of the Carlsberg sequence (serine 156, alanine 169, leucine 217) for that of the BPN′ enzyme (glutamate 156, glycine 169, tyrosine 217) by site-specific mutagenesis yielded a mutant of subtilisin BPN′ having a

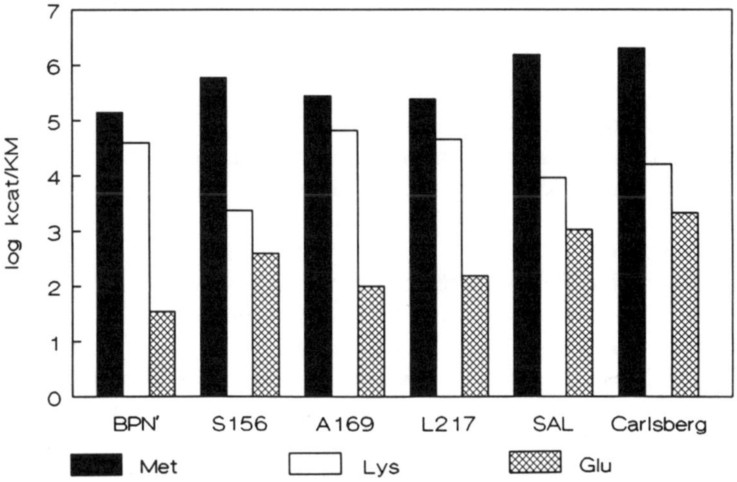

Figure 2. Comparison of substrate specificity profiles for wild-type subtilisins BPN′ and Carlsberg and also for the BPN′ mutants Ser156, Ala169, Leu217, and Ser156/Ala169/Leu217 (SAL). The log k_{cat}/K_M values are for substrates having the form *N*-succinyl-L-Ala-L-Ala-L-Pro-L-[X]-*p*-nitroanilide where the P1 amino acid, X, is methionine, lysine, or glutamate. Measurements were performed at 25 °C in 0.1 *M* Tris, pH 8.6.[54]

substrate specificity profile very similar to that of subtilisin Carlsberg (Figure 2). Furthermore, the substrate specificity profile shown in Figure 2 clearly illustrates the synergism of the three amino acid changes in recruiting the specificity of one enzyme into another. The substitution of serine for glutamate at position 156 exerts the dominant effect in altering specificity. This is not surprising since the change at 156 removes a negative charge from the S1 subsite which, as discussed above, can have a significant influence on specificity.

Using another, rather novel approach to the alteration of specificity, the active site structure of subtilisin may be changed in a manner such that a specific substrate can "assist" its own catalysis.[56] The imidazole side chain of histidine 64 acts both as a catalytic base and acid in the hydrolysis of peptide bonds. Model building of a substrate in the active site of subtilisin suggests that the imidazole side chain of a histidine at the P2 position in the substrate could be superimposed onto the imidazole of the catalytic histidine at position 64. Replacing histidine 64 with an alanine yields an enzyme 400 times more active on a synthetic substrate having histidine at the P2 position as opposed to alanine or glutamine at P2. The wild-type enzyme, on the other hand, is relatively nonspecific for substrates having different P2 amino acids. Although the alanine 64 mutant in comparison to the native enzyme is virtually inactive on most substrates, partial restoration of activity for substrates having histidine at P2 is clearly observed. Thus, "substrate-assisted" catalysis offers a new approach to designing highly selective proteases for uses such as peptide mapping or specific cleavage of heterologous fusion proteins. Recently, the specific cleavage of a fusion protein was reported for a variant of the alanine 64 enzyme which through a combination of protein engineering and substrate optimization exhibited a 20-fold increase in catalytic activity.[57]

So far we have seen how the analysis of substrate molecules built into the X-ray crystal structure of subtilisin BPN′ has been used to select sites for engineering the catalytic efficiency of the enzyme. However, in some cases it may not be possible to clearly determine from models which residues to modify. For example, the homologous subtilisin from *B. subtilis* (subtilisin SBT) is 85% identical in protein sequence to that of subtilisin BPN′. Yet, the kinetic properties of these two enzymes are quite different. The turnover rate, k_{cat}, for subtilisin SBT is seven-fold less than that of subtilisin BPN′ for a synthetic substrate (Table 5). The X-ray structures of the enzymes show no sequence differences within contact distance of a substrate molecule modeled in the active site. Hence, an alternative approach capable of identifying functionally important residues was required. Because of the high homology of the two subtilisins, the formation of hybrid proteins as described previously was a possible method for identifying the location of the residue responsible for the kinetic difference. The

Table 5. Comparison of Kinetic Parameters k_{cat}, K_M, and k_{cat}/K_M Measured for Wild-Types Subtilisin BPN′ and Subtilisin SBT as well as Position 224 Mutations in Subtilisin BPN′

Enzyme	k_{cat} (s^{-1})	K_M (M)	k_{cat}/K_M ($s^{-1}M^{-1}$)
Wild-type BPN′	50	$1.4 \cdot 10^{-4}$	$3.5 \cdot 10^{5}$
Alanine	46	$1.3 \cdot 10^{-4}$	$3.3 \cdot 10^{5}$
Threonine 224	16	$3 \cdot 10^{-4}$	$5.1 \cdot 10^{4}$
Valine 224	3	$3.3 \cdot 10^{-4}$	$8.2 \cdot 10^{3}$
Wild-type SBT	7	$3.2 \cdot 10^{-4}$	$2.2 \cdot 10^{4}$

[a] Kinetic measurements were performed at 25 °C in 0.1 *M* Tris, pH 8.6, using the substrate *N*-succinyl-L-Ala-L-Ala-L-Pro-L-Phe-*p*-nitroanilide.[58]

gene for subtilisin SBT up to codon 222 of the mature enzyme was joined in a plasmid near the 3′ end of the entire subtilisin BPN′ gene.[58] Subtilisin is synthesized in the cell as a precursor, preprosubtilisin,[26] which is processed autoproteolytically to the mature form as it is released from the cell membrane into the medium.[59] Hence, crossovers between the two genes could extend from within the prepro region of the coding sequence giving subtilisin BPN′ as the final gene product, to crossovers at or near codon 222 which would result in a hybrid protein having the first 80% of subtilisin SBT and the last 20% of subtilisin BPN′. Eight hybrids, which by hybridization studies were found to cover the range of possible crossovers, were selected for kinetic analysis. All of the hybrid enzymes resembled subtilisin BPN′ kinetically despite the fact that two of the hybrids contained approximately 80% of the subtilisin SBT sequence. Apparently, the residue responsible for the difference in kinetics between the wild-type enzymes is located after residue 222 in the terminal 20% of the protein.

There are nine sequence differences between the two subtilisins from residue 223 to the end of the molecule at 275. Except for residue 224, all of these differences can be seen in the X-ray structure to be on the surface of the enzyme away from the active site. Residue 224 is part of a central helix, one turn away from the catalytic serine at 221. To examine the possible influence of residue 224 on the kinetic activity of subtilisin, cassette mutagenesis was employed to introduce several mutations at this position in subtilisin BPN′.[58] The serine at position 224 in subtilisin BPN′ was replaced with threonine which is present at this site in subtilisin SBT. In addition, to determine whether the hydroxyl group on the side chains of serine and threonine is significant, the amino acids alanine and valine, which are homologous to serine and threonine, respectively, were also introduced at 224. The results of the kinetic analysis for the position 224 mutants are shown in Table 5. The substitution of threonine at position 224 in subtilisin BPN′ alters the kinetic properties of the enzyme to resemble those of subtilisin SBT indicating that the difference in the kinetic properties for the enzymes can be largely attributed to the sequence change at residue 224. The kinetics for the alanine substitution do not differ significantly from that of the native enzyme whereas the k_{cat} for the valine substitution is reduced considerably. Examination of the X-ray crystal structure of subtilisin BPN′ shows that although 224 does not contact bound substrate directly, it does make contact with alanine 152 which is in contact with the substrate. Modeling of a valine or threonine at 224 reveals a collision of the β-methyl group of these amino acids with alanine 152 which could result in a localized distortion in the vicinity of residue 224.[52] Thus, the analysis of hybrid subtilisins made possible the identification of a residue having previously unknown functional significance.

13.3.2. Alteration of pH Profile

Up to now the use of protein engineering in the modification of substrate specificity has been discussed at some length. However, as an example of the power and versatility of this technology, protein engineering also makes possible the systematic modification of the pH dependence of enzyme catalysis. This allows the catalytic activity of an enzyme with the required specificity to be optimized for the pH of an industrial process as well. The pH dependence of enzyme activity is often determined by the ionization constant of catalytic groups. For subtilisin BPN′, catalytic activity varies with pH according to the ionization of histidine 64 which at low ionic strength has a pK_a close to 7. During catalysis, histidine 64 acts as a general base by accepting a proton from serine 221 as it makes a nucleophilic attack on the substrate carbonyl carbon. Hence, the enzyme is most active at alkaline pH when histidine 64 is deprotonated. Studies involving the chemical modification of charged residues on the surface of chymotrypsin indicate that the pH activity profile of serine proteases may be altered by changing the surface charge of the enzyme.[60] Accordingly, the surface charge of subtilisin BPN′ has been modified via oligonucleotide-directed mutagenesis for the purpose of tailoring the pH dependence of the enzyme for catalysis.[61] Serine and lysine substitutions were made for the acidic residues aspartate 99 and glutamate 156 located on the surface of subtilisin. Although both of these residues are greater than 10 Å distant from histidine 64, downward shifts in the pK_a of this residue are seen for the mutant enzymes. At low ionic strength, the pK_a of histidine 64 decreases by 0.4 pH units for serine substitutions at either aspartate 99 or glutamate 156. The substitution of lysine at either of these same two sites causes the pK_a of histidine 64 to decrease by 0.6 pH units. The effect on the pK_a of histidine 64 is found to be cumulative for the double mutants Ser99/Ser156 and Lys99/Lys156. These mutants lower the pK_a by values of 0.6 and 1.0 pH units respectively. Based on studies using various types and concentrations of electrolytes the effect of these mutations on the pK_a appears to be maximized at ionic strengths of 0.1 *M* and below. Indeed, at an ionic strength of 1.0 *M*, the effect of the mutations on the ionization of histidine 64 is reduced considerably due to the shielding of electrostatic interreactions by the accumulation of counterions at charged residues on the surface of the enzyme.

The serine and lysine mutations at positions 99 and 156 in subtilisin BPN′ follow the convention that making the surface charge more positive will lower the pK_a of an ionizing group by destabilizing its protonated form. If the charge substitution is located less than 10 Å from the ionizing residue this rule will still apply. However, the pH dependence of enzyme activity can change in a direction different from that expected. Figure 3 compares the pH dependence of subtilisin BPN′ and a single mutant having the methionine at position 222 replaced with lysine. In this case, the addition of a positive charge drastically increases the pK_a apparent for k_{cat}/K_M.[62] With an apparent pK_a of 9 it appears that the catalytic activity of the enzyme is dependent on the protonation state of the lysine at 222 rather than that of histidine 64. The positively charged ε-amino group of the lysine should destabilize the formation of a positive charge on histidine 64 during catalysis as it accepts a proton from the catalytic serine 221. Hence, the mutant enzyme is most active at high alkaline pH where the

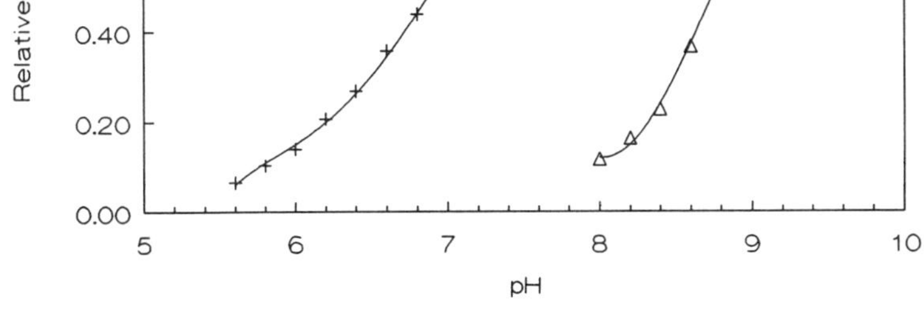

Figure 3. Comparison of the pH dependence of k_{cat}/K_M at 25 °C, ionic strength 0.1 *M*, for the hydrolysis of succinyl-L-Ala-L-Ala-L-Pro-L-Phe-*p*-nitroanilide by wild-type subtilisin BPN′ and a mutant enzyme having lysine substituted for methionine at position 222.[62]

lysine is no longer protonated. The narrow pH range where the lysine 222 mutant is active can be a useful feature for industrial processes which might require a self-regulating protease. In a weakly buffered reaction medium, the process of proteolysis will eventually lower the pH sufficiently to inactivate the lysine 222 mutant and effectively shut off catalysis.

13.3.3. Engineering Improved Stability

Just as important as the tailoring of substrate specificity and pH activity profile is the improvement of stability in enzymes by protein engineering. Site-specific mutagenesis has been successfully employed to increase the stability of subtilisin BPN′ to chemical oxidants.[28] The treatment of subtilisin with hydrogen peroxide leads to a rapid decrease in activity (Figure 4). This inactivation upon oxidation has been previously described and attributed to the generation of methionine sulfoxide at position 222.[63] Based upon the X-ray crystal structure of oxidized subtilisin BPN′ the dramatic decrease in activity does not seem to be due to a large distortion in the tertiary structure of the enzyme.[25] Rather, the activity loss can be attributed to destabilization of the transition state caused by the proximity of the methionine sulfoxide at position 222 to the oxyanion of the tetrahedral intermediate formed during catalysis. Hence, replacement of the methionine at 222 with a nonoxidizable amino acid seemed a plausible solution to the oxidative instability of subtilisin. For several reasons, however, predicting the optimal amino acid substitution at position 222 to

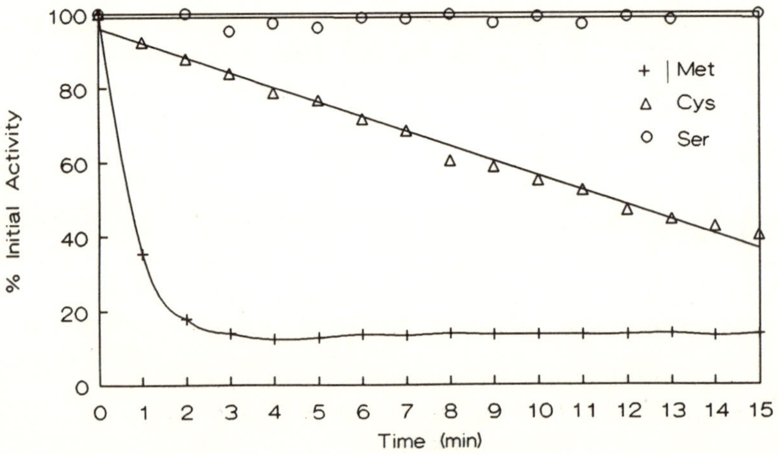

Figure 4. Time course of the effect of 1 M H_2O_2 on the activity of purified wild-type (methionine 222) and serine and cysteine substitutions at position 222 in subtilisin BPN′ (data from ref. 28).

enhance stability in oxidative conditions proved difficult. Methionine at position 222 is conserved in the known subtilisin sequences and is adjacent to the catalytic serine at position 221. Furthermore, analysis of the three-dimensional structure reveals methionine 222 to be closely packed with several residues forming a portion of the S1′ substrate binding site. As a result, cassette mutagenesis was employed to replace methionine at 222 with all nineteen of the coded amino acids.[28] Unexpectedly, all of the mutant enzymes were active (Table 6). Small amino acid substitutions (cysteine, alanine, serine, and glycine) resulted in higher enzyme activity than large or charged substitutions. Kinetic analysis of the mutant enzymes reveals a decrease in k_{cat} for all except the cysteine substitution. Although this mutant is not insensitive to oxidation it is evident in Figure 4 that the cysteine substitution at position 222 is oxidized at a much slower rate than methionine. Use of homologous exchange data would have incorrectly identified leucine, valine, or isoleucine as a good replacement for methionine.[64] Kinetic analysis of the leucine 222 mutant indicates, however, that although the K_M is not significantly altered, the k_{cat} is reduced tenfold (to the level of oxidized wild-type). Thus, selection of possible substitutions from homologous exchange data would appear to be of limited use for functionally significant sites in proteins.

Table 6. Relative Specific Activities of Position 222 Mutations in Subtilisin BPN′

Amino acid at position 222	Relative specific activity (%)
Cysteine	138
Methionine (wild-type)	100
Alanine	53
Serine	35
Glycine	30
Threonine	28
Asparagine	15
Proline	13
Leucine	12
Valine	9.3
Glutamine	7.2
Phenylalanine	4.9
Tryptohan	4.8
Aspartate	4.1
Tyrosine	4.0
Histidine	4.0
Glutamate	3.6
Isoleucine	2.2
Arginine	0.5
Lysine	0.3

The homologous exchange data base has proven useful, however, for an amino acid change at a less critical site in subtilisin BPN′. Treatment of the mutant glutamine 222 with 1 M H_2O_2 resulted in a detectable change in K_M although k_{cat} was unaffected (Table 7). This observation suggests that other residues prone to oxidation can lead to a slight structural alteration in the substrate-binding site when oxidized. Comparison of the highly refined three-dimensional structures of native and peroxide-oxidized enzymes shows that, in addition to position 222, methionines at positions 119 and 124 are also oxidized.[25] Unlike methionines 124 and 222 which are fully oxidized, however, the methionine at position 119 is only partially oxidized. Methionine 124 does not interact directly with a substrate molecule bound in the active site but rather makes contact with residues forming part of the substrate-binding region. Since the amino acid at position 124 is at a less functionally significant site than that of position 222, a suitable replacement for methionine 124 could be selected from the homologous exchange data base.[64] The kinetic data in Table 7 indicates that for a double mutant of subtilisin BPN′ having a leucine substitution at 124 and glutamine substitution at 222 both k_{cat} and K_M are unaffected by treatment of the enzyme with hydrogen peroxide.[65] Hence, with only two mutations the catalytic activity of subtilisin BPN′ can be made insensitive to peroxide oxidation.

In addition to identifying additional peroxide-sensitive residues, analysis of the crystal structure for the oxidized enzyme revealed two other

Table 7. Comparison of Kinetic Parameters k_{cat}, K_M, and k_{cat}/K_M Measured for Wild-Type and Mutant Subtilisins before and after Treatment with Hydrogen Peroxide[a,b]

Residue position		H_2O_2	k_{cat}	K_M	k_{cat}/K_M
124	222	(M)	(s^{-1})	(M)	($s^{-1}M^{-1}$)
Wild-type					
Met	Met	0	50	$1.4 \cdot 10^{-4}$	$3.6 \cdot 10^5$
Met	Met	1	14	$7.3 \cdot 10^{-4}$	$1.9 \cdot 10^4$
Mutant enzymes					
Met	Gln	0	8	$8.6 \cdot 10^{-4}$	$9 \cdot 10^3$
Met	Gln	1	8	$1.3 \cdot 10^{-3}$	$6.1 \cdot 10^3$
Leu	Gln	0	65	$1.3 \cdot 10^{-3}$	$5.1 \cdot 10^4$
Leu	Gln	1	64	$1.3 \cdot 10^{-3}$	$5.1 \cdot 10^4$

[a] Kinetic measurements were performed at 25 °C in 0.1 M Tris, pH 8.6, using the substrate N-succinyl-L-Ala-L-Ala-L-Pro-L-Phe-p-nitroanilide.[25,26]

[b] The double mutant Leu 124/Gln 222 also has a methionine at position 50 replaced with phenylalanine. However, because methionine 50 is not observed to be oxidized in the X-ray crystal structure of oxidized subtilisin BPN′ (42) the substitution at position 50 is not believed to be important for oxidative stability.

interesting observations.[25] First, the solvent accessibility of the five methionines in subtilisin BPN′ does not correlate with oxidation. The sulfur atom of methionine at position 50 is the most exposed to solvent and yet is not oxidized by hydrogen peroxide. By contrast, the sulfur atoms of the methionines which are oxidized are either completely inaccessible to solvent (positions 119 and 124) or only partially accessible (position 222). Furthermore, at the two positions where methionines are fully oxidized, 124 and 222, only one of the two possible stereoisomers is formed upon oxidation. The close packing of residues around the side chains of these methionines may impose steric constraints on the direction of attack by the sulfur atom on hydrogen peroxide.

The commercial utility of subtilisin is limited not only by its sensitivity to oxidation, but also by its tendency toward autolytic proteolysis. Environmental factors such as high temperatures, extreme pH, high levels of denaturants, and calcium chelation weaken the tertiary structure of subtilisin and increase its susceptibility to irreversible inactivation by autolysis. Two separate approaches have been utilized in attempting to enhance the autolytic stability of subtilisin. One approach involved the systematic use of computer modeling and site-specific mutagenesis to engineer a disulfide bridge into the structure of subtilisin.[66–69] This method had been previously used to enhance the stability of dihydrofolate reductase[70] and T4 lysozyme.[71] By contrast, the second approach involved the screening of random mutants of subtilisin exhibiting increased thermal stability.[40]

Disulfide bonds can serve to enhance the conformational stability of globular proteins.[72] Subtilisin, a compact globular protein consisting of a central β-sheet with six helices running antiparallel to the sheet, does not have any cysteine residues. Sites for introducing disulfides were selected by computer modeling based on the following criteria: 1) the mutations should be at least 10 Å from the catalytic serine to avoid possible deleterious effects on enzyme activity, 2) contact between the disulfide bond and nearby atoms should be minimal, and 3) energetically favorable geometries should be observed in the dihedral angles for the disulfide bond. A total of seven disulfide bonds have been engineered individually into subtilisin at different locations via site-directed mutagenesis.[69] The various locations of these disulfides include bridges between different secondary structures as well as linkage between neighboring or distant positions in the polypeptide chain. In addition, one of the disulfides was introduced at a site analogous to that of a disulfide present in proteinase K, a fungal protease containing two disulfides and sharing 37% sequence homology with subtilisin BPN′.[73] Another disulfide bond was designed to span a site occupied by calcium in the native structure.[74] The thermal stability of subtilisin has been shown to decrease substantially upon removal of this bound calcium ion.[75] Unlike T4 lysozyme and dihydrofolate reductase which required *in vitro* oxidation for the formation of disulfide bonds, the formation of disulfide bonds in subtilisin occurs *in vivo*. The activity and expression levels of the engineered enzymes were not significantly different from wild-type. However, in the presence of calcium, none of the disulfide enzymes were observed to be significantly more stable toward autolysis than native subtilisin BPN′, although in general they were more stable than their reduced counterparts or single cysteine mutants.

An alternative method of improving protein stability is the phenotypic screening of random mutants. This approach has been used with subtilisin BPN′ for the identification of mutants exhibiting enhanced thermal stability.[40] Using a simple plate assay screening procedure several thermal stable mutants were selected from a library of random mutants generated by treatment of the cloned gene with sodium bisulfite. One mutant in particular, having a serine substituted for asparagine at position 218, showed a fourfold increase in the half-life for inactivation at 65 °C compared to the native enzyme. To determine a structural basis for the positive effect of the serine 218 mutation on thermal stability, a highly refined X-ray crystal structure of the mutant enzyme was obtained.[40] The residue at position 218 is part of an irregular antiparallel β-pair known as a "β-bulge." Superposition of the structure of wild-type and mutant enzymes reveals closer packing of the two strands as the side chain of serine is much smaller than that of asparagine. In addition, although the number of hydrogen bonds between the two β-strands is the same for wild-type and mutant enzymes, it is apparent in the structure of the serine 218 mutant that several hydrogen bonds are strengthened by the shortening of bond length and improved geometry. The role of hydrogen bonding in the β-bulge in modifying thermal stability of subtilisin BPN′ was confirmed by introducing additional mutations at position 218 via site-specific mutagenesis.[40] The thermal stability of these mutants was found to rank in accordance with the relative hydrogen bonding capabilities of the different amino acids. Thus, we see that the phenotypic screening of random mutants is a valuable approach to increasing the thermal stability of a protein and that with further characterization the structural basis for the functional change may be revealed.

13.3.4. Engineering Reaction Specificity

The modification of functional properties in subtilisin by protein engineering extends beyond the examples of substrate specificity, pH profile, and stability described so far. The catalysis of synthetic reactions can also be modified. For example, one such type of reaction is the synthesis of peptides or ligation of peptide fragments.[76] This may be accomplished by proteases via a thermodynamically or a kinetically controlled reaction. Peptide synthesis by the former method involves altering the conditions for peptide bond hydrolysis so that the equilibrium for the reaction is reversed. In contrast, kinetically controlled peptide synthesis involves the competition between aminolysis and hydrolysis of the acylenzyme reaction intermediate of serine and thiol proteases. In general, the rate of product formation and total product yields are greater for kinetically controlled peptide synthesis. Enzymatic coupling of amino acids or peptide segments has the advantages of stereo-selective, racemization-free reactions which do not require side-chain protection. Already, proteases are being used commercially in the production of aspartame and human insulin.[77] One drawback of using proteases to catalyze peptide bond synthesis, however, is the potentially lower yields obtained due to proteolysis of the product. The problem of product degradation can be circumvented, however, by suppressing the amidase activity of the protease relative to esterase activity. This may be accomplished by the addition of water-miscible organic cosolvents to the reaction medium.[77] Under these conditions, the yield and product stability for peptide synthesis reactions is vastly improved. However, this approach to altering the relative amidase/esterase activity of proteases suffers from the possible disadvantages of poor enzyme stability, decreased selectivity, and high costs associated with organic waste disposal.

Alternatively, the reaction specificity of enzymes may be modified by altering their active-site structure directly either chemically or via site-specific mutagenesis. Chemical modification of catalytic residues in the active sites of proteases can alter the relative ratio of amidase/esterase activity. For example, methylation of the catalytic histidine in α-chymotrypsin renders the enzyme inactive on amide substrates and up to five orders of magnitude less active on ester substrates.[78] Peptide synthesis using the methylated α-chymotrypsin and activated ester substrates gave yields considerably higher than those obtained with the native enzyme. In similar fashion, the chemical conversion of the catalytic serine in subtilisin BPN′ to either cysteine or selenocysteine results in a significant decrease in amidase activity. Recently, thiolsubtilisin has been shown to synthesize small peptides in yields of up to 95% using an activated ester as the acyl donor.[79] Selenosubtilisin exhibits an additional property advantageous in peptide synthesis: the ratio of aminolysis to hydrolysis in selenosubtilisin is observed to be significantly higher than that of either thiolsubtilisin or native subtilisin.[80] Thus, the catalytic activity of selenosubtilisin has been effectively converted from hydrolysis to that of an acyl transferase.

In addition to transamination reactions, serine hydrolases have been observed to catalyze many other syntheses including enantioselective and regioselective transesterification.[81–84] These reactions can be achieved using either anhydrous or biphasic water-organic solvent reaction systems. Although the removal of water is sometimes required to prevent hydrolysis of activated ester substrates, it appears that alteration in the transesterification activity of an enzyme can also be achieved by specifically modifying the active-site structure. The ratio of transesterification to hydrolysis as well as the relative rate of transesterification have been increased for subtilisin BPN′ by introducing a single mutation at the catalytic center of the enzyme.[85] Substitution of methionine 222 with phenylalanine by cassette mutagenesis leads to a threefold increase over the native enzyme in the transesterification to hydrolysis ratio for the reaction of tyrosine ethyl ester with methanol. Compared to the native enzyme, the phenylalanine 222 mutant also exhibits an almost twofold increase in the rate of transesterification. The mutation at position 222 is located adjacent to the catalytic serine at 221 and also forms part of the S1′ subsite which interacts with the attacking nucleophile during deacylation. Substitutions at equivalent sites in other serine hydrolases might also be expected to alter nucleophile specificity.

13.4. CONCLUSION AND FUTURE OUTLOOK

Although our understanding of the relationship between protein structure and function is incomplete the success achieved in tailoring the substrate specificity, pH activity profile, oxidative and thermal stability, and biocatalytic properties of subtilisin by protein engineering clearly suggests that the ability to engineer useful properties in enzymes is now possible. In addition, the protein engineering of subtilisin has provided valuable information concerning the alteration of functional properties in proteins. Perhaps most important is the discovery that the active-site conformation of enzymes is sufficiently robust to accommodate extensive mutagenesis at functionally critical sites. This is evident from the finding that all of the nineteen mutations introduced at position 222 in the active site of subtilisin resulted in an active and stable enzyme. Of further significance is the finding that major alterations in the functional properties of enzymes can be obtained with only small structural changes. This has been repeatedly observed in the modification of structure in subtilisin. For example, mutations at a single site, position 222, have been shown to cause alterations in properties such as substrate specificity, nucleophile specificity, pH activity profile, and oxidative stability. A single mutation at position 218 and the resultant strengthening of a few hydrogen bonds substantially increased the thermal stability of subtilisin. Electrostatic interactions in the S1 subsite of subtilisin were found to have a dramatic effect on substrate specificity. Analysis of the alteration of substrate specificity also revealed the potential synergistic effect of multiple mutations on a functional property. This suggests that the systematic redesigning of the structure of a protein by introducing multiple mutations is indeed feasible. Finally, the identification of functionally important residues in subtilisin by the analysis of hybrid proteins as well as the phenotypic screening of random mutants will likely prove to be of considerable value to protein engineering in the near future. The successful use of these techniques to identify residues of catalytic and structural significance indicates that a precise knowledge of the three-dimensional structure of a protein is not mandatory in the modification of functional properties. Thus, it is evident from the study of subtilisin that the structure of enzymes can indeed be systematically redesigned to tailor functional properties for specific purposes.

As a result of these and other advances, protein engineering has stimulated a renaissance in the study of proteins. In recent years all areas of protein research, from protein folding to that of enzyme catalysis, have advanced rapidly as it is now possible to experimentally design new or modified proteins to study the relationship between protein structure and function. Eventually, the accumulation of structure–function data bases such as that generated for subtilisin will allow for the development and testing of computer algorithms useful in the design of proteins as well as the prediction of protein structure from a primary amino acid sequence. Hence, as our understanding of proteins improves in the coming years, our ability to engineer novel enzymes will surely improve as well. Based on the dramatic changes made in the functional properties of subtilisin by protein engineering, however, it appears that the design of engineered enzymes for the commercial production of specialty chemicals is already possible. Indeed, this goal is actively being pursued and may soon be achieved.

ACKNOWLEDGMENTS

The author would like to thank Jeannette Graycar, Helena Sagar, Randy Berka, Rick Bott, and David Estell for critical advice on this manuscript.

REFERENCES

1. C.-H. Wong, *Science 244*, 1145 (1989).
2. H. Yamada and S. Shimizu, *Angew. Chem. Int. Ed. Eng. 27*, 622 (1988).
3. G. M. Whitesides and C.-H. Wong, *Angew. Chem. Int. Ed. Eng. 24*, 617 (1985).
4. J. B. Jones, *Tetrahedron 42*, 3351 (1986).
5. R. Porter and S. Clark, eds., *Enzymes in Organic Synthesis*, Pitman, London (1985).
6. E. Polastro, A. Walker, and H. Teeuwen, *Bio/technology 7*, 1238 (1989).
7. R. Wetzel, *Protein Engineering 1*, 3 (1986).
8. R. J. Leatherbarrow and A. R. Fersht, *Protein Engineering 1*, 7 (1986).

9. D. L. Oxender and C. F. Fox, eds., *Protein Engineering*, Alan R. Liss, New York (1987).
10. J. R. Knowles, *Science 236*, 1252 (1987).
11. J. A. Wells and D. A. Estell, *TIBS 13*, 291 (1988).
12. W. F. Degrado, Z. R. Wasserman, and J. D. Lear, *Science 243*, 622 (1989).
13. G. D. Fasman, *TIBS 163*, 295 (1989).
14. A. Warshel and S. T. Russel, *Q. Rev. Biophys. 17*, 283 (1984).
15. A. R. Fersht and M. J. E. Sternberg, *Protein Engineering 2*, 527 (1989).
16. J. Kraut, *Ann. Rev. Biochem. 46*, 331 (1977).
17. A. Warshel, G. Naray-Szabo, F. Sussman, and J.-K. Hwang, *Biochemistry 28*, 3629 (1989).
18. M. Philipp and M. L. Bender, *Molecular and Cellular Biochemistry 51*, 5 (1983).
19. I. B. Svendsen, *Carlsberg Res. Commun. 41*, 237 (1976).
20. J. D. Robertus, J. Kraut, R. A. Alden, and J. J. Birktoft, *Biochemistry 11*, 4293 (1972).
21. T. L. Poulos, R. A. Alden, J. J. Birktoft, S. T. Freer, and J. Kraut, *J. Biol. Chem. 251*, 1097 (1976).
22. S. Hirono, H. Akagawa, Y. Mitsui, and Y. Iitika, *J. Mol. Biol. 178*, 389 (1984).
23. W. Bode, E. Papamokos, D. Musil, U. Seemueller, and H. Fritz, *EMBO J. 5*, 813 (1986).
24. C. A. McPhalen, I. Svendsen, I. Jonassen, and M. N. G. James, *Proc. Natl. Acad. Sci. USA 82*, 7242 (1985).
25. R. R. Bott, M. Ultsch, A. Kossiakoff, T. P. Graycar, B. Katz, and S. Power, *J. Biol. Chem. 263*, 7895 (1988).
26. J. A. Wells, E. Ferrari, D. J. Henner, D. A. Estell, and E. Y. Chen, *Nucleic Acids Res. 11*, 7911 (1983).
27. J. Yang, E. Ferrari, and D. J. Henner, *J. Bacteriol. 160*, 15 (1984).
28. D. A. Estell, T. P. Graycar, and J. A. Wells, *J. Biol. Chem. 260*, 6518 (1985).
29. M. V. Arbige and W. H. Pitcher, *TIBTECH 7*, 330 (1989).
30. R. E. Offord, *Protein Engineering 1*, 151 (1987).
31. D. Woo, I. Clark-Lewis, B. Chait, and S. Kent, *Protein Engineering 3*, 29 (1989).
32. A. Proudfoot, K. Rose, and C. Wallace, *J. Biol. Chem. 264*, 8764 (1989).
33. L. Polgar and M. L. Bender, *J. Am. Chem. Soc. 88*, 3153 (1966).
34. K. E. Neet and D. E. Koshland, Jr., *Proc. Natl. Acad. Sci. USA 56*, 1606 (1966).
35. E. T. Kaiser and D. S. Lawrence, *Science 226*, 505 (1984).
36. L. M. Bech and K. Breddam, *Carlsberg Res. Commun. 53*, 381 (1988).
37. M. J. Zoller and M. Smith, *Methods in Enzymology 100*, 468 (1983).
38. P. Carter, *Methods in Enzymology 154*, 382 (1987).
39. J. A. Wells, M. Vasser, and D. B. Powers, *Gene 34*, 315 (1985).
40. P. N. Bryan, M. L. Rollence, M. W. Pantoliano, J. Wood, B. C. Finzel, G. L. Gilliland, A. J. Howard, and T. L. Poulos, *Proteins: Structure, Function, and Genetics 1*, 326 (1986).
41. D. Botstein and D. Shortle, *Science 229*, 1193 (1985).
42. M. Smith, *Ann. Rev. Genet. 19*, 423 (1985).
43. S. S. Ner, D. B. Goodin, and M. Smith, *DNA 7*, 127 (1988).
44. D. W. Leung, E. Chen, and D. V. Goeddel, *Techniques 1*, 11 (1989).
45. R. Menzel, *Anal. Biochem. 181*, 40 (1989).
46. G. L. Gray, S. E. Mainzer, M. W. Rey, M. H. Lamsa, K. L. Kindle, C. Carmona, and C. Requadt, *J. Bacteriol. 166*, 635 (1986).
47. A. R. Fersht, *Enzyme Structure and Mechanism*, 2nd ed., Freeman, San Francisco (1985).
48. I. Schechter and A. Berger, *Biochem. Biophys. Res. Commun. 27*, 157 (1967).
49. D. A. Estell, T. P. Graycar, J. V. Miller, D. B. Powers, J. P. Burnier, P. G. Ng, and J. A. Wells, *Science 233*, 659 (1986).
50. R. Bott and M. Ultsch, in: *Fifth International Symposium on the Genetics of Industrial Microorganisms* (M. Alacevic, D. Hranueli, and Z. Toman, eds.), Pliva, Zagreb (1986).
51. J. A. Wells, D. B. Powers, R. R. Bott, T. P. Graycar, and D. A. Estell, *Proc. Natl. Acad. Sci. USA 84*, 1219 (1987).

52. R. Bott, personal communication.
53. R. Bott, M. Ultsch, J. Wells, D. Powers, D. Burdick, M. Struble, J. Burnier, D. Estell, J. Miller, T. Graycar, R. Adams, and S. Power, in: *Biotechnology in Agricultural Chemistry* (ACS Symposium Series No. 334, H. LeBaron, R. Mumma, R. Honeycutt, and J. Duesing, eds.), 139 (1987).
54. J. A. Wells, B. C. Cunningham, T. P. Graycar, and D. A. Estell, *Proc. Natl. Acad. Sci. USA 84*, 5167 (1987).
55. C. McPhalen, H. Schnebli, and M. James, *FEBS Lett. 188*, 55 (1985).
56. P. Carter and J. A. Wells, *Science 237*, 394 (1987).
57. P. Carter, B. Nilsson, J. P. Burnier, D. Burdick, and J. Wells, *Proteins: Structure, Function, and Genetics 6*, 240 (1989).
58. F. Graham, G. Gray, C. Carmona, E. Ferrari, and D. Estell, manuscript in preparation.
59. S. D. Power, R. M. Adams, and J. A. Wells, *Proc. Natl. Acad. Sci. USA 83*, 3096 (1986).
60. P. Valenzuela and M. L. Bender, *Biochim. Biophys. Acta 250*, 538 (1971).
61. A. J. Russell and A. R. Fersht, *Nature 328*, 496 (1987).
62. T. P. Graycar and D. A. Estell, unpublished results (1986).
63. C. E. Stauffer and D. Etson, *J. Biol. Chem. 244*, 5333 (1969).
64. M. O. Dayhoff, R. M. Schwartz, and B. C. Orcutt, in: *Atlas of Protein Sequence and Structure* (M. O. Dayhoff, ed.), Vol. 5, Supplement 3, 345 (1978), National Biomedical Research Foundation, Georgetown University Medical Center, Washington, D.C.
65. J. Wells, B. Cunningham, R. Bott, R. Adams, S. Power, T. Graycar, and D. Estell, unpublished results (1987).
66. J. A. Wells and P. B. Powers, *J. Biol. Chem. 261*, 6564 (1986).
67. B. A. Katz and A. Kossiakoff, *J. Biol. Chem. 261*, 15480 (1986).
68. M. W. Pantoliano, R. C. Ladner, P. N. Bryan, M. L. Rollence, J. F. Wood, and T. L. Poulos, *Biochemistry 26*, 2077 (1987).
69. C. Mitchinson and J. A. Wells, *Biochemistry 28*, 4807 (1989).
70. J. E. Villafranca, E. E. Howell, D. H. Voet, M. S. Strobel, R. C. Ogden, J. N. Ableson, and J. Kraut, *Science 222*, 782 (1983).
71. L. J. Perry and R. Wetzel, *Science 226*, 555 (1984).
72. R. Wetzel, *Trends Biochem. Sci. 12*, 478 (1987).
73. A. Pahler, A. Banarjee, J. K. Dattagapta, T. Fujiwara, K. Linder, G. P. Pal, D. Snick, G. Weber, and W. Saenger, *EMBO J. 3*, 1311 (1984).
74. J. Drenth, W. G. Hol, and J. Jansonius, *Eur. J. Biochem. 26*, 177 (1972).
75. G. Voordouw, C. Milo, and R. S. Roche, *Biochemistry 15*, 3716 (1976).
76. K. Morihara, *Trends Biotechnol. 5*, 164 (1987).
77. C. F. Barbas, III, J. R. Matos, J. B. West, and C.-H. Wong, *J. Am. Chem. Soc. 110*, 5162 (1988).
78. J. B. West, J. Scholten, N. J. Stolowich, J. L. Hogg, A. I. Scott, and C.-H. Wong, *J. Am. Chem. Soc. 110*, 3709 (1988).
79. T. Nakatsuka, T. Sasaki, and E. T. Kaiser, *J. Am. Chem. Soc. 109*, 3808 (1987).
80. Z.-P. Wu and D. Hilvert, *J. Am. Chem. Soc. 111*, 4513 (1989).
81. B. Cambou and A. M. Klibanov, *J. Am. Chem. Soc. 106*, 2687 (1984).
82. C.-S. Chen, S.-H. Wu, G. Girdaukus, and C. J. Sih, *J. Am. Chem. Soc. 109*, 2812 (1987).
83. S. Riva, J. Chopineau, A. P. G. Kieboom, and A. M. Klibanov, *J. Am. Chem. Soc. 110*, 584 (1988).
84. S.-H. Wu, L.-C. Lo, S.-T. Chen, and K.-T. Wang, *J. Org. Chem. 54*, 4220 (1989).
85. M. Arbige, D. Estell, M. Pepsin, and A. J. Poulose, European Patent Office Application No. 0260105.

14

Catalytic Antibodies

DAVID E. HANSEN

14.1. INTRODUCTION AND BACKGROUND

In December of 1986, groups led by Tramontano and Lerner[1] and by Schultz[2] independently demonstrated that antibodies can catalyze the hydrolysis of carboxylate esters and carbonates. In the short time since then, antibodies that catalyze a variety of additional reactions, including the sequence-specific hydrolysis of a peptide and a Claisen rearrangement, have been elicited. In this chapter, we will first discuss the key developments in the study of enzymatic catalysis that led to the successful isolation of these antibody catalysts and will comment on some earlier unsuccessful attempts toward this end. Then we will discuss the properties of the antibody catalysts isolated to date and will speculate on the future directions, including those involving protein engineering, that research in this area may follow.

14.1.1. The Search for New Protein Catalysts

Because enzymes are such powerful catalysts and because the range of reactions they catalyze is limited to those occurring in living organisms, the design and synthesis of new "enzymes" has long been a goal of biochemists. However, an incomplete understanding of protein folding and of the structure–function relationship of proteins has made the *de novo* construction of a new enzyme an unrealized goal.[3] Yet, from both a practical and theoretical standpoint, the generation of new enzymes is of fundamental impor-

DAVID E. HANSEN • Department of Chemistry, Amherst College, Amherst, MA 01002.

tance, as was realized almost a century ago by the legendary chemist Emil Fischer. In his 1902 Nobel Prize lecture on carbohydrate chemistry and enzymology, Fischer stated[4]:

> The examination of synthetic glucosides has shown that the action of the enzymes depends to a large extent on the geometrical structure of the molecule to be attacked, that the two must match like lock and key... To equal Nature here, the same means have to be applied, and I therefore foresee the day when physiological chemistry will not only make extensive use of natural enzymes as agents, but when it will also prepare synthetic [enzymes] for its purposes.

Indeed, synthetic organic chemists have begun to utilize enzymes to catalyze a variety of transformations, including, for example, the production of optically active molecules by the reduction of ketones with horse liver alcohol dehydrogenase and the selective hydrolysis of prochiral diesters with esterase.[5] Both of these enzymes handle a wide variety of substrates, and optically pure products are often obtained. Unfortunately, there are only very few nonselective enzymes. In an attempt to expand their reactivity, Kaiser's group has pioneered the creation of "semisynthetic enzymes," in which cofactors are covalently attached to enzyme active sites to yield new catalysts that have the binding characteristics of the enzyme but do the chemistry of the cofactor.[6] Ultimately, this approach too will be limited as only a small number of existing enzymes can be appropriately modified. Thus the search for truly new protein catalysts has continued unabated, and the recent demonstration that antibodies can have catalytic activity provides a significant step forward.

The recent isolation of antibody catalysts has followed directly from an understanding of enzymatic catalysis. Present thinking on the basis of enzyme catalysis began in the 1948 when Pauling, in a seminal statement,[7] suggested that Fischer's lock-and-key analogy must be modified: an enzyme active site should not match the substrate, it should rather be complementary to the transition state (or activated complex) of the reaction being catalyzed. Much recent work on enzyme mechanisms has supported Pauling's view[8] (indeed, as Jencks has recently reiterated, transition state stabilization "is required by the definitions of catalysis and of the transition state"[9]), and one may convincingly argue that most of the principles of enzyme catalysis are now well understood, even if the precise manifestation of these principles is not.[10] However, a concomitant understanding of protein folding has not, to date, been achieved. Nevertheless, in his 1948 article, Pauling suggested a strategy for the generation of new protein catalysts that did not require the ability to design a protein that would fold properly[7]:

> An enzyme has a structure closely similar to that found for antibodies, but with one important difference. Namely that the surface configuration of the enzyme

> is not so closely complementary to its specific substrate as is that of an antibody to its antigen, but instead is complementary to the activated complex.

Jencks, more directly, stated in 1969[11]:

> If complementarity between the active site and the transition state contributes significantly to enzymatic catalysis, it should be possible to synthesize an enzyme by constructing such an active site. One way to do this is to prepare an antibody to a haptenic group which resembles the transition state of a given reaction. The combining sites of such antibodies should be complementary to the transition state and should cause an acceleration by forcing bound substrate to resemble the transition state.

And, as Hood, Weissman, and Wood state in their book *Immunology*[12]:

> The antigen-binding sites of antibodies resemble the active sites of enzymes in several ways. Both involve multiple, weak noncovalent associations, including electrostatic, hydrogen bonding, and van der Waals interactions, which combine to give strong binding... The binding sites of both are predominantly nonpolar niches.

With the potential existence of at least 10^{10} different antibody molecules,[13] one may speculate that every antibody catalyst imaginable is already present in the existing repertoire of antibodies (just as every antibody needed to bind all antigens appears to be present). Potentially, huge rate enhancements could be effected as association constants as high as $10^{11}\ M^{-1}$ have been measured for antigen binding to antibodies,[14] which corresponds to a binding energy of over 16 kcal/mol at 298 °K. The challenge is therefore in selecting an antibody with the desired catalytic function from this huge repertoire. However, as stated by Jencks,[11] such a selection process might be at hand if, in a stable molecule, one could model the transition state for a reaction. In fact, the existence of such "transition-state analogs" had been realized almost twenty years prior to the generation of antibodies with catalytic activity.

14.1.2. Transition-State Analogs

In 1966, Jencks first recognized that a variety of enzyme inhibitors may be acting as transition-state analogs,[15] and in the early 1970s Wolfenden[16] and Lienhard[17] independently emphasized the potential for use of transition-state analogs as potent enzyme inhibitors. For example, as outlined in Figure 1A, the affinity of proline racemase for both pyrrole-2-carboxylic acid (I) and Δ-1-pyrolline-2-carboxylic acid (II) is 160 times greater than is the enzyme's affinity for either of the substrates, L-proline or D-proline.[18, 19] In the transition state for the enzyme-catalyzed racemization reaction, the α-carbon atom of proline and the substituents surrounding it are thought to become coplanar, and pyrrole-2-carboxylic acid and

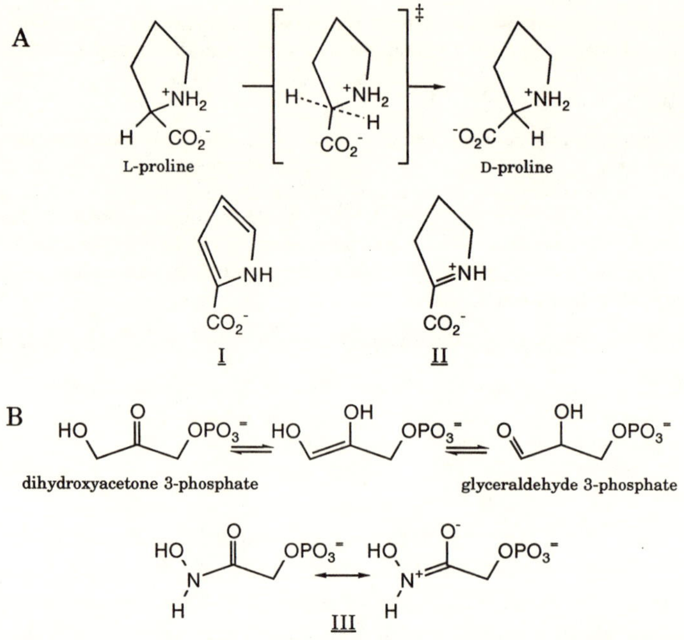

Figure 1. (A) The reaction catalyzed by proline racemase and the transition-state analogs pyrrole-2-carboxylic acid (I) and Δ-1-pyrolline-2-carboxylic acid (II). (B) The reaction catalyzed by triosephosphate isomerase and the transition-state analog 2-phosphoglycolohydroxamate (III).

Δ-1-pyrolline-2-carboxylic acid each mimic this geometry. Similarly, the transition-state analog 2-phosphoglycolohydroxamate (III) binds to triosephosphate isomerase approximately 250 times more tightly than does the substrate dihydroxyacetone 3-phosphate.[20] Because of the delocalized structure of the amide bond, 2-phosphoglycolohydroxamate is thought to mimic the transition states leading from either dihydroxyacetone 3-phosphate or glyceraldehyde 3-phosphate to the enediol phosphate intermediate as is outlined in Figure 1B. To date, nearly 100 examples of transition-state analogs have been reported, and they are among the best inhibitors of many enzymes.[21] Thus, prior to their use as immunogens to elicit antibodies with catalytic activity, transition-state analogs were well studied, and, as we shall see, known transition-state analogs have provided the inspiration for the isolation of a number of catalytic antibodies.

14.1.3. Earlier Attempts to Isolate Antibody Catalysts

During the 1970s, a number of reports describing the attempted production of catalytic antibodies appeared. In 1975 Raso and Stollar attempted to catalyze the condensation of pyridoxal phosphate and the amino acid tyrosine by raising antibodies against the multisubstrate transition-state analog IV (Figure 2A),[22] and in 1983 Summers and Johnson attempted to catalyze the enolization of a diarylacetophenone by raising antibodies against the α, β-unsaturated amide analog V (Figure 2B).[23] As low-molecular-weight molecules are by themselves not immunogenic,[24] both groups prepared transition-state analog-protein conjugates prior to immunization (the outlined arrow indicates the functional group that was utilized for the conjugation to the carrier protein, a convention that will be maintained throughout this review). The two attempts failed to produce antibody having the desired catalytic activities; however, Raso and Stollar

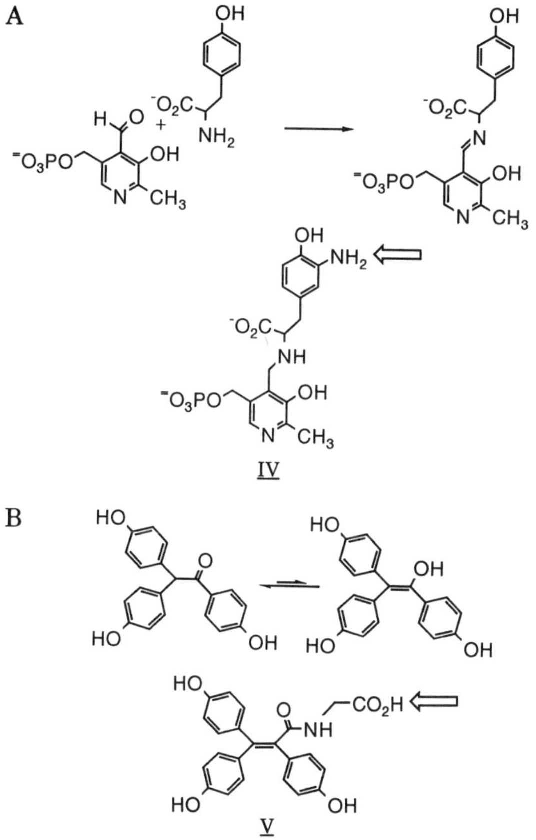

Figure 2. (A) The unsuccessful attempt by Raso and Stollar to elicit an antibody catalyst against the transition-state analog IV. (B) The unsuccessful attempt by Summers and Johnson to elicit an antibody catalyst against the transition-state analog V.

did report that their antibody apparently catalyzed a fivefold increase in the rate of a transamination reaction. It is important to note that these researchers had access only to polyclonal antibodies and thus did not explore the use of monoclonal antibodies.[25] During this period, Kohen *et al.*[26] published work on monoclonal antibodies raised against active ester derivatives. Although these antibodies enhanced the reaction of the ester haptens and related molecules, the "antibody-induced hydrolysis reaction was stoichiometric rather than catalytic." It is likely that reactive amino acid side chains in the antibody combining site are rapidly acylated by the ester "substrate," releasing an equivalent of chromaphoric product alcohol. However, the acyl-antibody intermediates produced are then unable to undergo further reaction.

14.2. THE ISOLATION OF ANTIBODY CATALYSTS

The recent successful work can be traced back to 1984 when Lerner again suggested eliciting antibodies against transition-state analogs.[27] Lerner wrote:

> The isolation of antibody catalysts is a wonderful possibility since one could fish in the immunological repertoire for any kind of enzyme so long as the substrate were sufficiently large to be immunogenic. The basic question is whether the diversity of the immune system is any match for the millions of years of molecular design which go into the evolution of an enzyme. I, for one, would bet on the diversity of the immune system.

Two years later, Lerner, working with Tramontano,[28] attempted to isolate antibodies that would catalyze the hydrolysis of a carboxylic ester by immunizing with a protein conjugate of the corresponding phosphonate. Phosphonates mimic the high-energy, tetrahedral intermediate in a base-catalyzed ester hydrolysis reaction as shown in Figure 3; by the Hammond postulate, the structure of the rate-determining transition state in an ester

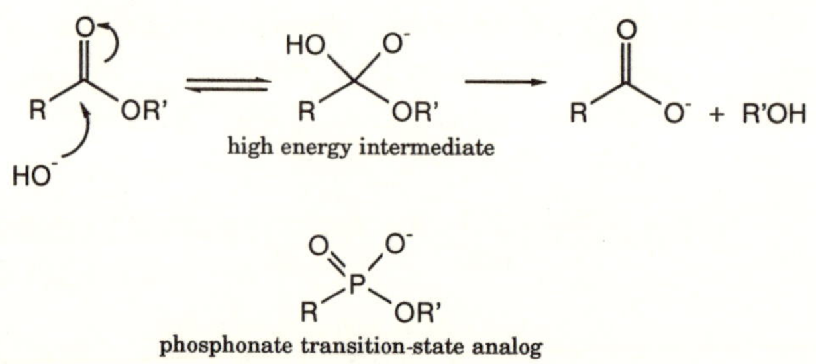

Figure 3. The mechanism of base-catalyzed ester hydrolysis and the corresponding phosphonate transition-state analog.

hydrolysis reaction will closely resemble that of this intermediate. The use of a phosphonate transition-state analog was suggested by the fact that phosphonamidate transition-state analogs are potent inhibitors of proteolytic enzymes.[29] Thus, evidence already existed that proteins that catalyze acyl-transfer reactions bind tightly to tetrahedral phosphorus derivatives in addition to the true transition states. In the present work,

VI

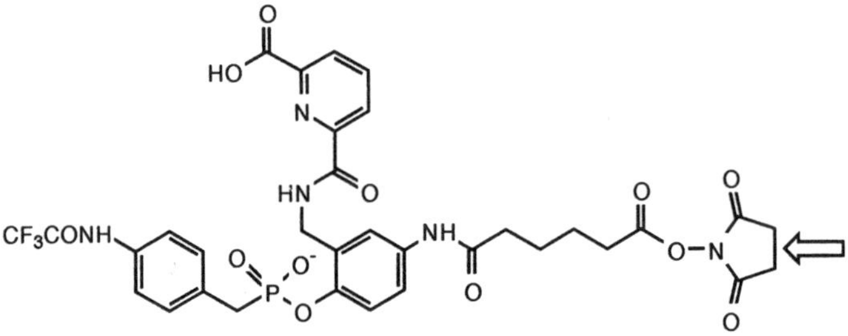

VII

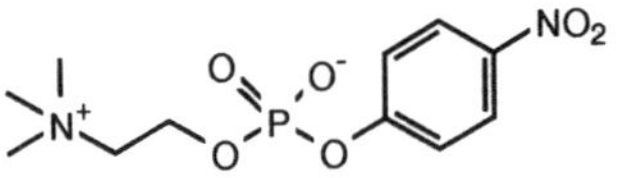

VIII

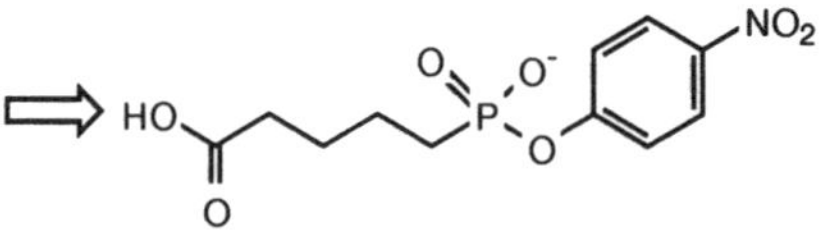

IX

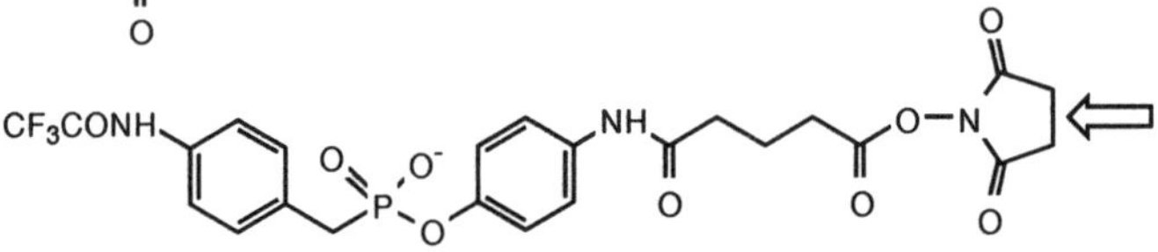

X

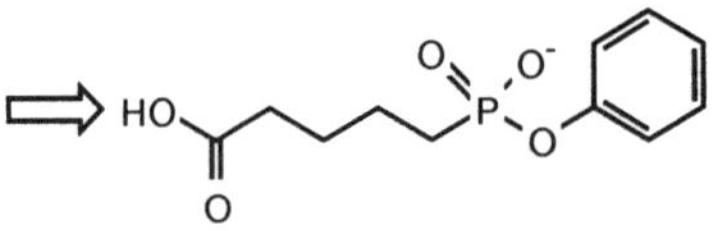

XI

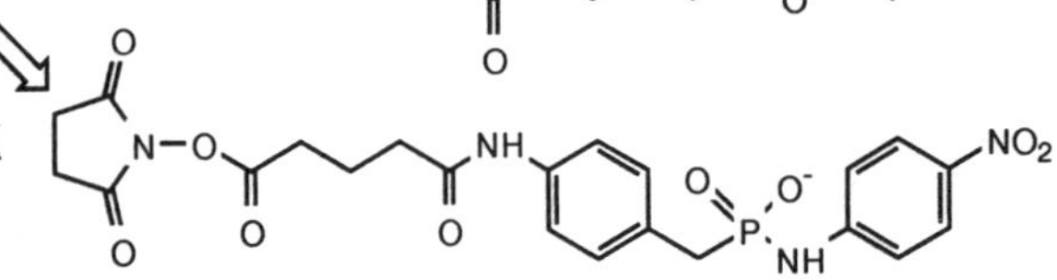

XII

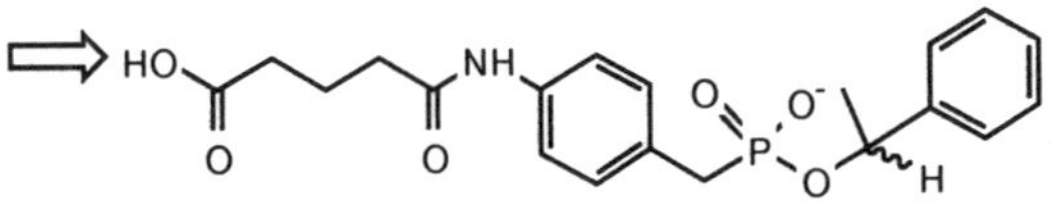

Figure 4. The tetrahedral phosphorus-containing transition-state analogs VI–XII utilized for the isolation of antibody hydrolases.

however, instead of using a transition-state analog to inhibit a known enzyme, a new "enzyme" would be created by raising antibodies against the transition-state analog. Specifically, Tramontano and Lerner raised monoclonal antibodies against the phosphonate hapten VI (Figure 4), which was conjugated to the carrier protein keyhole limpet haemocyanin (KLH) via its *N*-hydroxysuccinimide functionality. Approximately fifteen antibodies against VI were obtained, one of which, 6D4, was studied further. When monoclonal 6D4 was tested with the coumarin ester 6a (Figure 5), acylation of the antibody was observed, but turnover was not, analogous to the results obtained by Kohen *et al.* However, the acylation of 6D4 occurs at a rate at least 100-times that observed in Kohen's systems (where the antibodies were raised against the esters themselves, and not against a transition-state analog), and approximately 10^4-times that observed for the imidazole-catalyzed hydrolysis of 6a. Treatment of the acylated antibody with alkaline buffer or with hydroxylamine regenerated catalytic activity, presumably by chemically catalyzing the deacylation of the antibody.

In December of 1986, the Tramontano and Lerner[1] and the Schultz[2] papers were published. Tramontano and Lerner reported that when antibody 6D4 is tested with substrate 6b, which more closely resembles the phosphonate VI used for the immunization, catalytic activity is observed. Schultz, in a conceptually identical system, reported that the pre-existing murine myeloma protein MOPC167, which binds *p*-nitrophenylphosphorylcholine VII, catalyzes the hydrolysis of the carbonate 7. Antibodies 6D4 and MOPC167 each effect a rate acceleration of 1000-fold. In addition, each shows substrate specificity, exhibits saturation kinetics, and each is subject to competitive inhibition by the corresponding transition-state analog—an excellent indication that catalysis results from substrate binding in the antigen-combining site. (All of the catalytic antibodies isolated to date display these properties.) The respective transition-state analogs bind approximately 500-fold more tightly to these antibodies than do the substrates, a value that agrees well with the actual accelerations observed. (If one assumes that the substrate K_m is in fact equal to the substrate K_d, then a 1000-fold tighter binding of the analog would directly correspond to a 1000-fold rate acceleration.) Interestingly, Rao demonstrated that TEPC-15, another murine myeloma protein that binds phosphorylcholine analogs, does not catalyze the hydrolysis of the analogous choline *esters*[30]; however, Pollack and Schultz later demonstrated that the carbonate 7 is also a substrate for TEPC-15.[31]

To date, a variety of analogs have been employed to elicit antibodies with catalytic activity, but the use of tetrahedral phosphorus derivatives to isolate antibody hydrolases has been the predominate approach, and both the Schultz group, often in collaboration with scientists at IGEN, Inc., and

6a

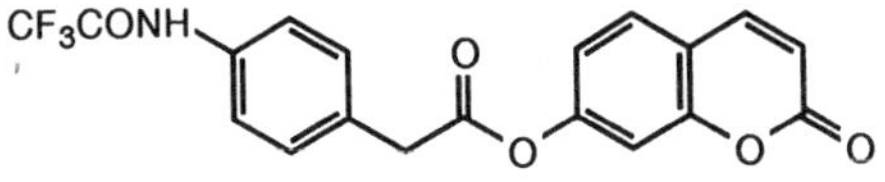

6b

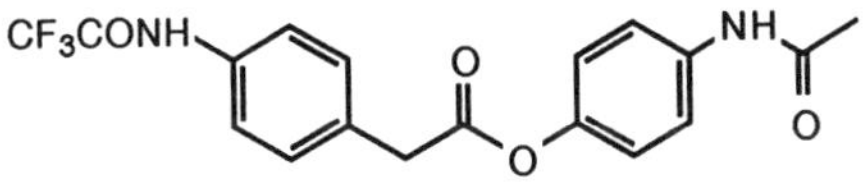

7

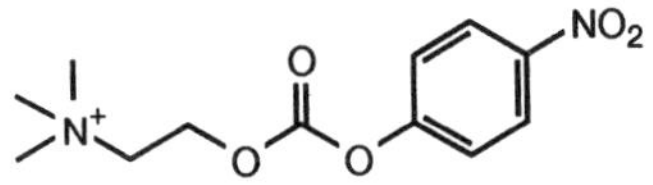

8

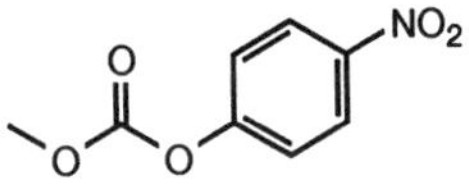

9

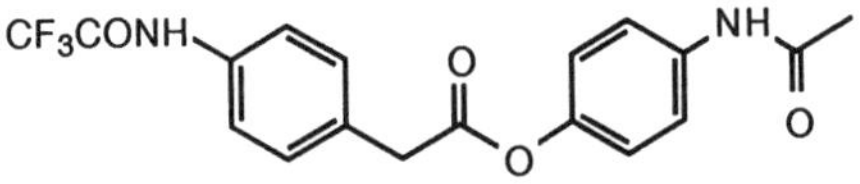

10

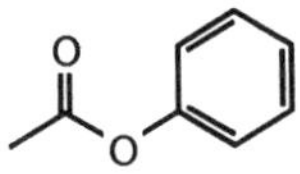

11

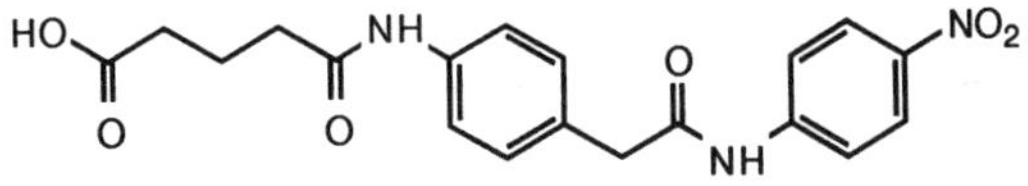

12

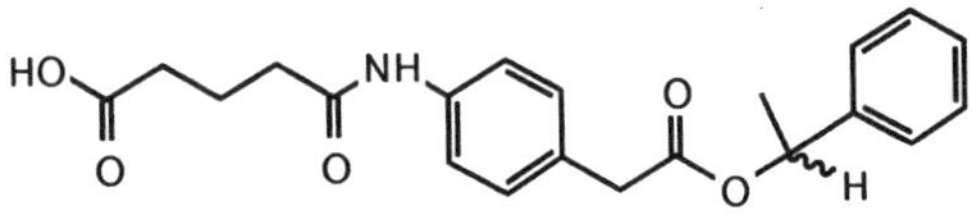

Figure 5. The corresponding substrates 6–12 for the antibody hydrolases raised against the analogs in Fig. 4.

the Tramontano and Lerner group have made significant contributions. We will now review the properties of these antibody hydrolases and will then discuss the other antibody catalysts that have been isolated to date. (As is evident by the groupings below, antibodies have been obtained that catalyze reactions corresponding to five of the six enzyme classes defined by the International Union of Biochemistry.[32])

14.2.1. Antibody Hydrolases

Soon after the initial work, Jacobs *et al.* demonstrated that at least one of the six antibodies raised against the phosphonate derivative VIII, which was conjugated to KLH through its carboxylate functionality with 1-(3-dimethylamino)propyl)-3-ethylcarbodiimide, catalyzes the hydrolysis of the carbonate 8.[33] (Virtually all of the analogs used to date have been conjugated through a free carboxylate or an *N*-hydroxysuccinimide functionality.) A rate acceleration of 810-fold was reported, and the reaction shows a first-order dependence on hydroxide-ion concentration. Overall, the properties of this antibody are almost identical to those of MOPC167, thus eliminating any concern over the generality of the MOPC167 result. Note, however, that hapten VIII is a phosphonate ester, while the substrate 8 is a carbonate, and thus the antibody clearly tolerates the substitution of the carbonate oxygen atom for the phosphonate methylene group. Whether this antibody will also hydrolyze the corresponding *p*-nitrophenyl ester was not discussed. Tramontano *et al.* then reported that five of twenty antibodies raised against the phosphonate IX show activity against the ester substrate 9,[34] and that one, 50D8, effects a rate acceleration of over a millionfold with some ester substrates. Shortly thereafter, Durfor *et al.* determined that five of thirteen antibodies raised against X catalyze the hydrolysis of phenyl acetate 10; one of these antibodies, 20G9, shows a rate enhancement of approximately 15,000-fold.[35] In addition, the authors showed that the antibody 20G9 functions as a catalyst in reverse micelles.

In an important step forward, Janda *et al.* then demonstrated that antibody NPN43C9, one of forty-four raised against the *p*-nitrophenyl phosphonamidate derivative XI, catalyzes the hydrolysis of the *p*-nitroanilide substrate 11.[36] This antibody effects a rate acceleration of 250,000-fold, and although the substrate 6 is an activated amide, NPN43C9 was still the first antibody isolated with amidase activity. The mode of catalysis is intriguing as the phosphonamidate transition-state analog XI apparently binds only 100-times more tightly to the antibody than does the substrate 11 (again, assuming substrate K_m equals K_d). Clearly to account for the observed catalytic acceleration, additional stabilizing interactions between the antibody and transition state must be realized. For example, the authors speculated that an amino acid present in the antigen-combining site may protonate the *p*-nitroanilide anion as it is expelled from the tetrahedral intermediate. The fortuitous presence of such an amino acid may explain why only one in forty-four of the antibodies obtained is catalytic, and why no antibodies capable of catalyzing the hydrolysis of unactivated amine bonds have ever been obtained by immunizing with phosphonamidate derivatives. Finally, Janda *et al.* demonstrated that antibodies raised against a chiral phosphonate derivative can catalyze the stereoselective hydrolysis of an unactivated ester.[37] Antibodies were raised against a racemic mixture of the phosphonate XII, and eighteen antibodies were obtained. Of these eighteen, nine catalyzed the hydrolysis of the (R)-isomer of 12, while two catalyzed the hydrolysis of the (S)-isomer, and rate accelerations of up to 80,000-fold were observed. Although a racemic mixture of the phosphonate XII was used for the immunizations, stereospecificity was observed in the monoclonals obtained as individual antibody molecules are induced by interaction with individual immunogen molecules—be they either the (R)- or the (S)-isomers. (Very recently, the Schultz group has also isolated antibodies that catalyze the stereospecific hydrolysis of unactivated esters.[38])

From the time of the first report that antibodies can have catalytic activity, the isolation of a sequence-specific protease has been of particular interest.[27] Such antibodies have potential for use in the treatment of diseases in which foreign structural or catalytic proteins are exposed to circulating antibodies. Furthermore, biochemical applications may include use as reagents for protein sequencing and for site-directed protein modification. As mentioned above, the obvious approach of eliciting antibodies against a phosphonamidate derivative has failed to yield antibodies that catalyze the hydrolysis of an *unactivated* amide bond. However, recently, Iverson and Lerner employed an entirely different strategy to isolate antibodies with sequence-specific peptidase activity.[39] Their strategy eschewed the traditional transition-state analog approach and instead concentrated on isolating antibodies that bind a peptide substrate juxtaposed with a catalytically potent metal ion. To accomplish this, mice were immunized with the Co(III) (triethylenediamine) exchange-inert complex XIII (Figure 6), and thirteen antibodies that bind both a chelated metal ion and a peptide derivative were isolated. That is, antibodies raised against a single molecule containing a peptide moiety covalently attached to a metal ion also bind a peptide and a metal ion together when presented with the individual pieces. Thus, remarkably, when these antibodies were tested against the peptide substrates 13a and 13b in the presence of an exchange-*labile* metal complex, catalytic activity was observed. The complexes of Zn(II), Fe(III), Ga(II), Cu(II), and Ni(II) are the most effective in mediating catalysis, and the point of cleavage of each the peptide substrates 13a and 13b is at the glycine-phenylalanine peptide bond as is indicated in Figure 6. Although the mechanistic details of catalysis by these antibodies have yet to be elucidated, the authors speculated that the peptide substrate may fold back upon itself so as to allow the reacting peptide bond to approach the metal ion. Whether the metal ion directly chelates the scissile carbonyl group thus facilitating attack of water, or instead delivers a hydroxide ion to the carbonyl group is currently unknown,

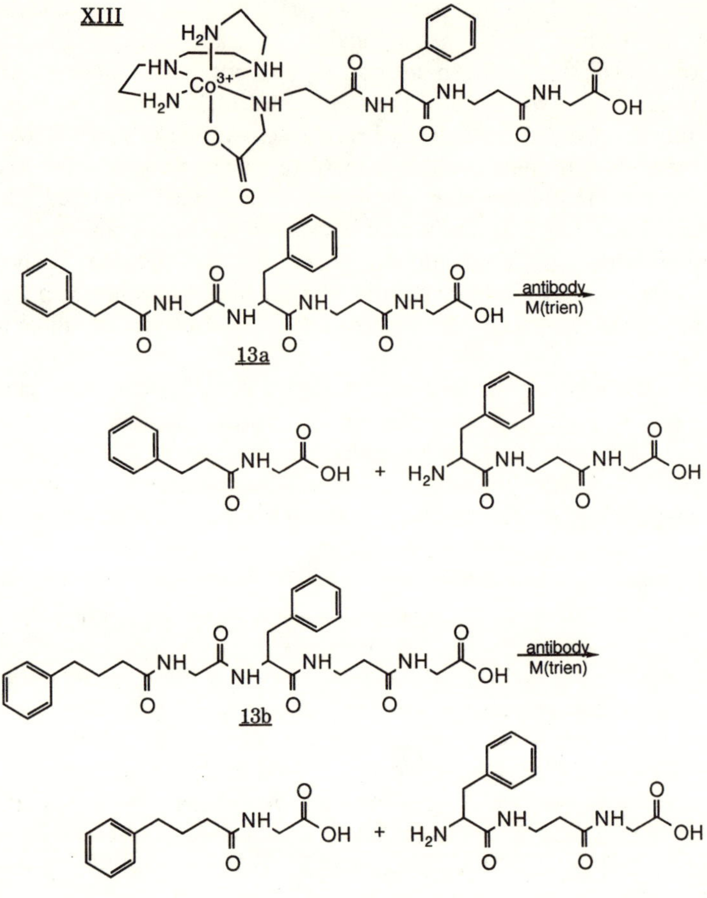

Figure 6. The analog XIII used to elicit sequence-specific antibody peptidases and the reactions catalyzed.

although, as the authors pointed out, metal ion delivery of a hydroxide ion is the more potent mode of catalysis. Given the ability of many antibodies raised against peptides to recognize analogous sequences in proteins,[40] the isolation of antibodies with sequence-specific *protease* activity appears well within reach.

The recent work of Paul *et al.* provides additional reason to believe that useful antibody proteases can be isolated. These workers isolated a pre-existing human autoantibody that catalyzes the sequence-specific cleavage of vasoactive intestinal peptide (VIP).[41] This antibody was isolated from a patient with an autoimmune disease and thus, obviously, was not elicited by immunization in the traditional sense. The anti-VIP antibody catalyzes the hydrolysis of VIP exclusively between glutamine$_{16}$ and methionine$_{17}$ and is the first example of a proteolytic antibody that catalyzes the cleavage of a peptide bond in the absence of added cofactors. In discussing the possible mechanism of the autoantibody-catalyzed hydrolysis, the authors stated:

> Amino acid residues within the antibody combining site may participate actively in the hydrolysis of this bond, in the same way that enzymes cleave peptide bonds. Alternatively, antibody binding to VIP could ... favor a peptide conformation that accelerates intramolecular nucleophile attack of the peptide bond carbonyl by the nitrogen in the side chain of glutamine.

The results of Paul *et al.* strongly suggest that sequence-specific antibody proteases that function in the absence of cofactors can be isolated by direct immunization with an appropriate transition-state analog.

14.2.2. Antibody Transferases

In collaboration with the Benkovic group, the Tramontano and Lerner group have also employed tetrahedral phosphate derivatives as immunogens for the isolation of antibodies that catalyze acyl-transferase reactions. Specifically, they demonstrated that monoclonal antibody 24B11, one of twenty-four raised against the cyclic phosphonate diester XIV, catalyzes both the intramolecular cyclization of the hydroxyester 14a to yield the lactone 14b,[42] and the intermolecular condensation of 1,4-phenylenediamine and 14b to yield the amide 14c (Figure 7A).[43] The intramolecular reaction proceeds with a rate acceleration of 787-fold, and only one of the two enantiomers of 14a is handled (as is expected as antibodies are chiral, optically pure molecules; to date, all catalytic antibodies acting on chiral substrates have shown great stereoselectivity). The intermolecular reaction proceeds with a modest rate acceleration of 16M (that is, the ratio of the first-order rate constant k_{cat} for the antibody-catalyzed reaction and the second-order rate constant k_{uncat} for the biomolecular reaction in the absence of antibody). Not surprisingly, the same enantiomer of the lactone formed in the cyclization reaction is utilized in the intermolecular amide-forming reaction.

Janda *et al.* then showed that a similar intermolecular reaction is catalyzed by an antibody raised against the phosphonamidate derivative XV (Figure 7B).[44] Out of fifty-five monoclonal antibodies raised against XV, one catalyzes the reaction of benzylamine and the ester 15a to yield the amide 15b, and an acceleration of 10.5M was observed. Interestingly, thirteen of the antibodies raised against XV catalyze the hydrolysis of the corresponding aryl ester, but none catalyze the hydrolysis of the aryl amide (cf. ref. 36 above).

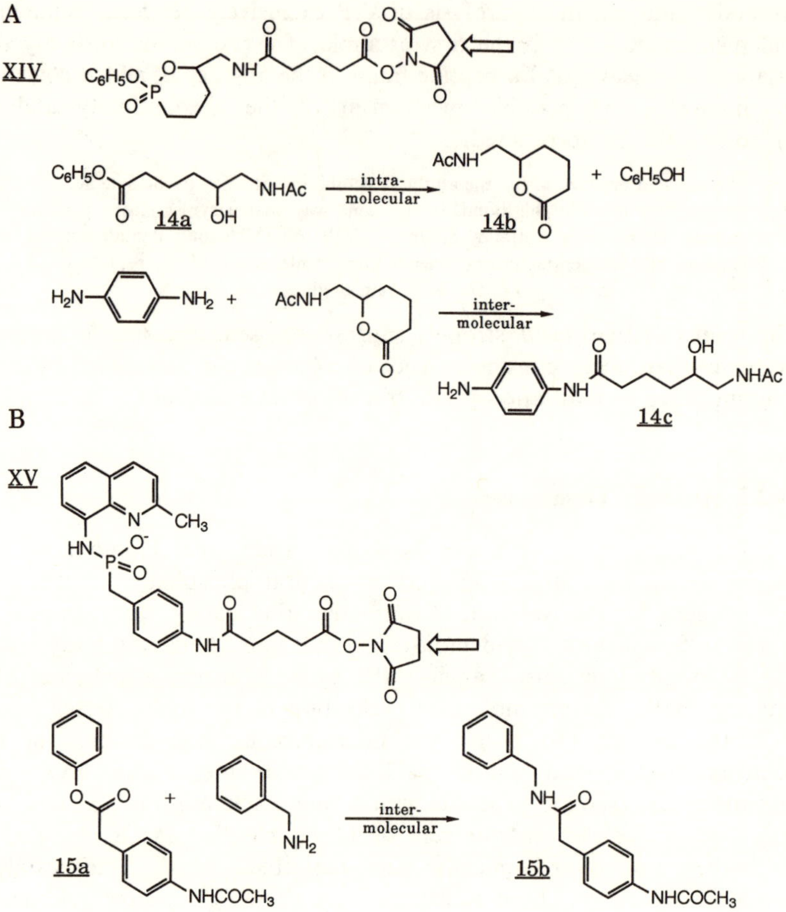

Figure 7. The analogs XIV and XV used to elicit antibody transferases and the reactions catalyzed.

Although the isolation of virtually all of the catalytic antibodies discussed so far relied on the use of tetrahedral phosphate transition-state analogs, more recently a variety of other analogs have successfully been employed as we will now discuss.

14.2.3. Antibody Isomerases

Groups led by Schultz[45] and by Hilvert[46] have independently obtained antibodies that effect the catalysis of the Claisen rearrangement of chorismate 16a to prephenate 16b (Figure 8). Both groups raised

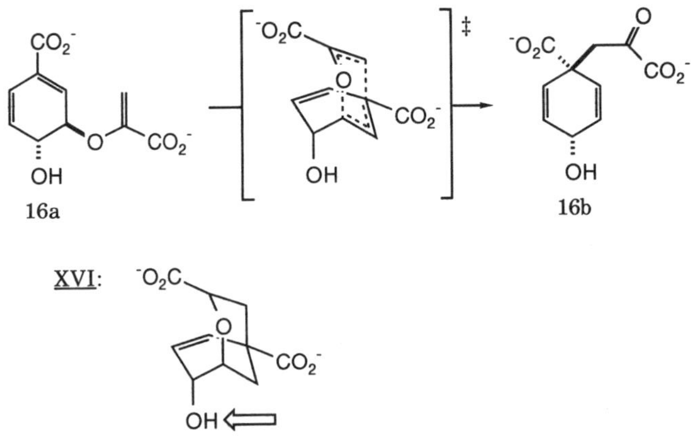

Figure 8. The Claisen rearrangement of chorismate to prephenate and the analog XVI used to elicit the antibody catalysts.

antibodies against the analog XVI, which, as Bartlett and Johnson had previously demonstrated, is a potent competitive inhibitor of the enzyme chorismate mutase.[47] Here then the strategy of utilizing a known enzyme inhibitor as the immunogen was again employed. The Schultz group assayed eight antibodies that bind XVI and determined that one is catalytic. This antibody effects a 10,000-fold acceleration of the rearrangement, but binds the transition-state analog only 100-times more tightly than the substrate. These data are provocative as catalysis is apparently effected primarily by "providing an environment complementary to the conformationally restricted transition state." The Hilvert group assayed fifteen antibodies against XI and showed again that one is catalytic. Their antibody effects an approximately 100-fold acceleration, and in this case, binds the analog XVI about 100-times more tightly than the substrate 17a. The Hilvert group has further demonstrated that their antibody stereospecifically catalyzes the rearrangement, preferentially utilizing (−)-chorismate as the substrate.[48]

14.2.4. Antibody Oxidoreductases

In an intriguing system, the Schultz/IGEN group has isolated antibodies that alter the redox potential of a flavin cofactor. These workers obtained six antibodies against the flavin XVIIa (Figure 9), one of which was studied in detail.[49] The reduction potential for free XVIIa, which is a planar molecule, is −206 mV, while that for the antibody-XVIIa complex is −342 mV. That is, when bound to the antibody, XVIIa is a far worse oxidizing agent than when free in solution; similarly XVIIb, which is

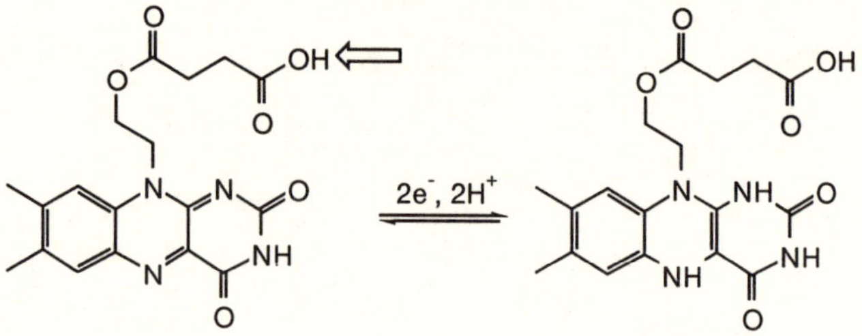

XVIIa XVIIb

Figure 9. The redox chemistry of the flavin XVIIa and 1,5-dihydroflavin XVIIb.

"butterfly"-shaped, is a far better reducing agent when bound than when free. (As the authors stated, from the difference in reduction potentials of free and bound flavin, one may determine that XVIIa is bound over 100,000 times more tightly by the antibody than is XVIIb.) Accordingly, the authors demonstrated that the antibody-XVIIb complex reduces the dye Safranine ($E = -289$ mV) and that free XVIIb does not. (Note that the transition-state analog strategy was not used in this system; instead, the antibody is used to modulate the properties of the hapten used in the immunization.) The authors concluded that "the use of antibody specificity to alter flavin redox potentials is the first step toward developing antibodies capable of using cofactors to catalyze stereoselective reductions."

14.2.5. Antibody Lyases

In the final class of antibody catalysts isolated to date, the Schultz/IGEN group has raised antibodies that effect the catalysis of two lyase reactions. (Again, for each of these systems, the traditional transition-state analog approach was not used.) In the first, antibodies were raised against the thymine dimer XVIII (Figure 10A).[50] As the authors stated:

> [We hoped] that an antibody combining site specific for a thymine dimer and containing an appropriately positioned sensitizer should act as a photoreactivating enzyme. It seemed reasonable to us that antibodies generated against the polarized system of a pyrimidine dimer might contain complementary tryptophan residue in the combining site.

Indeed, of the six antibodies against XVIII that were obtained, five catalyze its photocleavage. One, 15F1-3B1, was chosen for further study, and the k_{cat} for the reaction was determined to be 1.2 min^{-1}, which compares favorably with the k_{cat} of 3.5 min^{-1} for *E. coli* DNA photolyase and which is almost 1,000-times larger than the first-order rate constant for the unsensitized photocleavage. The quantum yield for the reaction at 300 nm was determined to be 0.08. (In a conceptually related system, Balan *et al.* have

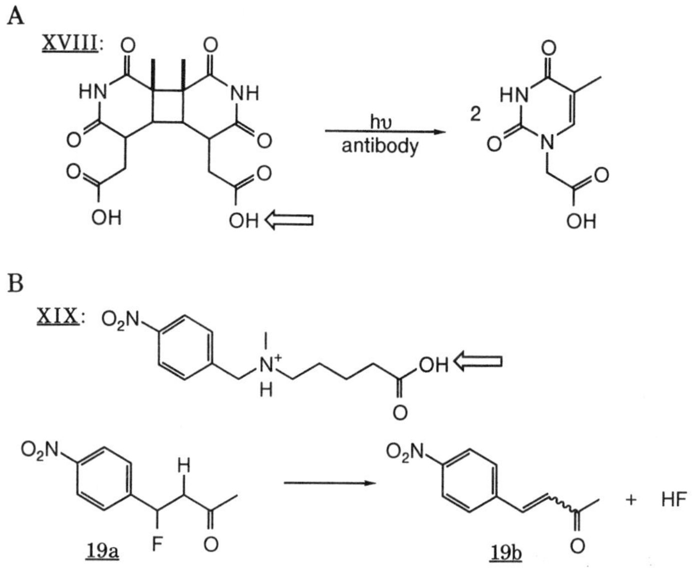

Figure 10. (A) The thymine dimer XVIII used to elicit antibody photolyases and the reaction catalyzed. (B) The analog XIX used to elicit antibodies that catalyze a β-elimination reaction.

reported that polyclonal antibodies raised against a photodimer catalyzes, in the presence of light, the stereoselective dimerization of two monomers.[51] However, the stereoselection is so small that direct interpretation of their data is difficult.)

In the second antibody-catalyzed lyase reaction, four of the six antibodies raised against the tertiary ammonium derivative XIX were found to catalyze the β-elimination of hydrogen fluoride from 19a to yield the unsaturated ketone 19b (Figure 10B).[52] Antibody 43D4-3D3 was characterized, and a rate acceleration of 88,000-fold was observed, as compared with the acetate-ion catalyzed reaction. Presumably, the positively charged ammonium group elicits a carboxylate group in the antigen-combining site which then acts as a base in the presence of the substrate. Furthermore, the hydrophobic environment of the antigen combining apparently raises the pK_a of this carboxylate group, enhancing its basicity. (The pH-rate profile for catalysis supports this conclusion.) At present, the stereochemical course of the elimination reaction is unknown, but the antibody apparently handles both enantiomers of 19a (which, in this case, is not surprising given the steric similarity of hydrogen and fluorine) and produces both the *cis*- and *trans*-isomers of the product olefin.

14.3. ENGINEERED ANTIBODY CATALYSTS

14.3.1. "Semisynthetic" Antibody Catalysts

Recently, as is outlined in Figure 11, the Schultz group has prepared a series of "semisynthetic" antibodies in which a nucleophilic group was chemically introduced in proximity with the antigen combining site of the pre-existing murine myeloma protein MOPC315.[53] MOPC315, which binds 2,4-dinitrophenyl groups, was treated with the derivative XX, which contains an electrophilic aldehyde functionality and a masked thiol functionality, followed by $NaCNBH_3$, and $lysine_{52}$ of the antibody heavy chain was labeled with 95% efficiency. Treatment of this labeled antibody with dithiothreitol (DTT) liberated the thiolated antibody MOPC315-I, which was tested for catalytic activity with a number of ester substrates; a rate acceleration of 60,000-fold was determined for the coumarin substrate 20, as compared with the DTT-catalyzed hydrolysis. Control experiments demonstrated that labeled antibody in which the thiol group remained masked was not catalytic.

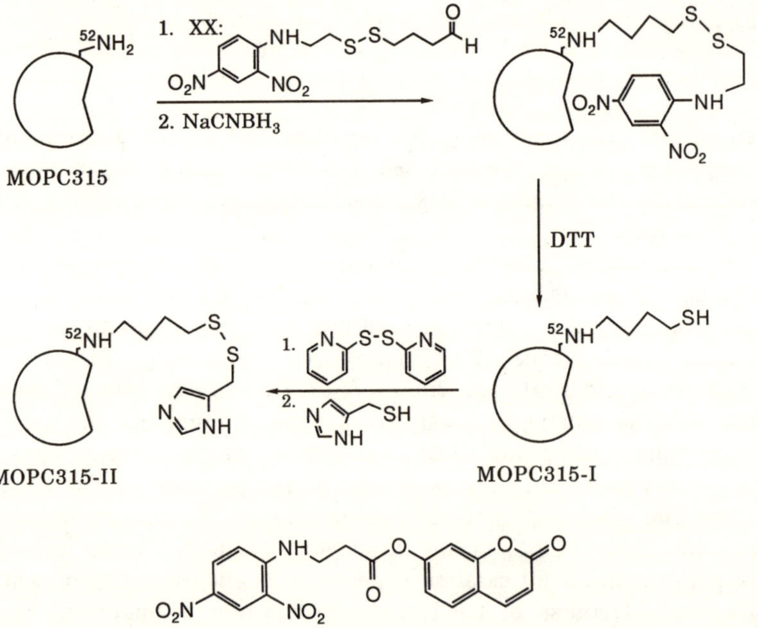

20

Figure 11. The chemistry used for the production of "semisynthetic" antibody catalysts.

In subsequent work, MOPC315-I, which had been treated with 2,2′-dithiopyridine for storage, was further modified by treatment with 4-mercaptomethylimidazole to yield the imidazole derivative, MOPC315-II.[54] MOPC315-II also catalyzes the hydrolysis of dinitrophenyl coumarin ester substrates, and a rate acceleration of approximately 1000-fold was observed with the ester 20, as compared with the reaction catalyzed by 4-methylimidazole. As the authors concluded:

> These experiments are a first step toward the development of selective catalysts which combine the high binding affinity and specificity of the immune system with the diverse, efficient catalytic groups available from synthetic chemistry.

14.3.2. Genetically Engineered Antibody Catalysts

To date, no reports have been published describing the use of protein engineering techniques to produce a catalytic from a noncatalytic antibody or to modify the properties of an existing catalytic antibody. However, such approaches are an obvious next step for the production of antibodies with more efficient catalytic properties and for the production of antibodies that catalyze reactions for which no good transition-state analogs exist. Not surprisingly then, the Schultz, Lerner, and Hilvert groups have all begun to work in this area.[55] Much as in the field of enzymology, two basic genetic approaches are possible: one, the use of "site-directed mutagenesis" techniques to introduce specific amino acid changes at precise points in an antibody structure, and two, the use of "random mutagenesis" techniques, followed by appropriate screening procedures, to generate virtually all possible advantageous mutations. Recent work by Knowles and co-workers on the enzyme triosephosphate isomerase provides an excellent example of the use of both of these techniques in the study of the catalytic properties of an enzyme.[56] Unlike for enzymes, however, techniques for the genetic manipulation of antibodies have only recently become available. The most basic of these is the ability to clone the gene for an antibody into a suitable host and to obtain a useful quantity of properly folded and expressed protein. During the past few years a number of groups have exploited mammalian cell lines to obtain small amounts of engineered antibodies; for example, Rees and co-workers have altered the affinity of the antilysozyme antibody Gloop2 by site-directed mutagenesis and have expressed the altered antibody in *Xenopus* oocytes.[57] A recent review by Morrison and Oi further discusses mammalian expression systems.[58]

However, because of the difficulty in obtaining large amounts of antibody from mammalian systems, much effort has been directed toward obtaining suitable bacterial and yeast expression systems. Although early attempts to express antibodies in bacteria and yeast were largely unsuccessful,[58] great strides have been made in the past two years. In 1988, Skerra and Pluckthun expressed the F_v region of MOPC603 in *E. coli*,[59] and Horwitz and co-workers expressed a chimeric mouse–human Fab fragment in *E. coli*[60] and in yeast.[61] Independent groups led by Bird[62] and by Huston[63] have also succeeded in expressing single-chain, tethered F_v analogs in *E. coli.* Fab fragments and the smaller F_v fragments and single-chain antibodies retain most, if not all, of the binding characteristics of intact antibodies (and presumably therefore will retain their catalytic properties as well). Thus the stage has been set to engineer antibody molecules quickly and cheaply with exactly the same means used for enzymes. In addition, computer modeling techniques have been developed for the study of antibody specificity, and such models may prove useful as a guide for engineering experiments.[64]

14.4. PROSPECTS FOR THE FUTURE

While it is clear from the above discussion that antibodies can effect the catalysis of a wide range of reactions, the mechanism of catalysis for most of the antibodies is far from clear. Indeed, even the expectations for catalysis differ—for example, Hilvert and Nared have commented, "Since strain and proximity are the principle catalytic effects antibodies are likely to impart..."[48] while Benkovic *et al.* have stated, "It seems reasonable based upon our present knowledge of antibody-catalyzed reactions to assume that [antibody catalysts] in general may eschew effects such as strain, distortion, or desolvation."[43] Thus, mechanistic information on the antibodies heretofore isolated is needed and will be essential for the creation of new strategies for the isolation of better antibody catalysts.

14.4.1. The Isolation of Additional Antibody Catalysts

Although a major goal to date of most of the researchers in the field has been the isolation of antibodies with sequence-specific protease activity, there exist many other important reactions for which antibody catalysts would be extremely useful. For example, antibodies with restriction exo- and endonuclease activity would be of great utility; however, their isolation by direct immunization will require a trigonal-bipyramidal transition-state analog, and few such suitable analogs exist. (Indeed, tetrahedral phosphorus derivatives may be the most effective transition-state analogs ever synthesized.) To isolate antibody catalysts for this reaction then, approaches other than immunization with a transition-state analog will most likely be necessary. Some of these approaches were discussed above. Of these, Iverson and Lerner's strategy of eliciting antibodies that bind a substrate in apposition with a potent catalytic cofactor may have the greatest generality. Much work in the field of physical organic chemistry has shown that simply holding a substrate and a catalyst together can lead to enormous rate enhancements. For example, as Kirby stated in 1980[65]:

> [Intramolecular reactions] are generally faster than the corresponding intermolecular processes, and are frequently so much faster that it is possible to observe those types of reactions involved in enzyme catalysis. Bimolecular reactions in water ... are frequently too slow to detect even under vigorous conditions. But when the catalytic and substrate groups are brought together in the same molecule such otherwise unreactive compounds may [react] under quite mild conditions.

A spectacular example of an intramolecular acceleration was recently published by Menger and Ladika who demonstrated that the amide XXI (Figure 12) hydrolyzes extraordinarily quickly under physiological conditions ($t_{1/2} = 8$ min).[66] The authors concluded:

> Note finally that enzymes [and thus presumably antibodies] need not summon esoteric mechanisms to split amides rapidly. If, for example, an enzyme positioned one of its carboxyls adjacent to an amide substrate with the geometry given [by the amide XXI], little additional catalytic power would be required.

As was amply demonstrated by Iverson and Lerner,[39] similar accelerations can occur when an antibody is used to bring together the catalyst and substrate, and this approach may well yield numerous antibody catalysts, including antibody exo- and endonucleases.

14.4.2. Protein Engineering of Antibody Catalysts

The use of protein engineering techniques may provide an alternative strategy for the isolation of new antibody catalysts. Given the recent successes in expressing antibody F_v and Fab fragments in *E. coli* discussed above, the protein engineering of antibody molecules should now be far more straightforward. Nevertheless, given the history of enzyme engineering, one must ask whether new and improved antibodies will, in fact, be obtained by this approach. Even when beginning with a well characterized enzyme, "rationally" improving its catalytic properties or altering its sub-

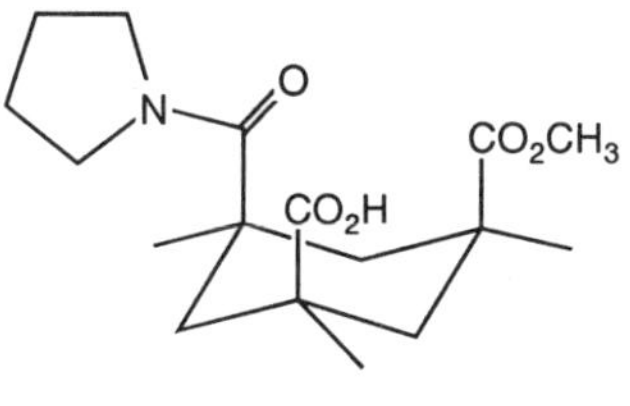

XXI

Figure 12. The amide derivative XXI.

strate specificity has not been an easy task. (Of course, given the catalytic efficiency of many enzymes, increasing catalytic activity is certainly a challenge and in some cases is, in fact, impossible.[67]) In addition, for those enzymes that have been successfully engineered, the engineers usually have made use of pre-existing biological information. For example, the catalytic activity of phospholipase A_2 has been improved using information obtained from prior chemical modification and X-ray crystallographic studies,[68] and the substrate specificity of lactate dehydrogenase has been altered by making mutations based upon the structural similarities between lactate dehydrogenase and malate dehydrogenase.[69] Furthermore, experiments with altered enzymes have shown that a molecular understanding of most mutations is difficult. For example, removal of an "essential" residue can in fact have only a small effect,[70] while, in other instances, removal of a single hydrogen bond can have an enormous effect.[71] As Gerlt stated in 1987[72]:

> [The] primary lesson that is being repeatedly learned by those already using site-directed mutagenesis is that the...most important aspect to success in its utilization is not the ability to generate and produce mutant proteins but to perform the structural characterizations of the mutant proteins so that their chemical and kinetic properties can be definitively understood. Much of the data discussed in this review cannot be definitively explained in the absence of crystal structures for essentially each and every mutant protein.

Thus even when mutant enzymes are in hand, understanding their properties, in spite of decades of work on enzymatic catalysis, has been difficult. Thus, it will be interesting to see if antibody molecules that possess little or no catalytic activity and for which there is very likely no crystal structure data can be engineered onto mighty catalysts. Given the above difficulties, it is not surprising that many in the field are concentrating on altering antibodies of the phosphorylcholine binding family, of which there are in excess of twenty-five examples. Extensive sequence data are available for these antibodies,[73] and the crystal structure of the Fab fragment of one, MOPC603, has been solved.[74] However, as mechanistic information becomes available on the existing antibodies and more is learned about altered enzymes, protein engineering methods may ultimately yield new antibody catalysts. At present, random mutagenesis techniques, including the induction of catalytic activity by growth of host organisms under selective pressure, would seem to provide the best prospects.

In his Nobel Prize lecture, Fischer suggested that one day scientists may be able to "equal Nature" in their preparation of synthetic enzymes.[4] Although in just three years spectacular examples of antibody catalysis have been demonstrated, it remains to be seen whether antibodies can be isolated that will match the enormous rate accelerations effected by enzymes and whether they will catalyze an equally wide range of reactions. (Of course, it seems only fair to posit that all of the antibody catalysts obtained by direct immunization are in fact Nature's, and that the preparation of a true "synthetic enzyme" still remains to be achieved.) In addition, it will be of great interest to see if an antibody catalyst, or indeed if any nonenzymic catalyst, will effectively function using a catalytic mechanism that Nature herself did not uncover during the long process of evolution.

NOTE ADDED IN PROOF

Recently a number of additional publications have appeared. Examples of an antibody-catalyzed redox reaction,[75] an antibody-catalyzed Diels–Alder reaction,[76] and an intriguing antibody-catalyzed hydrolysis reaction have been reported.[77] Baldwin and Schultz have reported the use of site-directed mutagenesis to generate an antibody esterase.[78] The antibody 6D4[1] has been expressed in tobacco plants.[79] Finally, in two extremely important papers, Ward *et al.*[80] and Huse *et al.*[81] have independently reported the expression of mouse antibody libraries (V_H domains and Fab fragments, respectively) in *E. coli.*

REFERENCES

1. A. Tramontano, K. D. Janda, and R. A. Lerner, *Science 234*, 1566 (1986).
2. S. J. Pollack, J. W. Jacobs, and P. G. Schultz, *Science 234*, 1570 (1986).
3. P. S. Kim, *Protein Engineering 2*, 249 (1988).
4. *Nobel Lectures Chemistry 1901-1921*, p. 34, Elsevier, New York (1966).
5. G. M. Whitesides and C.-H. Wong, *Aldrichimica Acta 16*, 27 (1983).
6. E. T. Kaiser and D. S. Lawrence, *Science 226*, 505 (1984).
7. L. Pauling, *Am. Sci. 36*, 51 (1948).
8. A. Fersht, *Enzyme Structure and Mechanism*, 2nd ed., W. H. Freeman, New York (1985).
9. W. P. Jencks, *Symposia on Quantitative Biology 52*, 65 (1987).
10. J. Kraut, *Science 242*, 533 (1988).
11. W. P. Jencks, *Catalysis in Chemistry and Enzymology*, p. 268, McGraw Hill, New York (1969).
12. L. E. Hood, I. L. Weissman, and W. B. Wood, *Immunology*, p. 206, Benjamin/Cummings, Melno Park, California (1978).
13. S. Tonegawa, *Sci. Am. 253*, 122 (1985).
14. L. E. Hood, I. L. Weissman, and W. B. Wood, *Immunology*, p. 158, Benjamin/Cummings, Melno Park, California (1978).
15. W. P. Jencks, in: *Current Aspects of Biochemical Energetics* (E. P. Kennedy, ed.), p. 273, Academic Press, New York (1966).
16. R. Wolfenden, *Acc. Chem. Res. 5*, 10 (1972).
17. G. E. Lienhard, *Science 180*, 149 (1973).
18. G. J. Cardinale and R. H. Abeles, *Biochemistry 7*, 3970 (1968).
19. M. V. Keenan and W. L. Alworth, *Biochem. Biophys. Res. Commun. 57*, 500 (1974).
20. K. D. Collins, *J. Biol. Chem. 249*, 136 (1974).

21. R. Wolfenden and L. Frick, in: *Enzyme Mechanisms* (M. I. Page and A. Williams, eds.), Royal Society of Chemistry Press, London (1987).
22. V. Raso and B. D. Stollar, *Biochemistry 14*, 584, 591 (1975).
23. R. Summers, *Catalytic Principles of Enzyme Chemistry*, Ph.D. Thesis, Harvard University (1983).
24. B. F. Erlanger, *Methods Enz. 70*, 85 (1980).
25. G. Kohler and C. Milstein, *Nature 256*, 495 (1975).
26. F. Kohen, J. B. Kim, H. R. Lindner, Z. Eshhar, and B. Green, *FEBS Lett. 111*, 427 (1980).
27. R. A. Lerner, *Adv. Immunology 36*, 1 (1984).
28. A. Tramontano, K. D. Janda, and R. A. Lerner, *Proc. Natl. Acad. Sci. USA 83*, 6736 (1986).
29. N. E. Jacobson and P. A. Bartlett, *J. Am. Chem. Soc. 103*, 654 (1981).
30. G. C. Rao, *Enzyme Models Based on Boronic Acids and on a Monoclonal Antibody*, Ph.D. Thesis, City University of New York (1987).
31. S. J. Pollack and P. G. Schultz, *Symposia on Quantitative Biology 52*, 97 (1987).
32. A. L. Lehninger, *Biochemistry*, p. 210, Worth, New York (1982).
33. J. Jacobs, P. G. Schultz, R. Sugasawara, and M. Powell, *J. Am. Chem. Soc. 109*, 2174 (1987).
34. A. Tramontano, A. A. Ammann, and R. A. Lerner, *J. Am. Chem. Soc. 110*, 2282 (1988).
35. C. N. Durfor, R. J. Bolin, R. Sugasawara, R. J. Massey, J. W. Jacobs, and P. G. Schultz, *J. Am. Chem. Soc. 110*, 8713 (1988).
36. K. D. Janda, D. Schloeder, S. J. Benkovic, and R. A. Lerner, *Science 241*, 1188 (1988).
37. K. D. Janda, S. J. Benkovic, and R. A. Lerner, *Science 244*, 437 (1989).
38. S. J. Pollack, P. Hsiun, and P. G. Schultz, *J. Am. Chem. Soc. 111*, 5961 (1989).
39. B. L. Iverson and R. A. Lerner, *Science 243*, 1184 (1989).
40. H. J. Dyson, R. A. Lerner, and P. E. Wright, *Annu. Rev. Biophys. Biophys. Chem. 17*, 305 (1988).
41. S. Paul, D. J. Volle, C. M. Beach, D. R. Johnson, M. J. Powell, and R. J. Massey, *Science 244*, 1158 (1989).
42. A. D. Napper, S. J. Benkovic, A. Tramontano, and R. A. Lerner, *Science 237*, 1041 (1987).
43. S. J. Benkovic, A. D. Napper, and R. A. Lerner, *Proc. Natl. Acad. Sci. USA 85*, 5355 (1988).
44. K. D. Janda, R. A. Lerner, and A. Tramontano, *J. Am. Chem. Soc. 110*, 4835 (1988).
45. D. Y. Jackson, J. W. Jacobs, R. Sugasawara, S. H. Reich, P. A. Bartlett, and P. G. Schultz, *J. Am. Chem. Soc. 110*, 4841 (1988).
46. D. Hilvert, S. H. Carpenter, K. D. Nared, and M.-T. M. Auditor, *Proc. Natl. Acad. Sci. USA 85*, 4953 (1988).
47. P. A. Bartlett and C. R. Johnson, *J. Am. Chem. Soc. 107*, 7792 (1985).
48. D. Hilvert and K. D. Nared, *J. Am. Chem. Soc. 110*, 5593 (1988).
49. K. M. Shokat, C. J. Leumann, R. Sugasawara, and P. G. Schultz, *Angew. Chem. Int. Ed. Engl. 27*, 1172 (1988).
50. A. G. Cochran, R. Sugasawara, and P. G. Schultz, *J. Am. Chem. Soc. 110*, 7888 (1988).
51. A. Balan, B. P. Doctor, B. S. Green, M. Torten, and H. Ziffer, *J. C. S. Chem. Commun.*, 106 (1988).
52. K. M. Shokat, C. J. Leumann, R. Sugasawara, and P. G. Schultz, *Nature 338*, 269 (1989).
53. S. J. Pollack, G. R. Nakayama, and P. G. Schultz, *Science 242*, 1038 (1988).
54. S. J. Pollack and P. G. Schultz, *J. Am. Chem. Soc. 111*, 1929 (1989).
55. P. G. Schultz, R. A. Lerner, and D. Hilvert, private communication.
56. J. D. Hermes, S. C. Blacklw, and J. R. Knowles, *Symposia on Quantitative Biology 52*, 597 (1987).
57. S. Roberts, J. C. Cheetham, and A. R. Rees, *Nature 328*, 731 (1987).

58. S. L. Morrison and V. T. Oi, *Adv. Immunol. 44*, 65 (1989).
59. A. Skerra and A. Pluckthun, *Science 240*, 1038 (1988).
60. M. Better, C. P. Chang, R. R. Robinson, and A. H. Horwitz, *Science 240*, 1041 (1988).
61. A. H. Horwitz, C. P. Chang, M. Better, K. E. Hellstrom, and R. R. Robinson, *Proc. Natl. Acad. Sci. USA 85*, 8678 (1988).
62. R. E. Bird, K. D. Hardman, J. W. Jacobson, S. Johnson, B. M. Kaufman, S.-M. Lee, T. Lee, S. H. Pope, G. S. Riordan, and M. Whitlow, *Science 242*, 423 (1988).
63. J. S. Huston, D. Levinson, M. Mudgett-Hunter, M.-S. Tai, J. Novotny, M. N. Margolies, R. J. Ridge, R. E. Bruccoleri, E. Haber, R. Crea, and H. Oppermann, *Proc. Natl. Acad. Sci. USA 85*, 5879 (1988).
64. A. R. Rees and P. de la Paz, *TIBS 11*, 144 (1986).
65. A. J. Kirby, *Adv. Phys. Org. Chem. 17*, 183 (1980).
66. F. M. Menger and M. Ladika, *J. Am. Chem. Soc. 110*, 6794 (1988).
67. A. Fersht, *Enzyme Structure and Mechanism*, 2nd ed., pp. 147–152, W. H. Freeman, New York (1985).
68. O. P. Kuipers, M. M. G. M. Thunnissen, P. de Geus, B. W. Dijkstra, J. Drenth, H. M. Verheij, and G. H. de Hass, *Science 244*, 82 (1989).
69. H. M. Wilks, K. W. Hart, R. Feeney, C. D. Dunn, H. Muirhead, W. N. Chia, D. A. Barstow, T. Atkinson, A. R. Clarke, and J. J. Holbrook, *Science 242*, 1541 (1988).
70. D. Hilvert, S. J. Gardell, W. J. Rutter, and E. T. Kaiser, *J. Am. Chem. Soc. 108*, 5298 (1986).
71. W. M. Kati and R. Wolfenden, *Science 243*, 1591 (1989).
72. J. A. Gerlt, *Chem. Rev. 87*, 1079 (1987).
73. P. J. Gearhart, N. D. Johnson, R. Douglas, and L. Hood, *Nature 291*, 29 (1981).
74. Y. Satow, G. H. Cohen, E. A. Padlan, and D. R. Davies, *J. Mol. Biol. 190*, 593 (1986).
75. N. Janjic and A. Tramontano, *J. Am. Chem. Soc. 111*, 9109 (1989).
76. D. Hilvert, K. W. Hill, K. D. Nared, and M.-T. M. Auditor, *J. Am. Chem. Soc. 111*, 9261 (1989).
77. K. D. Janda, M. I. Weinhouse, D. M. Schloeder, R. A. Lerner, and S. J. Benkovic, *J. Am. Chem. Soc. 112*, 1274 (1990).
78. E. Baldwin, and P. G. Schultz, *Science 245*, 1104 (1989).
79. A. Hiatt, R. Cafferkey, and K. Bowdish, *Nature 342*, 76 (1989).
80. E. S. Ward, D. Gussow, A. D. Griffiths, P. T. Jones, and G. Winter, *Nature 341*, 544 (1989).
81. W. D. Huse, L. Sastry, S. A. Iverson, A. S. Kang, M. Alting-Mees, D. R. Burton, S. J. Benkovic, and R. A. Lerner, *Science 246*, 1275 (1989).

15

Enzymatic Catalysis in Bioseparations

DAVID G. RETHWISCH and JONATHAN S. DORDICK

15.1. INTRODUCTION

Organic acids and alcohols often constitute the active components of biologically functional molecules including pharmaceuticals, sweeteners, insectides, herbicides, and hormones.[1–5] The large-scale production of these molecules generally requires efficient and selective methods of purification.[6] Presently, individual peptides are separated from fermentation broths, blood serum, or chemical liquors by reverse phase or ion-exchange high-performance liquid chromatography (HPLC).[7] Similarly, racemic mixtures of organic compounds are separated (i.e., optically resolved) into their optically active, biologically functional isomers using chiral-based chromatographic supports,[8] or by reaction of the racemate with an optically active co-reactant to form diasteriomers which are more easily separated.[9] While these methods have proved to be effective for laboratory-scale separations, they suffer from several limitations for commercial-scale applications. Chromatographic methods are expensive, solvent-intensive, and cannot be run in a continuous fashion. Diasteriomeric separations often require multiple recrystallization steps and consume expensive and often toxic coreactants. Thus, a need has developed for high-volume, low-cost processes for separations of organic acids and alcohols.

DAVID G. RETHWISCH and JONATHAN S. DORDICK • Department of Chemical and Biochemical Engineering, University of Iowa, Iowa City, Iowa 52242.

Enzymes display a high degree of both regio- and stereospecificity. Recently, there has been a great deal of interest in using this capability of enzymes in the discrimination step of separations. This has proved effective in performing both regio- and stereoselective separations. The objective of this review is to highlight the primary approaches which are currently being used and indicate where possible the future directions in the field. In the sections that follow the methods in which enzymes have been used in separations will be reviewed. This review will include single-phase systems (aqueous or nonaqueous), two-phase systems (organic/aqueous), and three-phase liquid-membrane systems (aqueous/organic/aqueous). Each of these will be discussed in detail below.

15.2. SINGLE-PHASE SYSTEMS

The simplest approach conceptually for using the selectivity of enzymes for separations is the single-phase system. In traditional separations, molecules are differentiated based on differences in physical properties such as boiling point or solubility. In a single-phase system, the enzyme is generally used to modify (often reversibly) the target substrate in such a way that its physical properties become significantly different from those of the remaining materials. The desired product is then isolated by conventional methods. Consider the resolution of a racemic mixture. The physical properties of the optical isomers are generally indistinguishable. However, enzymes can readily distinguish between the two isomers, and only one isomer is modified. Hydrolases (e.g., lipases, esterases, and acylases) are the enzymes most commonly used for separations because they are commercially available at reasonable cost, require no cofactors, are useful with a wide range of substrates, and many are stable in both aqueous and organic solvents.[10] Specific examples are given below for both aqueous and nonaqueous systems.

15.2.1. Aqueous Systems

Until recently, enzymes were thought to be viable only in aqueous systems. Therefore, it is not surprising that the first uses of enzymes for separations were in water. A good example is the commercial resolution of amino acids developed by Uzuki *et al.*[11] In the first step of this process, racemic amino acids are chemically acetylated to yield the racemic *N*-acetyl amide of the amino acid. Treatment of the *N*-acetyl amino acid with aminoacylase in an aqueous solution yields free L-amino acid plus the D-*N*-acetyl amino acid derivative[11] (Eq. (1)).

$$(\text{D, L})-R_1-\underset{}{\overset{\displaystyle NH-\overset{\displaystyle O}{\overset{\|}{C}}-CH_3}{\overset{|}{C}H}}-COOH + H_2O \longrightarrow \text{L}-R_1-\overset{\displaystyle NH_2}{\overset{|}{C}H}-COOH$$

$$+ CH_3\overset{\displaystyle O}{\overset{\|}{C}}OH + \text{D}-R_1-\overset{\displaystyle NH-\overset{\displaystyle O}{\overset{\|}{C}}-CH_3}{\overset{|}{C}H}-COOH \qquad (1)$$

Similarly, one can chemically prepare a mixture of D and L methyl esters from a racemic mixture of amino acids. The L-isomer can then be stereoselectively hydrolyzed to yield the L-amino acid. This approach has been commercialized using the enzyme chymotrypsin for the resolution of L-phenylalanine.[12] In both of the reaction sequences discussed above, the residual ester or amide was separated from the desired amino acid via solvent extraction.

Lipases have also been used in the stereoselective hydrolysis of racemic esters.[13, 14] Miyano and co-workers[15] carried out stereoselective hydrolysis of racemic binaphthols using porcine pancreatic lipase. Optical purities (as measured by enantiomeric excess) exceeded 95%.

15.2.2. Water-Miscible Cosolvents

While aqueous systems have been successful for a number of applications, they suffer from a serious drawback. For large-volume separations, these systems are only effective for substrates which are water-soluble. However, many commercially important species (pharmaceuticals, pesticides, etc.) are only sparingly soluble in water. In addition, the synthesis of many compounds is thermodynamically unfavorable. For example, in the synthesis of esters and amides water is a product; therefore, in aqueous solutions equilibrium favors the starting reagents. To overcome these problems people have used water-organic mixtures with a high concentration of a water-miscible organic cosolvent such as methanol, ethanol, acetone, DMF, and DMSO.[16, 17] However, in practice these systems have limited application. The miscible solvents tend to reduce enzyme activity, selectivity, or stability. In addition, the aqueous/organic solvent system is seldom hydrophobic enough to significantly increase the solubility of lipophilic substrates. Bielski and Freed[18] used a water/methanol system for the α-chymotrypsin catalyzed hydrolysis of *N*-acetyl-L-tryptophan ethyl ester. Chymotrypsin was rapidly denatured by this solvent system at room temperature. However, the enzyme was stable at concentrations as high as 80% methanol at $-33\,^{\circ}C$, though 24 h were required to observe a measurable conversion of the ester.

15.2.3. Nonaqueous (Organic) Media

Recent studies indicate that many enzymes are stable and active in nonaqueous systems.[10] This suggests that the solubility problems discussed above are better handled in single-phase nonaqueous systems. Indeed, the use of enzymes as catalysts for separations including optical resolutions has been extended to nonaqueous systems. While selective hydrolysis of esters is used in aqueous systems selective esterifications, transesterifications, and aminolysis are used in organic media. Consider the reactions in Eq. (2).

$$(\text{D, L})\text{-R-}\overset{\text{O}}{\overset{\|}{\text{C}}}\text{-OH} + \text{R}'\text{-OH} \underset{\text{hydrolysis}}{\overset{\text{esterification}}{\rightleftarrows}} \text{L-R-}\overset{\text{O}}{\overset{\|}{\text{C}}}\text{-O-R}' + H_2O + \text{D-R-}\overset{\text{O}}{\overset{\|}{\text{C}}}\text{-OH} \quad (2)$$

The procedure is straightforward: racemic alcohols or amines are esterified or transesterified with an achiral carboxylic acid or ester, respectively, to produce a chiral ester or amide that can be isolated from the starting materials via conventional chemical/physical separations. Racemic acids can be resolved by reacting with an achiral alcohol or amine to form a chiral ester or amide, respectively. The purified ester is then chemically hydrolyzed to give the desired optically pure product.

The lipase from *Candida cylindracea* has become a versatile catalyst for the resolution of racemic carboxylic acids and alcohols. For example, (±)-menthol can be resolved by the enzyme-catalyzed esterification of the (−) isomer with lauric acid in heptane to give optical purities in excess of 95%.[19] The (+) isomer can be recovered as free menthol. Similar resolutions were performed with the same lipase entrapped in polyurethane gels, using 5-phenylvalerate in isooctane.[20] Gram-scale resolutions were afforded. Lipase from *C. cylindracea* has been effective in the preparative resolution of 2-halocarboxylic acids.[21, 22] Gram quantities of 2-chloro- and 2-bromopropionic acids with optical purities in excess of 95% were prepared. Racemic aromatic acids have also been resolved in organic media.[23] The most important of these was 2-(4-chlorophenoxy)propionic acid, the R-isomer having use as a potent herbicide.[24]

Porcine pancreatic lipase has proved effective in the optical resolution of racemic alcohols. For example, 2-octanol, 2-dodecanol, and *sec*-phenethanol were resolved by transesterification with trichloroethyl butyrate.[22] Only the R-isomer reacted with the enzyme affording an R-based ester and optical resolution. The residual S-alcohol could be recovered easily by distillation. Acylation can be performed using lipases in the presence of an anhydride.[25] Inactivation due to functional group acylation was not evident. In another example of the power of enzyme-catalyzed optical resolution of alcohols, stereospecific esterification of the hydroxyl moiety of amino-alcohols is feasible by blocking the amine of 2-amino-alcohols.[26] The blocked amine is unable to act as the nucleophile and esters form in the absence of amides.

In addition to aliphatic alcohols, optical resolution of organometallic alcohols and aliphatic amines has been carried out. In the former case, racemic ferrocenylethanol was resolved in toluene using porcine pancreatic lipase.[27] In the latter case, subtilisin in 3-methyl-3-pentanol was shown to resolve racemic amines with enantiomeric excess above 80%.[28] Solvent polarity was vital in determining the efficiency of optical separations. Non-polar solvents such as toluene and cyclohexane did not provide a medium capable of sustaining enzymatic resolution of the amines. This is similar to the relaxation in enantioselectivity of various proteolytic enzymes afforded by hydrophobic solvents for the transesterification of amino acid derivatives (see above). Enzyme-catalyzed resolution of racemic mixtures is summarized in Table 1.

Thus, by careful selection of the enzyme and solvent (aqueous or organic), a wide range of organic acids and alcohols can be separated and purified using a single-phase system. However, several additional processing steps must be included to carry out a separation. For example, in an optical resolution, a three-step sequence is required: (1) starting with the racemic acid (alcohol) the racemic esters are prepared using traditional (nonselective) chemistry; (2) one enantiomer is hydrolyzed using an enzyme; (3) the unreacted ester (in aqueous systems) or the acid produced (in organic systems) are extracted. In general, it is desirable to reduce the number of processing steps required. One step can be removed by going to a biphasic system as discussed below.

15.3. MULTIPHASE SYSTEMS

15.3.1. Biphasic Separations

Enzyme-assisted separations of sparingly water-soluble substrates can also be carried out using multiphase systems.[33, 34] For example, in "biphasic" systems, the enzyme is dissolved in the aqueous phase and the substrate (and often products) is in a separate, immiscible organic phase. Either the aqueous phase can be the continuous phase, with the organic phase dispersed, or the aqueous phase can be dispersed in a continuous organic phase. In a variant of this process the enzyme can be supported on a third, solid phase.[35] Indeed, Cambou and Klibanov[36] demonstrated that in a biphasic system, lipase from *C. cylindracea* was able to readily resolve the R-acid from a racemic mixture of the methyl ester of (R, S)-2-

Table 1. Representative Summary of Optical Resolutions Employing Enzymes in Monophasic Organic Solvents

Racemate	Acyl donor or acceptor	Enzyme	Solvent	Isomer reacted	Yield[a]	EE[b]	Reference
Menthol	Lauric acid	Lipase from *C. cylindracea*	Heptane	(–)	45	95	19
2-Decanol	*n*-Butyl alcohol	Lipase from *M. miehei*	Hexane	R	40	87	29
2-Chloropropionic acid	*n*-Butyl alcohol	Lipase from *C. cylindracea*	Hexane	R	42	95	21
2-Bromopropionic acid	*n*-Butyl alcohol	Lipase from *C. cylindracea*	Hexane	R	45	96	21
2-Octanol	Trichloroethylbutyrate	Lipase from porcine pancreas	Diethyl ether	R	47	95	21
2-Methyl-2,4-pentanediol	Trichloroethylbutyrate	Lipase from porcine pancreas	Heptane	R	48	87	21
Butyl-*O*-formylmandelate	*n*-Butyl alcohol	Lipase from *C. cylindracea*	Isopropyl ether	R	39	75	30
2-Octanol	Dodecanoic acid	Lipase from porcine pancreas	Heptane	R	43	97	31
2-(4-Chloro)-propionic acid	Cyclohexanol	Lipase from *C. cylindracea*	Isooctane	R	40	98	23
Menthol	Lauric acid	Lipase from *C. cylindracea*	Isooctane	R	40	99	23
2-*N*-Alk-aminopropanol[c]	Ethyl acetate	Lipase from porcine pancreas	Ethyl acetate	R	32[d]	92	26
2-*N*-Alk-aminopropanol	Ethyl acetate	Pancreatin	Ethyl acetate	R	25[d]	95	26
2-*N*-Alk-aminopropanol	Ethyl acetate	Lipase from *Pseudomonas* sp.	Ethyl acetate	R	14[d]	85	26
1,1-Binaphtyl-2,2′-diol	Vinyl acetate	Lipase from *Pseudomonas* sp.	Isopropyl ether	R	53	95	32
2-*O*-Benzylglycerol	Isopropenyl acetate	Lipase from *Pseudomonas* sp.	Chloroform	S	36	96	27
Ferrocenylethanol	Vinyl propionate	Lipase from *Pseudomonas* sp.	Toluene	S	30	84	27
α-Methyl benzylamine	Trifluoroethylbutyrate	Subtilisin	*s*-Methyl-pentanol	S	35	85	28

[a] Unless otherwise stated, this represents percent of reacting isomer.
[b] Enantiomeric excess.
[c] Alk = alkoxycarbonyl.
[d] Isolated yield of recovery of 2-aminopropanol following alkaline hydrolysis.

(*p*-chlorophenoxy)-propionic acid. The solid-phase, supported enzyme, and the organic phase (the methyl ester) were dispersed in a continuous aqueous phase. Using this technique, the R-acid was produced at 97% purity in 85% yield with an optical purity of 96%.[36]

In general, biphasic systems have been found to have little industrial utility. Cambou and Klibanov[36] indicated that pH control was difficult to achieve in this system. Also, due to the emulsion, continuous operation and scale-up are difficult, and the emulsion must be broken to recover the product.

15.3.2. Two-Phase Membrane Systems

Sepracor, Inc. has recently developed a membrane reactor system for large-scale stereoselective hydrolysis of racemic amino acid esters resulting in purification of the desired L-amino acids.[36–40] A primary advantage of such a system over a biphasic system is increased stability and easier scale-up and continuous operation. In this system, a racemic mixture of amino acids is chemically esterified and fed in an organic solvent to the membrane reactor. In the reactor an enzyme such as aminoacylase is supported on a porous polymer membrane (see Figure 1). This enzyme catalyzes the hydrolysis of the L-amino acid ester. The resulting L-amino acid partitions into an aqueous recovery stream while the hydrophobic D-amino ester is retained in the organic phase. Optical resolution has been demonstrated with this system for 2-chloropropionic acid, Naproxen, and Ibuprofen using lipase from *C. cylindracea* and for *N*-benzoyl-tyrosine and *N*-acetyl-tyrosine using chymotrypsin.[38]

As discussed above, an advantage of the Sepracor system is that the stereospecific reaction and the extraction of the product occur in a single processing step; however, a second process vessel is required for the initial formation of the ester.

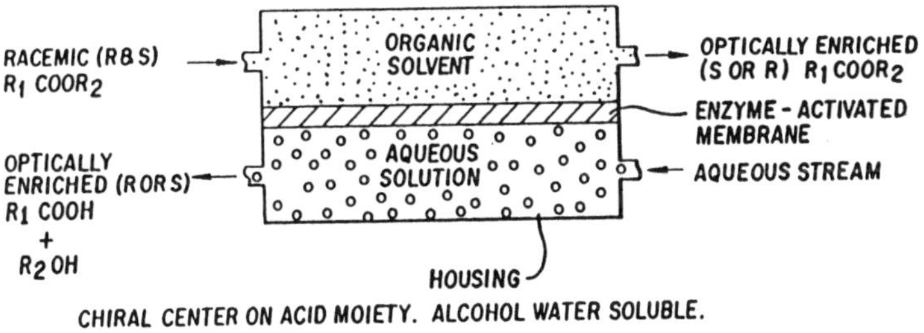

Figure 1. Schematic of extractive membrane reactor.[38]

15.3.3. Liquid Membranes

While the use of enzymes either in aqueous or organic media permits regio- and stereospecific purifications, the processes which have been developed require multistep procedures, which increases both the capital and operating costs of the separation.[40, 41] In addition, they are solvent-intensive. This solvent must either be recycled or disposed of, further adding to the operational expense and associated health risks. A possible way to reduce unneccessary steps is the combination of the selectivity of enzymes with a liquid membrane system. For example, carbonic anhydrase has been used in a variety of liquid membranes for transport of CO_2.[42, 43] Thus, the selectivity of enzymes is coupled with liquid membranes to provide a single-step method to selectively separate and purify organic acids.[44]

This experimental strategy uses facilitated transport of a desired organic acid through a liquid membrane. This was carried out by Rethwisch *et al.*[44] for the lipase-catalyzed esterification of an organic acid with a hydrophobic alcohol contained in an organic liquid membrane (see Figure 2). The resulting ester partitions into the organic phase or is hydrolyzed to the parent acid. Once in the organic phase, the ester diffuses

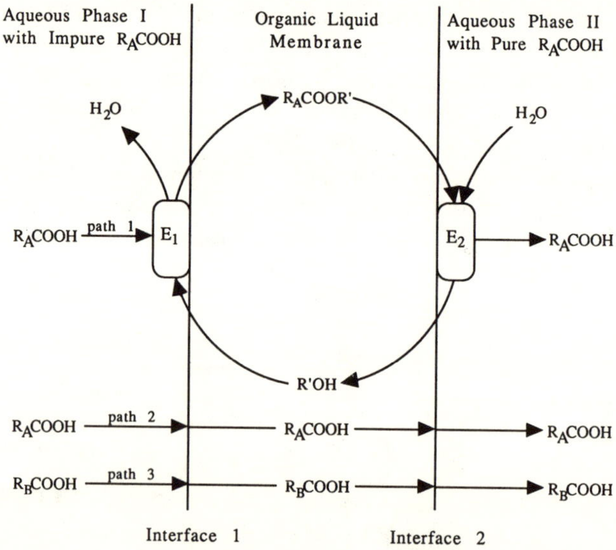

Figure 2. Schematic of enzyme-facilitated liquid membrane transport for organic acid separations.[44] E_1 and E_2 represent the enzymes in aqueous phase I and aqueous phase II, respectively. Path 1 represents the enzyme-facilitated transport of the desired organic acid. Paths 2 and 3 represent nonfacilitated transport of both organic acids.

across the membrane where a second lipase catalyzes ester hydrolysis into the alcohol and the parent acid. If the enzyme-facilitated pathway is significantly faster than transport of the organic acids through the membrane, this will result in selective purification of the desired acid. The system employed for these studies is shown in Figure 3.

The results for the transport of 2-phenoxypropionic acid (PPA) are shown in Figure 4. 2-Phenoxypropionic acid was chosen as a model organic acid because it belongs to a known class of substrates of commercial lipase-catalyzed esterifications,[21–23] is easily analyzed by reverse-phase HPLC and has the correct degree of hydrophobicity such that in its anionic form it resides almost solely in the aqueous phase while in its esterified form it partitions favorably into the organic liquid membrane.

Lipase from *C. cylindracea* (CCL) was used as the esterification catalyst in aqueous phase I, while lipase from porcine pancreas (PPL) was used as the hydrolysis catalyst in aqueous phase II. Lipase from *C. cylindracea* has been shown to be an effective esterification catalyst.[21–23] Lipase from porcine pancreas is a well-known hydrolytic catalyst,[45] as well as transesterification catalyst, yet is unable to catalyze esterification reactions.[21] Hence reverse esterification and transport from aqueous phase II to phase I is eliminated.

The organic liquid membrane phase consisted of isooctane supplemented with $1M$ *n*-butyl alcohol. Isooctane was selected as the liquid membrane phase because it is immiscible with and less dense than water, provides for a favorable partitioning of the resulting ester, and has a low

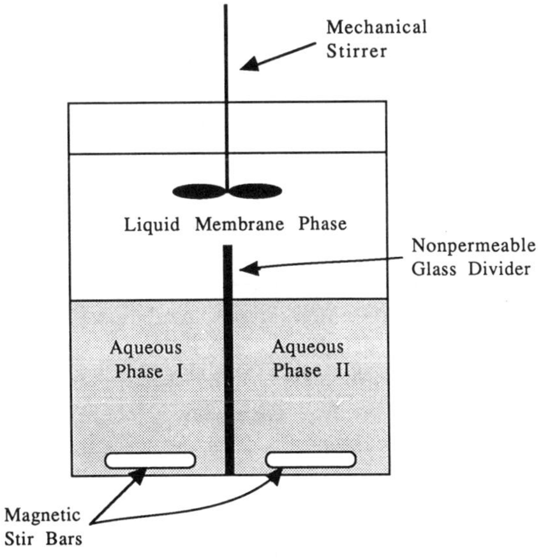

Figure 3. Enzyme-facilitated membrane apparatus.[44]

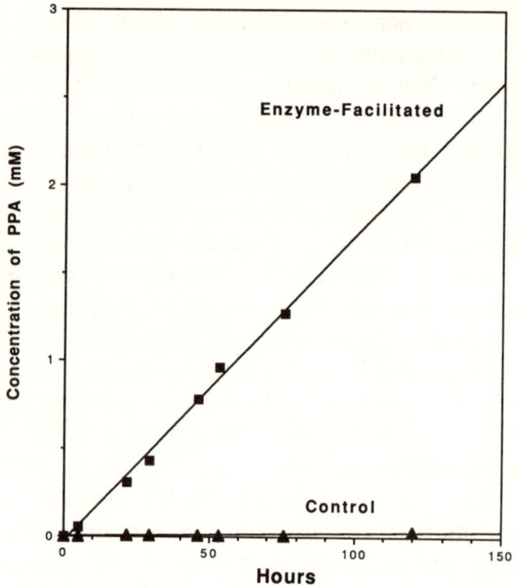

Figure 4. Enzyme-facilitated transport of 2-phenoxypropionic acid (PPA) across an organic liquid membrane.[44] (■) enzyme-facilitated; (▲) control, no enzyme. Aqueous phase I initially consisted of 50 m*M* NaCl, pH 6.3, 10 m*M* PPA, and 30 mg/mL CCL, while aqueous phase II initially consisted of 50 m*M* NaCl, pH 6.3, and 20 mg/mL PPL.

vapor pressure that limits evaporation during the separation.[44] It may be expected that the CCL-catalyzed esterification of organic acids occurs through the acid form of the substrate. Hence, it is desirable to operate at the lowest pH possible without significant loss of enzyme activity. In this case, pH 6.3 fulfilled both requirements.[44]

A control performed by omitting the enzymes indicates that the addition of enzymes resulted in a 70-fold increase in the amount of organic acid transported through the liquid membrane. Specifically, after 120 h, 2.05 m*M* PPA was measured in aqueous phase II in the enzyme-assisted case, while less than 0.03 m*M* PPA was transported in the control. Furthermore, the ester, butyl-2-phenoxypropionate was observed in the isooctane phase only in the presence of the enzymes, while no PPA was observed in the organic phase, indicating that the species transporting through the liquid membrane is the butyl ester of the 2-phenoxypropionic acid. An increase in the PPA concentration to 25 and 100 m*M* resulted in 40 and 32% transport, respectively, after ten days of operation.[44]

In addition to PPA, the transport of other organic acids was studied (Table 2). Phenylacetic acid and 2-(4-chlorophenoxy)propionic acid were transported across the liquid membrane with efficiencies similar to those obtained for PPA. Furthermore, *n*-octyl alcohol was capable of replacing

Table 2. Enzyme-Facilitated Transport Rates of Organic Acids through an Organic Liquid Membrane[a]

Enzyme[a] (30 mg/mL)	Organic acid (10 mM)	Alcohol (1 M)	Initial rate of transport (μmol/L-h)
CCL	PPA	*n*-Butyl alcohol	60
CCL	PPA	*n*-Octyl alcohol	28
CCL	4-Cl-PPA[b]	*n*-Butyl alcohol	67
CCL	Phenylacetic	*n*-Butyl alcohol	123
CCL	Mandelic	*n*-Butyl alcohol	0
PPL	PPA	*n*-Butyl alcohol	0
Lipase G	PPA	*n*-Butyl alcohol	5

[a] Enzymes were obtained from commercial suppliers and were used without pretreatment. Lipase G is from *Penicillium* sp.
[b] 2-(4-Chlorophenoxy)propionic acid.

n-butyl alcohol and lipase from *Penicillium* sp. (lipase G) could replace CCL, although in the latter substitution the rate of transport was significantly reduced. No transport was observed with PPL employed as the esterification catalyst.

The enzyme-assisted liquid membrane was also shown to be selective. Mandelic acid could not be transported across the liquid membrane. This selectivity was induced by the great selectivity afforded by CCL. In independent experiments it was found that CCL could not catalyze the esterification of mandelic acid with butanol in either toluene or hexane under conditions analogous to organic acid esterifications in nearly anhydrous media.[20] To demonstrate selective separation a solution of 10 mM 2-phenoxypropionic acid and 10 mM mandelic acid was prepared in aqueous phase I (all other conditions as described above). After 170 h, 3.3 mM PPA was transported across the liquid membrane with the total exclusion of mandelic acid (<0.01 mM detected in aqueous phase II). Hence, ca. 100% selectivity was achieved, and this selectivity was imparted by the enzyme.[44]

In summary, the liquid membrane-based separation technique is attractive for the following reasons: (1) the separation occurs in a one-stage process; (2) both the feed streams and product streams are aqueous, thus reducing solvent usage.

It is desirable to have a small partition coefficient, P_i, for the acid to make it insoluble in the organic phase. The alcohol, however, should have a large value of P_i so that it remains primarily in the organic phase. The ester should also have a large partition coefficient so that the resulting lactic acid ester is concentrated in the organic phase.

Thus, coupling enzymes with liquid membranes provides a simple, single-step method to selectively separate and purify organic acids through a facilitative transport mechanism. This methodology could be expanded to include optical resolutions and purifications as well as alternate membrane geometries. It can be envisioned that by varying the enzyme used, it may be possible to control membrane selectivity for specific separations.

15.4. FUTURE DIRECTIONS

While two-phase membrane systems have been commercially developed, the industrial use of enzymes for separations in general has not become widespread. The areas best suited to commercial development are enzymes in organic solvents and membrane systems. In addition, enzymes will generally be cost-effective only in separations where traditional separation techniques are ineffective and the unique selectivity of enzymes is required (e.g., resolution of optical isomers). The following improvements are required to increase commercial application and acceptance of this technology:

1) Increased separation speeds and selectivities.
2) Reduced solvent usage.
3) Increased enzyme utilization and recovery.
4) Additional reactor (separator) configurations.
5) Application to a larger number of systems (development of a data base).

For example, liquid membrane systems (Figure 3) can be implemented in a contained liquid membrane device (membrane contactor).[46] This would reduce the diffusion resistance in the organic phase and increase the surface area to volume ratio (increasing throughput). In addition, the enzymes could be immobilized on the porous polymeric membranes which separate the aqueous from the organic phases in these systems, thereby increasing enzyme utilization and recovery. Thus, with enhancements such as those indicated above, enzymes will gain increased importance and usage in bioseparations.

REFERENCES

1. P. E. Hansen and B. A. Morgan, in: *The Peptides*: *Analysis*, *Synthesis*, *Biology*, (S. Udenfriend and J. Meinhofer, eds.), Academic Press, Orlando (1984).
2. W. H. Pirkle and E. M. Doherty, *J. Am. Chem. Soc. 111*, 4113 (1989).
3. E. T. Kaiser and F. J. Kezdy, *Science 226*, 249 (1984).
4. S. Wolfe, A. L. Demain, S. E. Jensen, and D. W. S. Westlake, *Science 226*, 1386 (1984).
5. A. R. Macrae, *JAOCS 60*, 243A (1983).

6. E. E. Bjurstrom, *Chem. Eng. 92*, 142 (1985).
7. J. M. Cooney, *Bio/technology 2*, 41 (1984).
8. J. F. Bunnett and J. L. Marks, *J. Amer. Chem. Soc. 74*, 5893 (1952).
9. J. Gal, *LC-GC 5*, 106 (1987).
10. J. S. Dordick, *Enzyme Microb. Technol. 11*, 194 (1989).
11. T. Uzuki, M. Takahashi, H. Yokomichi, and Y. Hawahara, US Patent 3,907,638 (1974).
12. C. Waundry, in: *Enzymes as Catalysts in Organic Synthesis* (M. P. Schneider, ed.), NATO ASI series, D. Reidel Publishing, Dordrecht, Holland (1986).
13. W. J. Marshek and M. Miyano, *Biochim. Biophys. Acta 316*, 363 (1973).
14. W. E. Ladner and G. M. Whitesides, *J. Am. Chem. Soc. 106*, 7250 (1984).
15. S. Miyano, K. Kawahara, Y. Inoue, and H. Hashimoto, *Chem. Lett.*, 355 (1987).
16. A. Tanaka and S. Fukui, in: *Enzymes and Immobilized Cells in Biotechnology*, (A. I. Laskin, ed.), Benjamin/Cummings Publishing, Menlo Park, CA (1985).
17. A. Klibanov, G. P. Samokhin, K. Martinek, and I. V. Berezin, *Biotechnol. Bioeng. 19*, 1351 (1977).
18. B. H. J. Bielski and S. Freed, *Biochim. Biophys. Acta 89*, 314 (1964).
19. G. Langrand, J. Baratti, G. Buono, and C. Triantaphylides, *Tetrahedron Lett. 27*, 29 (1986).
20. K. Yokozeki, S. Yamanaka, K. Takinami, Y. Hirose, A. Tanaka, K. Sonomoto, and S. Fukui, *Eur. J. Appl. Microbiol. Biotechnol. 14*, 1 (1982).
21. G. Kirchner, M. P. Scollar, and A. M. Klibanov, *J. Am. Chem. Soc. 107*, 7072 (1985).
22. A. M. Klibanov and G. Kirchner, US Patent 4,601,987 (1986).
23. C.-S. Chen, S.-H. Wu, G. Girdaukas, and C. J. Sih, *J. Am. Chem. Soc. 109*, 2812 (1987).
24. R. Cremlyn, *Pesticides, Preparation and Mode of Action*, p. 143, Wiley, Chichester (1978).
25. D. Bianchi, P. Cesti, and E. Battistel, *J. Org. Chem. 53*, 5531 (1988).
26. F. Francalanci, P. Cesti, W. Cabri, D. Bianchi, T. Martinengo, and M. Foa, *J. Org. Chem. 52*, 5079 (1987).
27. Y.-F. Wang, J. J. Lalonde, M. Momongan, D. E. Bergbreiter, and C.-H. Wong, *J. Am. Chem. Soc. 110*, 7200 (1988).
28. H. Kitaguchi, P. A. Fitzpatrick, J. E. Huber, and A. M. Klibanov, *J. Am. Chem. Soc. 111*, 3094 (1989).
29. P. E. Sonnet, *J. Org. Chem. 52*, 3477 (1987).
30. H. S. Bevinakatti and R. V. Newadkar, *Biotechnol. Lett. 11*, 785 (1989).
31. D. Gerlach and P. Schreier, *Biocatalysis 2*, 257 (1989).
32. M. Inagaki, J. Hiratake, T. Nishioka, and J. Oda, *Agric. Biol. Chem. 53*, 1879 (1989).
33. M. D. Lilly, *J. Chem. Tech. Biotechnol. 32*, 162 (1982).
34. G. Carrea, *Trends Biotechnol. 2*, 102 (1984).
35. L. E. S. Brink and J. Tramper, *Biotech. Bioeng. 27*, 1258 (1985).
36. B. Cambou and A. M. Klibanov, *Biotech. Bioeng. 26*, 1449 (1984).
37. R. L. Bratzler, *Proc. World Biotech. Report, Vol. 1*, p. 23 (1987).
38. S. L. Matson, US Patent 4,800,162 (1989).
39. S. L. Matson, US Patent 4,795,704 (1989).
40. S. L. Matson and J. A. Quinn, *Ann. N.Y. Acad. Sci. 469*, 152 (1986).
41. I. Chibata, T. Tosa, and I. Takata, *Trends Biotechnol. 1*, 9 (1983).
42. T. L. Donaldson and J. A. Quinn, *Chem. Eng. Sci. 30*, 103 (1975).
43. S. R. Suchdeo and J. S. Schultz, *Biochim. Biophys. Acta 352*, 412 (1974).
44. D. G. Rethwisch, G. Yi, A. Subramanian, and J. S. Dordick, *J. Am. Chem. Soc. 112*, 1649 (1990).
45. B. Bjorgstrom and H. L. Brockman, eds., *Lipases*, Elsevier, Amsterdam (1984).
46. A. Sengupta, R. Basu, and K. K. Sirkar, *AiChE Journal 34*, 1698 (1988).

Index